Harvard-Yenching Institute Studies XXVI

JAPANESE STUDIES
of
MODERN CHINA

A Bibliographical Guide to Historical and Social-Science Research on the 19th and 20th Centuries

JOHN KING FAIRBANK
Harvard University

MASATAKA BANNO
University of Tokyo

SUMIKO YAMAMOTO
International Christian University

Harvard University Press
Cambridge, Massachusetts
1971

First edition, Charles E. Tuttle Company, 1955

Reissued 1971, Harvard University Press

Distributed in Great Britain by Oxford University Press, London

Library of Congress Catalog Card Number 74-134948

SBN 674-47249-7

Printed in the United States of America

FOR

WILMA AND RYŌKO

PREFACE TO 1971 REISSUE

This volume, completed in mid-1953, is reissued in 1971 without revision because it seems to have stood the test of time: in the intervening seventeen years the corpus of Japanese writings on China which it describes has become of greater, not less, interest to Western researchers on China. There are more of them today and they are more often able to read Japanese. Study of modern China has not diminished in importance, nor has the Japanese contribution to it.

A supplementary volume, dealing in the same fashion with Japanese studies published since 1953, is in preparation.

In the present reissue we have corrected a few errors of reference or dating and several readings of Japanese names. The principal correction we have *not* made is to substitute Sutō for Sudō as the correct transliteration of the name of Professor Sutō Yoshiyuki; our reason is simply that his name appears in these pages so frequently as to make textual correction more difficult than this specific notice. We would also call attention to the fact, with reference to item 1.2.1, **Shinkoku gyōseihō,** that a reprint of the whole seven volumes in a smaller format was issued by Daian 大安 in 1966–67 together with a separate new index volume of 107 pages prepared by Professor Yamane Yukio 山根幸夫 (Daian, 1967), which is much more useful than the original index of 1914.

Finally, we should like to explain how the two authors of 1953 have grown to three authors on our title page of 1971. This actually goes back to 1954 after our manuscript had been sent to the printer and was in process of indexing. Professor Sumiko Yamamoto generously consented to take responsibility for the final completion of the indexing, both the "general index" of all persons, titles, and subject headings and the "character index of authors' names." This involved her in a degree of scholarly effort that we had not adequately foreseen. The indexes have proved to be a strategic and essential part of the book for reasons indicated on page 321. They required knowledge, skill and care commensurate with the other parts. *Ergo*, Professor Yamamoto became a co-author and we are happy to be able now to acknowledge the fact.

For the opportunity to reprint we are indebted to Dr. Glen W. Baxter and the Harvard-Yenching Institute.

JOHN KING FAIRBANK
MASATAKA BANNO

CONTENTS

CONTENTS

CONTENTS

INTRODUCTION

This volume describes more than one thousand Japanese books and articles, which constitute the main body of Japanese research on modern China. This on-the-spot reconnaissance, by an American and a Japanese working with the advice of Japanese specialists, is designed to assist Americans and other non-Japanese in Chinese studies. By "modern China" we mean the 19th and 20th centuries, although we have pursued some topics farther back and have avoided ephemeral outpourings on the latest developments.

The Use of Japanese Studies of China in America

By an historical accident, American studies of modern China (as distinct from those of traditional China) have largely ignored Japanese materials. The first generation of American specialists on modern China to receive systematic sinological training, in the early 1930's, naturally studied in China rather than Japan, where the political climate was unfavorable to American research on the controversial China question. Much of the Japanese research noted in this volume, in any case, has been done since that time, on the basis of intensified wartime interest and contact. The startling reversal of the American position vis-à-vis China and Japan since 1945 now makes American scholarship dependent on Japanese almost as much as on Chinese scholarship, for stimulation, fresh evidence, disagreement, or expert guidance in the study of the modern century and a half of Chinese history. At all events, this volume should make it evident that much can be learned about China in the West through the rather neglected channel of Japanese studies.

The First Phase of Japanese Research

Through this same channel, much can also be learned about modern Japan. Research concerning continental problems began in Japan with a stage of fact-finding, under government or semi-official leadership, and particularly under the stimulus of the Sino-Japanese War of 1894-95 which also touched off the revolutionary movement in the Chinese empire. In the same period when Japanese expansionists and frustrated liberals were aiding the plots of Sun Yat-sen, Japanese administrators were inaugurating research programs on the current Chinese scene. The Tōa Dōbun Shoin or T'ung-wen College was set up as a training and research center in Shanghai in 1901 (see items no. 4. 6. 1–4). Scholarly study of the administrative institutions of the Ch'ing empire was instituted in the same year by the Taiwan Government-general (see item no. 1. 2. 1). After the Russo-Japanese War, the South Manchurian Railway Company eventually set up its productive research department in Dairen.

These field research programs were generally conducted by men with a veritable passion for compilation. They used not only the compendia of Ch'ing scholarship—the gazetteers and imperially sponsored collections—but also the reports of Japanese observers who roamed notebook in hand over the length and breadth of the country. The massive compilations presided over by Negishi Tadashi and others at the Tōa Dōbun Shoin (see 7. 2. 1 et seq.) were based on the annual field trips and reports of Japanese students in training there for professional service in business, foreign service, the armed forces, journalism or academic life. While these deposits of an earlier generation are no more than raw material for the modern social scientist, they nevertheless provide an independent view of contemporary China, more modern than that of the traditional Chinese compilers of the day, and rich in data.

Meanwhile the scholars of the imperial universities at Tōkyō and Kyōto were led by events increasingly to appraise the current Chinese scene. Professor Naitō Torajirō of Kyōto worked out a major interpretation of modern Chinese history which still has wide influence (see 1. 4. 1 et seq.). Professor Hattori Unokichi of Tōkyō, as a young man in Peking, produced an important study of Ch'ing government (1. 2. 2). In time men who had studied China in the research centers at Shanghai and Dairen returned to professorial posts in Japan.

The Japanese and American Conceptions of China

Out of this scholarly study, in the context of Japan's modern development, there emerged a Japanese idea of China which was considerably different from the American idea of China. This is a fascinating contrast, which awaits research. Possibly the early Western missionary, aware of his own dedication to good works and convinced of human perfectibility, fell to some degree under the spell of the great Confucian doctrines of scholar-government and rule-by-virtue, and so tried to see the best in a society toward which his motives were generous. Possibly the Japanese student interpreters, journalists, and merchants, coming later upon the scene and seeking the advancement of their own national welfare and prestige, were more readily conscious of modern China's sordid state, when compared with the legendary grandeur of her past as studied in Japan. At any rate, Japanese histories of China, like those which Inaba Iwakichi prepared for the Japanese army during the first years of the Republic (see 2. 1. 3 et seq.), emphasized the phenomenon of barbarian conquest, at the very time when Western missionary historians were inclined to feel that China had "always absorbed her conquerors." Again, while the often a-historical Americans were responding optimistically to modern China's republican and democratic stirrings, Dr. Naitō at the same time was locating the birth of " modern " China in the Sung period (from which one might easily infer that little progress had since been made). Because of such contrasts of judgment and interpretation, we suggest, the United States and Japan more easily diverged in their approaches to the Chinese Republic.

Having both now suffered a grievous ejection from the Chinese mainland, these two overseas trading nations, sharing a similar loss, may perhaps acquire a common understanding of it. Yet the American and Japanese ideas of China are still far from similar.

By the late 1920s, the Japanese compilers were beginning to meet frustration in their assiduous effort to master the China problem intellectually through the accumulation of facts about it. Perhaps because a plethora of facts had been accumulated, the sweeping generalizations of academic Marxism were more readily welcomed as a means of giving coherence to the data which had been so voluminously published about modern China. This ushered in a new phase of Japanese studies of China in the 1930s, which is only now being succeeded by a newer and more up-to-date effort at independent scholarly interpretation. This Japanese background, however, confronts the American student with a problem of discrimination and judgment.

The Problem of Marxism

It has long since ceased to be fashionable for historians to say, as some once did, that their function is to "let the facts speak for themselves." On the contrary, it is now accepted that historians must present both facts and theories in their effort to make the past understandable to the contemporary reader. Data and interpretation are thus the cognate components of modern history writing, shading into each other at every point and mixed by each historian according to his own taste and sense of reality. It is generally agreed among the academic fraternity of non-totalitarian countries, however, that the historian must pick his own individual way between the devil of false theory and the deep sea of distorted fact. If he presents only data, he is a mere compiler or at best an old-style chronicler. If he lets theory run away with him, on the other hand, he has become a dreamer, or an ideological and semantic acrobat, or a reverent commentator on dogma.

The essential objection to doctrinaire history writing, particularly the Marxist type, thus lies in its unrealism. Man cannot save himself by constructing fancy statements of how history must have been, or must be, on an a priori and dogmatic basis; he must get closer to reality if he is to master it. To survive above the totalitarian level, modern society must receive the benefit of individual minds thinking creatively; for the insight into social problems which is prerequisite to solving them can be achieved only in the thought processes of individuals who are free to think.

Approaching the problem of Marxist historical theory in Japan from this point of view, we are at once impressed with two aspects of it, visible at opposite ends of a spectrum. At one extreme is the party-line indoctrination or propaganda which is admired by communists and their easy-going followers as "scientific" historical truth, but which in reality is hardly more than a tool of the communist dictatorships in Russia or China, used by them purposefully for the ideological enlistment and intended absorption of the individual.

At the other end of the spectrum is a different picture—the large and varied body of ideas more or less vaguely associated with the name of Karl Marx, which form part of the intellectual heritage received by the Western world from the nineteenth century. Much of what is now included in "Marxism" was at one time in vogue among the pioneer social scientists of Europe. At any rate, it would be a gross error to assume that an attempt to analyze tenancy relations or class structure, for example, is a "Marxist" operation, merely because Marx and Engels attempted it. In our revulsion against the thought-controlled states which idolize Marx and Marxism, we will gain nothing by becoming anti-historical and denying the bearded father of the totalitarian cult his individual place in the history of social thought. Many systematic thinkers have made their contributions partly in response to his stimulus.

We are confronted then, as are the historical fraternity in Japan, with a mixed and puzzling situation, in which the interesting intellectual challenge of broad Marxist conceptions like that of the "Asiatic mode of production" is intermixed with party-line stuff which may run from cynical propaganda by disciplined agents to mere wish-fulfillment on the part of undiscriminating enthusiasts.

From the survey described in this volume, several further impressions emerge. One is that genuine historical scholarship, the creative addition to understanding of our past and so of our present, can be achieved only by individual minds. The masters of the historical craft in Japan as elsewhere are non-dogmatic searchers after truths which are not yet fully known. The leading social scientists are similarly men with open minds who regard their interpretations as malleable in response to new hypotheses, tests, and verifications.

A second impression, however, is that several of the broad ideas of Marxism have taken deep root in Japan's intellectual soil, for reasons about which it is interesting to speculate. Possibly the recent experience of a transition from feudalism to capitalism of a certain sort in Japan, which was comparable in some ways to the experience which lay behind Marx and his generation in Europe, lent credence to a theory which emphasized feudalism and capitalism. Another factor may have been the rather tragic intellectual history of modern Japan. In the early Meiji period liberal historians like Fukuzawa Yukichi had had a comprehensive view of "civilization" and "civil society". But under the gradually consolidated Emperor system, with its pseudo-constitutionalism and the resulting absorption of the civil society into the state, the political and intellectual climate became quite intolerant both of a free and critical development of social science and of a comprehensive and causal understanding of history. Social scientists could hardly speak other than as conscious or unconscious ideologues of the status quo. Anything smelling of radicalism was drastically suppressed. Chances were rare for any pragmatic and problem-solving approach to reality, and historians either became devoted to assiduous fact-finding, based on the elaborate handling of primary sources, in defense of the barely established tradition of modern historiography; or followed the a-political approach of intellectual and cultural studies, if they did not actually cling to the traditionalistic-moralistic view of history.

In these circumstances, the breakdown of the comprehensive sociology of Spencer and Comte in the late 19th century and the resulting compartmentalization of the social sciences, had peculiar repercussions in Japan: there set in an extreme and sectionalistic division into separate fields or disciplines, without any serious attempt at a comprehensive grasp of the various aspects or factors of social life. Scholars indulged in unrealistic and doctrinaire speculation, each entrenched in his citadel, concerning theories and systems imported from abroad such as neo-Kantian methodology, formal sociology, pure economics, or pure jurisprudence.

Into this intellectual climate, which lacked an effective alternative body of theory and suffered from the recurrent economic instability that culminated in the post-1929 depression, came Marxism with its comprehensive approach and keen sense of catastrophe. By this time, moreover, Marxism had been developed into Marxism-Leninism, which stressed the "contradictions" within capitalism and the "inevitability" of expansive imperialism in a way that appealed to many Japanese.

Its enthusiasm and cynicism were also something quite new and interesting to Japanese intellectuals, who felt guilty over their utter impotence to confront social problems. It is therefore intelligible that the Marxist way of thinking was taken up not as a result of individual study but rather as an esoteric doctrine, and even with a sense of martyrdom.

Post-war conditions in Japan, especially in the earlier stage of the occupation, have made possible a rather free and open discussion of Marxism and as a result its impact has been (somewhat paradoxically) diluted and interfused with other approaches, in the arena of scholarly pressures and counter-pressures; yet the fact remains that it has a wide currency. Scholars in the industrialized Western world sometimes forget that the conditions of an agrarian-based and economically undeveloped peasant-village society like China, subject to the powerful political and economic influence of foreign powers, offer fertile ground for a Marxist-Leninist interpretation of modern history. In Japan, in particular, the fact that Marxism-Leninism came under militarist attack gave it the opportunity to be the ideology of protest among poverty-stricken intellectuals and frustrated youth. The political and social developments in China, finally, from the 1920s made it almost imperative for the Japanese to follow and study, critically or not, the communist movement on the Asiatic mainland, and the recent spectacle of China in the process of going communist has undoubtedly exerted a considerable influence upon intellectuals in a country so close, historically and strategically, as Japan.

Whatever may be the reasons for the vogue of Marxism in Japan, there can be no doubt about the strength of its influence in the scholarly world there. This fact in turn has created a special problem for us in compiling this volume, and presents the reader with a similar problem in using it. How are we to describe and evaluate the element of Marxism in our effort to describe Japanese studies of modern China?

First, we would emphasize that recognition that a work exists is not necessarily equivalent to recommending it. In other words, we have included some party-line writings, if only as examples, and a great many works which seem sympathetic, if they do not actually resort, to the Marxist or Marxist-Leninist or Stalinist approach in one way or another. Secondly, we assume no responsibility to warn the reader as to how far a work may present a party-line view, since we lack ideological Geiger counters by which to make the test. We have, however, been rather free with our impressions in many cases, and have usually preferred to indicate items which seem to us doctrinaire, dogmatic or over-schematized in the well-known pattern. At the same time, American readers will do well to recognize that many Marxist formulations, if taken as hypotheses or problems for investigation, may be valid, important and respectable subjects of honest academic research.

Whether a Marxist-seeming work is or is not a party-line product must be determined in each case. To meet this problem, our main device has been to assemble in certain sections a representative collection of writings concerning a disputed topic or interpretation—in particular, section 1.4 concerning the periodization of Chinese history, section 1.5 on the nature of Chinese society (including the questions of "Oriental society" and the "Asiatic mode of production"), and section 8.2 on the nature of the Chinese village community. In each case the reader can see for himself, if he looks at the original items, the major positions which specialists and polemicists have worked out in these matters of interpretation.

Our personal view is a dual one—(1) that Marxism, taken in the broad sense of the term, has been a powerful stimulus to Japanese scholarship and (2) that among Japanese scholars of superior competence, the balance has swung distinctly against the party-line form of Marxism, which has been and is being increasingly rejected.

Some Major Research Agencies and Their Productions

A large proportion of Japanese research on modern China has been produced by official or semi-official agencies like the SMR, the Foreign Office, or the Tōa Kenkyūjo. Much of this research has not been published through commercial publishing houses. Instead, it has been printed (or in wartime often mimeographed), sometimes classified, and distributed to a restricted list of recipients. This practice puts the ordinary librarian and civilian researcher under a handicap, even though copies of many such works no doubt eventually percolate into collections where they can be used, or even come onto the second-hand book market. To compound the difficulty, bibliographies of such works have seldom been compiled or published. Lists like item 9. 2. 9 for the SMR materials are hard to find.

On the other hand, the researcher and librarian outside Japan may be consoled by the fact that these unpublished productions, judging by what we have seen, are of uneven quality, often printed purely for office use (sometimes as first drafts which were never revised), and not always up to the standard or in the form desired by publishers. Some of them correspond, perhaps, to background research studies written by American scholars in the agencies of wartime Washington and since declassified. But the comparison is not very apt, for in the different institutional context of Japan entire research commissions of the best brains obtainable were maintained by government agencies for years on end, before the war.

One may expect, therefore, to come across in Japanese libraries a considerable array of rather miscellaneous and unforeseen research studies on aspects of modern China. Many are numbered as parts of one or another series or labelled secret or confidential, but otherwise unidentified. From those that we have seen, we have selected the more substantial for inclusion in this present volume, without attempting a complete coverage. Attention may be called to the following agencies which produced such materials. (We omit the Tōa Dōbun Shoin at Shanghai, whose products were mainly published in periodicals or through commercial firms, or have been noted by us individually.)

1)   The SMR research bureau at Dairen (called Chōsabu 調査部 or Chōsaka 課) originated in the vision of the first head of the SMR, Gotō Shimpei. It developed another research center at Tōkyō (Tōa Keizai Chōsakyoku 東亞經濟調査局) and subsequently research sections in the SMR branch offices at Harbin (1923), Tientsin (1935) and Shanghai (1936). Altogether these research offices, as well as other sections of the SMR, put out a tremendous amount of research and field reports, documentation, bibliography and translated materials. **Mantetsu chōsa geppō** (item 9. 9. 7) and other SMR publications, however, provide a key to a good deal of it.

2)   The Government-general of Taiwan, as noted above, also under the original stimulus of Gotō Shimpei when he was civil governor, maintained a temporary research commission of note and a regular research bureau. Whereas the SMR concerned itself primarily with Manchuria and North China, this smaller-scale

work focussed more upon Taiwan, South China and also Southeast Asia.

3) The Foreign Office or Ministry of Foreign Affairs (Gaimushō) in Tōkyō maintained a Bureau of Research and Documentation (recently it has been broken up and the parts distributed among the area desks in a manner which may be of wry interest to researchers in Washington). Both it and the Bureau of Commercial Affairs (now called Economic Affairs Bureau) used consular reports and library materials to turn out an important flow of research studies on contemporary China, much of which is now hard to find.

4) During the war the Ministry for Greater East Asia (Daitōashō 大東亞省) and especially its Asia Development Board (Kōain 興亞院) produced many reports, which included substantial results of field investigation in China.

5) One of the most extensive bodies of literature on China was turned out under the auspices of the Tōa Kenkyūjo 東亞研究所, a wartime agency which had both private and governmental sponsorship. It mobilized a powerful staff and enlisted the collaboration of professorial groups at the imperial universities of Tōkyō and Kyōto and elsewhere. Among several large projects of scholarly value which this agency and its sub-sections undertook, we have noted those on foreign investments in China (items 7. 18. 1–4), alien rule over China (3. 3. 1–5), and North China rural society and economy (see especially 8. 1. 1; this last in fruitful cooperation with the SMR research staff).

Further data on these and other agencies may be found through our General Index. Since the war, periodic lists of publications of Japanese government agencies have been compiled and made available by the National Diet Library in Tōkyō.

Problems of Bibliographical Methodology

While we have examined all the materials here described, except for a dozen items noted as not seen, it need hardly be remarked that we have not read them all. Neither are we expert in all the fields of learning represented. Mindful of our own limitations, we have sought guidance from a number of more competent specialists, as indicated in our Acknowledgments.

In forming our own impressions, however, we have sought to err on the side of enthusiasm. This is a voyage of discovery, not a final accounting. In looking for materials of probable interest to Western researchers, we have frequently devoted as much space to an article as to a book, in the effort to indicate its significance or content. As a result, books and articles tend to be described on two different levels of intensity, more space, in proportion to their length, being devoted to articles. We have also noted recent book reviews, as indications of scholarly opinion. Where we may have made less or more than a just appraisal, we regret it sincerely; but this survey does not pretend to be more than a survey, nor can it substitute for the reader's reading the original works.

In order to make our romanizations as meaningful as possible, we have capitalized names, like *Shindai* for the Ch'ing dynasty; and hyphenated combinations which are so in English, like *Nichi-Ro* for Russo-Japanese. Confusion overtakes us, however, in the case of Sino-Japanese, *Nisshi*, which we let stand as mute witness of lexicographic defeat. The same is true of *Nisshin*, Japanese-Ch'ing, or *Nikka*, Sino-Japanese.

Quotation marks have two uses: to surround romanizations of all articles, and to indicate translations of titles which are copied from the original Japanese publications, where English translations are often supplied. Chinese names transliterated from Japanese are hyphenated as though from Chinese: Shō Kai-seki for Chiang Kai-shek, Mō Taku-tō for Mao Tse-tung.

On the subject of Japanese scholars' personal or given names, we confine our sentiments to the "Character Index of Authors' Names" at the back (see "Note").

Acknowledgments

We have received invaluable help and guidance from many persons specialised in the study of China. None of them can be regarded as the source of inadequacies which may appear below, and they have contributed in such different degrees and on such various topics, that we can do no more than list their names with this expression of thanks: Abe Takeo, Amano Motonosuke, Etō Shinkichi, Fujieda Akira, Hatada Takashi, Hori Toshikazu, Ichiko Chūzō, Iwamura Shinobu, Kobori Iwao, Makino Tatsumi, Maruyama Masao, Matsumoto Yoshimi, Miyazaki Ichisada, Muramatsu Yūji, Nakamura Tadayuki, Niida Noboru, Ono Shinobu, Onogawa Hidemi, Otake Fumio, Saeki Yūichi, Shiga Masatoshi, Ueda Toshio, Wada Sei, Yamamoto Sumiko, Yamamoto Tatsurō.

We feel a special obligation to Professor Ueda, who gave us not only helpful comment but also office space and the run of his private library. Most of our debts, in fact, have involved the borrowing of books, and we are similarly indebted to the librarians and staffs of the Tōyō Bunka Kenkyūjo of Tōkyō University, where our research was conducted, and to those of the Jimbunkagaku Kenkyūjo of Kyōto University which we visited in May 1953. The manuscript was typed by Imaki Haruko and the index prepared by Yamamoto Sumiko, to both of whom we are greatly indebted. We wish to thank Professor S. Elisséeff, Director, and the Harvard-Yenching Institute, for moral and financial support of this project; and the John Simon Guggenheim Memorial Foundation and the Institute for Oriental Culture (Tōyō Bunka Kenkyūjo) of Tōkyō University, respectively, for making possible the allocation of our time to it.

As collaborators, however, our greatest debt is to each other, without whose assistance this work would not have been possible.

Tokyo, July 10, 1953

JOHN KING FAIRBANK
MASATAKA BANNO

JAPANESE STUDIES

of

MODERN CHINA

1. GENERAL WORKS

Note: On the theory that this volume is for reading as well as reference, we have organized it to proceed from the general to the particular and (to a lesser extent, since it is not the same thing) from the elementary to the complex. This first section accordingly presents survey works in modern history, government, and foreign relations. Further general historical works will be found in sec. 2.1 (e.g., the survey by Inaba Iwakichi, item 2.1.3), on the history of the Ch'ing period, and in sec. 7.1, on economic history. Our judgment being fallible, the reader is reminded that all works in this volume, no matter where we have placed them, may be found through the index at the back.

1.1 SURVEYS OF MODERN HISTORY

Note: We begin with textbooks and surveys for the general reader, some of which may have value for the Western beginner. Among them, item 1.1.3 is perhaps the most fresh and interesting. The last five items are collections of rather miscellaneous essays. Note that item 9.1.1 (Wada, **Chūgokushi gaisetsu**) is a compact survey by a major historian, though even more noteworthy for bibliographical purposes. These "surveys of modern history" merge almost imperceptibly into the "surveys of Ch'ing history" in sec. 2.1 below. Note the companion survey volumes edited by Wada (2.1.13) and by Ishida (5.1.8), and the series of articles and booklets issued by various groups or publishers (9.8.1-2-4-7-9).

1.1.1 OSHIBUCHI Hajime 鴛淵一, TAMURA Jitsuzō 田村實造, MITAMURA Taisuke 三田村泰助, HANEDA Akira 羽田明, joint authors, **Dokusai seiji no jidai** 獨裁政治の時代 (The era of despotic government), vol. 3 of **Kyōdai Tōyōshi** 京大東洋史, Sōgensha, 1952, 195 pp., map.

> This text book (vol. 3 of 5 vols.) presents an attractively condensed summary of Chinese history and all aspects of Chinese society from post-T'ang to the beginning of the 19th century in 153 pp., with charts, dates, bibliography and index attached. The resultant highly generalized and sophisticated account should be of particular interest to Western students.

1.1.2 OTAKE Fumio 小竹文夫, MIYAZAKI Ichisada 宮崎市定, SAEKI Tomi 佐伯富, joint authors, **Tōa no kindaika** 東亞の近代化 (The moderni-

3

zation of East Asia), vol. 4 of **Kyōdai Tōyōshi** 京大東洋史, Sōgensha, 1952, 167 pp., maps.

In addition Continuing the Kyōto University text of Oriental history, this streamlined summary recounts the intrusion of the West, the resultant cultural mixing, and the various developments of modern China up to date in 134 pp., with charts, dates, bibliography and index attached. See review of the first 4 volumes of **Kyōdai Tōyōshi** by Tanikawa Michio 谷川道雄 in **Tōyōshi kenkyū**, 11.5–6 (July 1952), 489–492.

1.1.3 NIIDA Noboru 仁井田陞, NOHARA Shirō 野原四郎, MATSUMOTO Yoshimi 松本善海, and MASUI Tsuneo 増井經夫, **Sekai no rekishi: Tōyō** 世界の歴史・東洋 (World history: the Orient), Mainichi Shimbunsha 毎日新聞社, 1949 (6th printing, 1951, 391 pp., illus.); also, revised ed., 1952, 380 pp.

In addition to sections on India and Islam, this volume of a popular series is mainly devoted to China (pp. 2–291)—ancient, medieval, and modern. The ancient period sees the eventual creation of the unified empire, its decline, and revival under Sui and early T'ang. Medieval China, bureaucratic and "inclining toward feudalism" 封建主義への傾斜, then succumbs to the dynasties of conquest—Chin, Yuan, Ch'ing. This leaves the modern period, that of a "People's society" 人民の社會, to develop under the stimulus of Western contact (pp. 244–288). This easy-to-read survey by well-informed authors necessarily embodies a good deal of recent interpretation (especially the section on medieval China written by Professor Matsumoto), much of which should be of interest to scholars elsewhere. The revised edition recasts somewhat the treatment of ancient and medieval China. The book closes with a separate essay, "Tōyō to wa nanika" 東洋とは何か (What is the Orient?), by Professor Niida, which provides a provocative problem-analysis, especially critical of the stationary theories accepted by some Westerners.

1.1.4 ICHIMURA Sanjirō 市村瓚次郎, **Tōyōshi tō** 東洋史統 (A general history of the Orient), vol. 4, Fuzambō, 1950, 977+37 pp. (Vol. 1, 1939, 867 pp.; vol. 2, 1940, 729 pp.; vol. 3, 1943, 832 pp.)

This massive volume concludes a comprehensive history of East Asia produced in the grand manner. It centers about China under the Ch'ing but rounds out the picture with sections on Central Asia, India, the European expansion, Southeast and Northeast Asia, as the following chapter headings indicate: (1) Rise and flourishing of the Ch'ing dynasty, 1–81. (2) Rise and fall of the Mogul empire, eastward expansion of Westerners and their transmission of Western culture, 82–175. (3) Ch'ing relations with Mongolia-Central Asia, and with the countries of Southeast Asia; Liu-ch'iu and Korea, 176–245. (4) Culture of the Ch'ing period, beginning of dynastic decline, 246–385. (5) Brief account of the British in India, the Opium War, 386–450. (6) Domestic rebellions and foreign invasions under the Ch'ing, 451–544. (7) Central Asia, British-Russian-Chinese rivalry, 545–598. (8) The countries of Southeast Asia, 599–662. (9) Korea, Liu-ch'iu and Taiwan; Japanese-Chinese-Russian relations, 663–770. (10) Advance of the European-American powers in East Asia and the state of China; Russo-Japanese War, Japanese annexation of Korea, 771–885. (11) Chinese revolution and political situation of the Chinese Republic, changing circumstances at home and abroad, 886–977.

Dr. Ichimura was a scholar particularly of early Chinese history. He was able to correct the proofs of this last volume before his death in 1947 at the age of 84, and it was published with the aid of Dr. Wada Sei and others. Although lacking footnotes and bibliography as such, the volume presents a tremendous number of proper names, identifications, dates and source references in the text, which reflect the author's wide grasp of subject matter (his first survey of Chinese history was published in six volumes before he was 30). Marginal headings make up somewhat for the lack of an index. This is a trustworthy, factual, narrative history, meticulously prepared, rather than a work of interpretation or comparative study, and treats its topics as units, without stressing their causal interconnections. See review by Ichiko Chūzō 市古宙三 in **Shigaku zasshi**, 60. 1 (Jan. 1951), 63–68.

1.1.5 URA Ren'ichi 浦廉一, ed., **Tōyō kinseishi** 東洋近世史 (History of East Asia in the modern period), vol. 1, Heibonsha, 1939 (5th printing, 1941, 536 pp., map, illus.).

This interestingly-illustrated survey text, written by twelve scholars, opens with a 150-page summary of the Ch'ing regime in China (by Ura Ren'ichi) and moves on to the arrival of Europeans (by Masui Tsuneo 増井經夫), the opium war and the growth of foreign trade (by Miyazaki Sonoji 宮崎其二) and its effect on China's feudal society (by Yamauchi Shizuo 山內靜夫). A chapter on the special "feudal" character of this society (by Numata Tomoo 沼田鞆雄) is followed by two on the growth of capitalism (by Sano Riichi 佐野利一 and Hashimoto Isamu 橋本勇) and another two on the scholarship and literature of the Ch'ing period (by Nakayama Hachirō 中山八郎) and on the earlier introduction of Western science by the Jesuits (by Fujino Takeshi 藤野彪). Final chapters are on Manchuria (by Kawakubo Teirō 川久保悌郎), Russian encroachment (by Momose Hiromu 百瀬弘) and the British in India (34 pp., by Suzuki Tomohide 鈴木朝英). While produced in the form of an attractive textbook, this volume embodies a high degree of historical interpretation.

1.1.6 MATSUI Hitoshi 松井等, ed., **Tōyō kinseishi** 東洋近世史 (History of East Asia in the modern period), vol. 2, Heibonsha, 1939 (4th printing, 1941, 510 pp., maps, illus.).

Published as a companion volume to the one edited by Ura Ren'ichi (1.1.5), this work is written by seven scholars and treats the modern century under the more conventional topics of political history. The first 100 pages survey the period from the Opium War to the 1930s (by Matsui Hitoshi) and the following chapters take up such topics as Sino-British relations (by Satō Masashi 佐藤正志), the Taiping Rebellion (by Nohara Shirō 野原四郎), the revolutionary movement (by Sano Riichi 佐野利一), railroads and loans (by Momose Hiromu 百瀬弘), modern international relations in peripheral areas (Korea and Manchuria by Nohara Shirō, Mongolia and Tibet by Yoshida Kin'ichi 吉田金一). Two concluding chapters (by Suzuki Tomohide 鈴木朝英) are on India (50 pp.) and Indo-China (30 pp.). Well illustrated.

1.1.7 MATSUI Hitoshi 松井等, **Manshū minzoku seisuijidai** 満洲民族盛衰時代 (The period of the rise and fall of the Manchu people), and **Shin-Shina jidai** 新支那時代 (The period of the new China), in **Tōyōshi kōza** 東洋史講座 (Lectures on Oriental history), Yūzankaku 雄山閣, 1930,

vols. 8 and 9 (Dai yonki zempen, kōhen 第四期前編，後編), 596 pp.

Of these two volumes in a broad popular series (now out of date), vol. 8 recounts the history of the Ch'ing dynasty, emphasizing its relations with and tribulations at the hands of the West. Vol. 9 narrates events from the 1890 s down to the Nanking Government, similarly stressing foreign relations (and touching on India) but with some attention to the new thought (ch. 27) and warlord politics (ch. 28).

1.1.8 MATSUI Hitoshi 松井等, **Tōyōshi gaisetsu** 東洋史概說 (A survey of Oriental history), Kyōritsusha 共立社, 1930 (11th printing, 1941, 357 pp., illus.).

A well-balanced textbook widely used in the 1930 s, based on lectures delivered at 9 or more Tōkyō colleges and universities. Its coverage, from ancient to modern times and including India, is preeminently simple and factual. By the same author are two narrower survey pamphlets: **Shina minzoku** 支那民族 (The Chinese people), in **Iwanami kōza Tōyō shichō** (item 9.8.4), ser. 10, 1935, 45 pp.; and **Shina gendai shisō** 支那現代思想 (Contemporary Chinese thought), in ibid., ser. 5, 1934, 43 pp., a topical survey by types of thought.

1.1.9 OTAKE Fumio 小竹文夫, **Gendai Shinashi** 現代支那史 (Modern Chinese history), Kōbundō, 1940, 154 pp. (Kyōyō bunko 教養文庫, 58).

This vest-pocket volume by a leading professor of Tōa Dōbun Shoin at Shanghai provides a neat popular summary of the old society, its revolutionary stirrings, the Republic, KMT vs. CCP, Manchoukuo and the war since 1937. It presents the social interpretation of a careful scholar widely experienced in Chinese economic history and textual studies.

1.1.10 NIWA Masayoshi 丹羽正義, **Gendai Shina no seikaku** 現代支那 の性格 (The character of modern China), Kōbundō, 1942, 177 pp. (Kyōyō bunko 教養文庫, 48).

Acknowledging the inspiration over two decades of Professors Naitō Torajirō and Yano Jin'ichi, the author of this vest-pocket survey provides a quick wartime look at China's past foreign relations with the West and with Japan, leading to the collapse of the Ch'ing and the difficulties of the Republic. This approach in the Yano tradition provides a possibly interesting contrast to Otake Fumio's **Gendai Shinashi** (1.1.9) in the same series.

1.1.11 HIRANO Yoshitarō 平野義太郎, **Dai-Ajiashugi no rekishiteki kiso** 大アジア主義の歴史的基礎 (Historical foundations of the Great Asia principle), Kawade Shobō, 1945, 410 pp.

This somewhat camouflaged wartime volume with no title page and·a mimeographed publication statement is a series of sweeping essays on modern China, beginning with the stimulus of Meiji Japan for the revolutionary movement, the 1911 Revolution, China's village structure, its community organization and morality, and various studies and theories about it, East and West. A great many topics are touched upon (e.g. the early industrialist, Chang Chien 張謇, pp. 105-113; Quesnay on China's despotism, pp. 304-319), often with a good deal of emphasis. The author had participated in the SMR North China field research and urges the importance of the institutions of village cohesion 共同體 (pp. 137 ff.), apparently under the persistent influence of the Marxist theory of the Asiatic mode of production. Note also Mr. Hirano's **Nōgyō mondai to tochi henkaku** 農業問題と土地變革 (The agrarian problem and land reform), Nippon Hyōronsha, 1948, 440 pp. He is now head of the Chūgoku Kenkyūjo.

1.1.12 OZAKI Shōtarō 尾崎庄太郎, **Chūgoku nōson shakai no kindaika katei** 中國農村社會の近代化過程 (The process of modernization of the Chinese rural society), **Shakai kōseishi taikai** (item 9.8.7), series 7, 1950, 84 pp.

A conventional schematic account of the collapse of the Ch'ing feudal system and manor system (sic), and the subsequent bourgeois revolution, including the Taiping Rebellion as a "peasant war of liberation," etc., etc., down to the people's revolution. The author is now at the Chūgoku Kenkyūjo.

1.1.13 SHIRATORI Kurakichi 白鳥庫吉, "Tōzai kōshōshijō yori mitaru yūboku minzoku" 東西交涉史上より觀たる遊牧民族 (The pastoral peoples as viewed in the history of East-West relations), pp. 1–41 in Shigakukai, ed., **Tōzai kōshō shiron** (9.8.11).

This essay by a former leading sinologue of Tōkyō University pointed out the great influence exercised by the nomadic peoples, especially the Mongols, on Chinese culture and society. Though mainly concerned with the earlier periods, this article is noted here as a typical expression of Professor Shiratori's view that the history of Asia has been that of pressures and counter-pressures between the Northern nomads and the Southern settled or civilized peoples, in a process further complicated by East-West contacts.

1.1.14 WADA Sei 和田清, **Shina** 支那 (China), in **Iwanami kōza Tōyō shichō** (9.8.4), in 2 vols., series 13 (1935) and 18 (1936), total 199 pp.

This chronological survey of Chinese social and cultural history, in the second volume (which comprises parts 3 and 4 of the subject matter listed in the contents), summarizes major developments from the Five Dynasties to late Ch'ing. Some useful bibliographical references are given but the rate of speed seems a bit rapid, and this survey is in any case superseded by Dr. Wada's two volumes of 15 years later, also prepared for Iwanami, **Chūgokushi gaisetsu** (9.1.1). Reviewed by Konuma Tadashi 小沼正 in **Rekishigaku kenkyū**, 7.1, whole number 39 (Jan. 1937), 106–109; and by Kishibe Shigeo 岸邊成雄 in **Shigaku zasshi**, 48.1 (Jan. 1937), 129–133.

1.1.15 WADA Sei 和田清, "Rekishijō yori mitaru Shina minzoku no hatten" 歷史上より見たる支那民族の發展 (The development of the Chinese people as viewed historically), 45 pp., pub. in vol. 12 of **Tōyōshi kōza** 東洋史講座 by Yūzankaku 雄山閣, 1931.

This broad essay considers Chinese ethnic expansion toward the south during the Ch'ing period (from p. 30) in the light of preceding eras. As a whole, the essay emphasizes the uninterrupted and continuing process of Chinese expansion, political, economic and cultural, and provides a vivid and stimulating picture of the broad sweep of Chinese history from ancient down to contemporary times. Dr. Wada is a professor emeritus of Tōkyō University, now teaching at Nippon Daigaku, Tōkyō.

1.1.16 WADA Sei 和田清, **Tōashi ronsō** 東亞史論藪 (Collected essays on East Asian history), Seikatsusha, 1942, 579 pp., maps.

Dr. Wada groups these pieces under China (10 items, to p. 208), Japan (2), Manchuria-Mongolia (6, pp. 241–420), and finally "peoples" and "historical writings". Those on China range through the Ming and Chʻing periods, touching both highly specific topics and some very broad ones, e.g. general essays on the Ming and Chʻing periods. The following three articles in particular should be of interest for the modern period:

1. "Ri Kō-shō to sono jidai" 李鴻章とその時代 (Li Hung-chang and his times), pp. 139–155.—A transcript of a lecture with a selected bibliography, originally published in **Tōa mondai** 東亞問題, 3.1 (Jan. 1941).
2. "Shina kansei hattatsushijō no santokushoku" 支那官制發達史上の三特色 (Three characteristics of the historical development of political organization in China), pp. 156–168.—Originally published as the prefatory chapter of Wada Sei, ed., **Shina kansei hattatsushi** 支那官制發達史, vol. 1, Chūō Daigaku 中央大學, 1942.
3. "Kōtō rokujūshi ton no mondai ni tsuite" 江東六十四屯の問題について (On the problem of the Chiang-tung liu-shih-ssu tʻun), pp. 380–420, with a map.—Originally published in Shigakukai, ed., **Tōzai kōshō shiron** 東西交渉史論 (Treatises on the history of East-West relations), Fuzambō, 1939, 2 vols., 1410 pp. (item 9.8.11). This is a scholarly article on Sino-Russian conflicts concerning the Manchu bannermen colonies on the east bank of the Amur River, which remained under Chʻing jurisdiction after the Treaties of Aigun (1858) and of Peking (1860). See review by Mimura Keikichi 三村啓吉, in **Shigaku zasshi**, 54.12 (Dec. 1943), 1413–1419.

1.1.17 HATTORI Unokichi 服部宇之吉, **Shina kenkyū** 支那研究 (Researches on China), Kyōbunsha 京文社, enlarged ed. 1926, 670 pp. (1st ed. 1916).

A book consisting of 25 essays on various topics like Chinese political thought, Han Fei-tzu 韓非子, constitutional preparations, Chinese national character, the moral crisis, effect of Japanese culture, pacifism, geomancy, etc., by a leading scholar of Chinese studies in Japan. While Dr. Hattori ranges widely over the field of classical learning, his early experience in China enables him to write penetrating essays on "The duties of local officials in China" (pp. 10–49, which describes the inner working of a yamen), and "An inside view of the first revolution" (pp. 210–243, which appraises the elements on the scene in 1911).

1.1.18 IZUSHI Yoshihiko 出石誠彦, **Tōyō kinseishi kenkyū** 東洋近世史研究 (Researches in the modern history of East Asia), Taikandō 大観堂, 1944, 387 pp.

Thirteen scholarly essays on various aspects of Chinese and Manchurian history and Japanese relations thereto—including the adoption of Western educational systems, the 1898 reforms, Sun's Three Principles, British and Russian policy, etc. Most of these are well-balanced survey articles, though sometimes tinged with patriotic sentiments. An essay on Kʻang Yu-wei (pp. 179–218) is a scholarly pioneer analysis of Kʻang's memorials, perhaps overly sympathetic.

1.1.19 YANO Jinʻichi 矢野仁一, **Kindai Shina no seiji oyobi bunka** 近

8

代支那の政治及び文化 (Politics and culture of modern China), Idea Shoin イデア書院, 1926, 420 pp.

These twelve miscellaneous essays include a lengthy study of tea in China (pp. 218–319), and others on opium, firearms, bandits, extraterritoriality, abolition of the unequal treaties, and Sino-Japanese relations, as well as more general essays on Chinese politics and society. Some sources are noted in the text. For further studies by Dr. Yano, see 1.3.3–5, 2.1.11–12, 2.2.1 and 4.7.1–3.

1.1.20 YANO Jin'ichi 矢野仁一, **Gendai Shina gairon** 現代支那概論 (A general discussion of contemporary China), Meguro Shoten 目黒書店, 1936, 2 vols.—**Ugoku Shina** 動く支那 (China in motion), 304 pp.; **Ugokazaru Shina** 動かざる支那 (Unmoving China), 308 pp.

The first of these volumes of historical essays (some of which were previously published in **Gaikō jihō** 外交時報 and other journals) dwells mainly on traditional elements now in transition: frontier problems in Outer and Inner Mongolia, Tibet, Sinkiang and Yunnan and their historical background, bandits and secret societies, and North China as a separate region. The second volume is even more diverse, but deals with the persistence of traditional things in seemingly modern developments: the stationariness of Chinese society, republicanism vs. monarchy, extraterritoriality, *wang-tao* 王道 (especially in the puppet East Hopei "government" of Yin Ju-keng 殷汝耕), Japanese policy, Russian aid, etc. Seen in retrospect these popular articles retain rather uncertain historical value, except as they may indicate certain misappraisals which later influenced Japanese continental policy. See review by Ono 小野 in **Shirin**, 21.3 (July 1936), 659–662.

1.2 SURVEYS OF GOVERNMENT

Note: Item 1.2.1 is one of the monuments of Japanese study of modern China but primarily a reference work. Item 1.2.2, on the other hand, has many of the same merits but in briefer compass. The remaining items begin a long parade of Japanese compilations, arranged chronologically, which is continued below in sec. 7.2. Each of them in its day, whether labelled political or economic, sought to capture China and the China problem within the printed page. Our examination would indicate that these heavy tomes contain much raw material for the historian.

1.2.1 Rinji Taiwan Kyūkan Chōsakai 臨時臺灣舊慣調査會 (Temporary commission of the Taiwan Government-general for the study of old Chinese customs), **Shinkoku gyōseihō** 清國行政法 (Administrative laws of the Ch'ing Dynasty), 6 *kan* 卷 in 7 *satsu* 冊 with one separate volume as general index, Tōkyō and Kōbe: 1910–14.

This basic study, the fruit of ten years expert research on the **Ta-Ch'ing hui-tien** 大清會典 (Collected statutes of the Ch'ing dynasty) and related sources, has been relatively unknown and unused by American specialists on China.

The first *kan* published in 1905 in one volume, was revised and reissued in 1914 (2 vols., 302 and 326 pp.). It constitutes a general survey, from the viewpoint of modern legal science, of the Ch'ing administrative structure. There is a 14-page bibliography. The succeeding treatises are as follows: *kan* 2 and 3 (Kōbe 1910, 528 and 534 pp.), on domestic administration; *kan* 4 (Tōkyō 1911, 462 pp.) on domestic and military administration; *kan* 5 (Tōkyō 1911, 338 pp.), on judicial and fiscal administration; *kan* 6 (Tōkyō 1913, 394 pp.), on fiscal administration. A separate volume of 162 pp. was added as a general index (Tōkyō 1914), but this index is not very thorough.

This work originated in the efforts which began after Taiwan became Japanese territory to conduct research on Chinese institutions which would be useful for the colonial administration. The result was the establishment, in 1901, of the Rinji Taiwan Kyūkan Chōsakai as a research commission. The First Section (*Dai-ichi-bu* 第一部) was for study in the sphere of institutions and the Second Section (*Dai-ni-bu* 第二部) for study in the fields of agriculture, industry and commerce. **Shinkoku gyōseihō** was produced as a report of the First Section. Eagerly encouraged by Gotō Shimpei 後藤新平, then Civil Governor of Taiwan and president of the commission, Oda Yorozu 織田萬, professor of public law in Kyōto Imperial University, took general charge of the research. Kano Naoki 狩野直喜, an authority in the study of Chinese literature and history, worked as his associate. There were also three other able scholars who participated as assistant researchers: Asai Torao 淺井虎夫, Higashikawa Tokuji 東川德治, both specialists in Chinese legal history, and Katō Shigeshi 加藤繁, an expert in Chinese economic history. Influence of this basic research may be seen in the later works of these participants (noted elsewhere in this volume). The result of their labors was a modern-minded application of legal concepts to the traditional forms and procedures of Chinese government, taking account of late-Ch'ing reforms, quoting both Chinese and Japanese and also Western reference works, and touching on informal as well as statutory institutions. This was, in effect, to put the structure and functions of the Ch'ing regime through an analytic sieve, using the concepts of early 20th century Japanese and German law and politics. This job should now be done again from a more advanced and sophisticated viewpoint.

1.2.2 HATTORI Unokichi 服部宇之吉, **Shinkoku tsūkō** 清國通考 (A general account of the Ch'ing government), Sanseidō 三省堂, 1905, 2 vols., 166+204 pp.

These two thin volumes give a remarkably penetrating picture of the imperial administration of China in its last years. Vol. 1 describes the historical development and functions of the *Chün-chi-ch'u* 軍機處 or Grand Council (including a valuable summary of its procedure), the *Nei-ko* 內閣 or Grand Secretariat, the Censorate, and the Hanlin Academy (pp. 1–75), and then describes the examination system at length, concluding with an appraisal of its social influence and the current changes in it. Vol. 2 analyzes the administrative organization of the Ch'ing central government, taking into account both its historical origins and the new agencies and changes made by post-Boxer reforms. After analyzing the policy-forming, executive and supervisory organs in general, the author then deals with the organization and personnel of the various Boards (at this time the new Foreign Office and Ministry of Commerce, in addition to the Six Boards); the selection and promotion of officials; the yamen clerks and other underlings, together with the officials' personal advisers (*mu-yu* 幕友). Vol. 2 from p. 100 gives a history of the Six Boards and provincial administration from Ch'in to Ch'ing. The great value of these volumes lies in their analysis of the actual procedures of the imperial regime from a modern Japanese point of view. The author was first sent to China by the Ministry of Education in 1899, in time to participate in the Boxer seige (see his diary, **Pekin rōjō nikki**, 2.7.11), and later returned to teach in Peking for six years, until 1909. Subsequently he became the leading specialist on the Chinese classics at Tōkyō Imperial University and dean of the Faculty of Letters, as well as head of the Tōhō Bunka Gakuin. See also his essays on China, 1.1.17.

1.2.3 INOUE (NARAHARA) Nobumasa 井上 (楢原) 陳政, for the Ōkurashō 大藏省 (Ministry of Finance), **Uiki tsūsan** 禹域通纂 (A general compendium on China), 1888, 2 vols., 1220 and 774+353 pp., illus., many tables (stamped "not for sale" 禁賣買).

This grandfather of modern Japanese gazetteers of China was based on the author's assiduous observations during 6 years travel in the provinces and treaty ports during 1882–1887, much of it in the company of the former Chinese minister to Japan, Ho Ju-chang 何如璋. The author had been ordered to begin his Chinese studies at the Tōkyō Chinese legation in 1878, where Huang Tsun-hsien 黃遵憲 was counsellor. Returning to China with Ho, and sent by the Ministry of Finance, he found occasion to visit almost all the major provinces and centers, and was present at Foochow in 1885 when the French navy succeeded in anchoring peacefully near the Chinese navy and then somewhat unexpectedly sank it. (Ho was removed as Minister of the Navy). The 72-page table of contents would indicate that the compiler went through Peking with the **Collected Statutes** in hand. Vol. 1 deals mainly with the central and provincial administration, including public finance; vol. 2 touches on miscellaneous items including the official postal service, *pao-chia*, communications, gentry, *mu-yu* 幕友, yamen clerks 胥吏, education, the military establishment in many aspects (pp. 384–574), trade and the ports, water conservancy and waterways, etc. The last 87 pages are on sericulture, symptomatic of Japan's special concern with this topic. This early eye-witness potpourri should have considerable interest historically. It is full of succint institutional and historical explanations, together with the author's personal observations and opinions as well as translations of current discussions by Chinese officials and scholars. The author (1862–1900) later studied in England, entered the foreign service, and was in Peking at the time of the Boxer rising, where he died on July 23 from tetanus caused by bullet wounds. See the biography in **Tai-Shi kaikoroku** (4.2.10), 2.244–247.

1.2.4 Shinkoku Chūtongun Shireibu 清國駐屯軍司令部 (Headquarters of the army stationed in China), comp., **Pekin shi** 北京誌 (A gazetteer of Peking), Hakubunkan, 1908, 926 pp., maps, photos.

Produced under military orders, mainly in 1905, by a research group which included Hattori Unokichi 服部宇之吉 as chief editor, with Yano Jin'ichi and a dozen others as collaborators, this volume compiles useful knowledge on the history, geographic circumstances, palaces, temples, yamen, population, Manchu aristocracy, bannermen, etc., etc. (see contents, 44 pp.) of the Ch'ing capital. Perhaps its greatest value lies in the accounts of the Ch'ing official system and its functioning (pp. 111–164), the legal system (165–239), military (240–268), education (269–325), police (326–386) and corresponding economic aspects in the last years of the dynasty—a period for which comparable Chinese gazetteers are not numerous. Both Peking as a city and the imperial government are dealt with. The companion volume on Tientsin (7.2.12) is largely economic in subject matter.

1.2.5 ŌMURA Kin'ichi 大村欣一, for the Shanghai Tōa Dōbun Shoin, **Shina seiji chiri shi** 支那政治地理誌 (Political and geographic gazetteer of China), Maruzen, vol. 1, 1913, 968 pp.; vol. 2, 1915, 1025+21 pp., many maps.

This large compendium brings together a vast amount of geographic data on place names, river conservancy, canals, steamer routes, and the like, but will be of greater interest perhaps for its data on government (vol. 1, from p. 382), especially local government (pp. 458–494, 520–539). Late Ch'ing central government receipts and expenditures are dealt with in detail (581–663), as well as the main categories of taxation. The Republican government structure is then described (915–68), its loans, taxes, currency (in vol. 2), and the banks, railroads and communications of China just after the Revolution.

1.2.6 BABA Kuwatarō 馬場鍬太郎, **Shina keizai chiri shi, seido zempen** 支那經濟地理誌制度全編 (Gazetteer of Chinese economics and geography, volume on institutions, complete), Shanghai: Uiki gakkai 禹域學會, 1928 (4th printing, 1933, 1464 pp.).

This volume is in continuation of an earlier publication, **Shina keizai chiri shi, kōtsū zempen** 交通全編 (ibid., volume on communications, complete), 1922, 1406 pp. Both were produced by a professor who was at the Tōa Dōbun Shoin from 1916 and who in this volume sought to up-date earlier works like **Shinkoku gyōseihō** (1.2.1), although he alone was hardly as competent as the compilers of that work. As outlined in his table of contents (66 pp.), under (1) "Administration" 行政 he deals with all aspects of both the Ch'ing and the Republican governmental systems, central and local; under (2) "Treatises on various important institutions", he surveys the legal system, extrality, police, army, education, religion, and races; and under (3) "Fiscal and special commercial institutions," he also goes into great detail in many directions, including likin and currency. Similarly the volume on communications of 1922, above-noted, provides a vast body of data on water transport and railroads, and also on roads and posts.

1.3 SURVEYS OF FOREIGN RELATIONS

Note: Except for the first two items, these works are largely outdated; after the first seven items, they are arranged chronologically by date of issue and form a rather unimpressive list, possibly because comprehensive surveys of China's international relations have seldom been attempted by first-class scholars. The outstanding Japanese works in diplomatic history, by men like Tabohashi Kiyoshi, e. g. item 4.4.3, and Yano Jin'ichi, e. g. item 4.7.1, will be found in sec. 4, particularly 4.4. Major works on the treaty system are in sec. 3.7. Thus this section consists mainly of left-overs. Items 1.3.2 (Momose and Numata) and 1.3.1 (Ueda, ed.)

will be found of value, but the later items are largely dead wood, listed here (contrary to our practice in other sections) only for their historical interest. For recent articles which break new ground in certain fields of diplomatic history, see sec. 2, particularly 2. 3.

1.3.1 UEDA Toshio 植田捷雄, ed., **Gendai Chūgoku o meguru sekai no gaikō** 現代中國を繞る世界の外交 (Contemporary China in world diplomacy), Nomura Shoten 野村書店, 1951, 330 pp.

This recent symposium of research studies on modern China's foreign relations contains the following articles:

1) Banno Masataka 坂野正高, "Dai-ichiji taisen kara Go-sanjū made, kokken kaifuku undōshi oboegaki" 第一次大戰から五・卅まで——國權回復運動史覺書 (From World War I to the May 30th movement, notes on the history of the rights recovery movement), pp. 1–67. From the 21 Demands to about 1926, with an analysis of the various social classes concerned with the movement.

2) Ueda Toshio, "Taishō yonen Nikka nijūikka jōyaku to Manshū jihen, tokuni jōyaku no kōryoku o chūshin to shite" 大正四年日華二十一箇條約と滿州事變——特に條約の效力を中心として (The Sino-Japanese treaties of 1915 following the Twenty-one Demands, and the Manchurian incident, with special reference to the validity of the treaties), 69–105. An examination of the legal and diplomatic relationships.

3) Ichimata Masao 一又正雄, "Kokusai remmei ni okeru Manshūjihen oyobi Shanhaijiken shori no gaikan" 國際聯盟における滿州事變および上海事件處理の概觀 (A survey of the disposal in the League of Nations of the Manchurian and Shanghai incidents), 107–183. Details of the negotiations in and under the League, 1931–33.

4) Uchida Naosaku 内田直作, "Zai-Ka Eikoku shōsha no gaikōjō no katsudō, sono dentōshugiteki seikaku" 在華英國商社の外交上の活動——その傳統主義的性格 (The activity regarding foreign policy of British trading companies in China, with special reference to their traditionalistic character), 185–234. Considers the East India Co., private firms, China Association, etc.

5) Irie Keishirō 入江啓四郎, "Ka-So yūkō dōmei jōyaku ron" 華ソ友好同盟條約論 (On the Sino-Soviet treaty of amity and alliance), 235–300. Analysis of the various aspects of the 1945 treaty.

6) Ōhira Zengo 大平善梧, "Beikoku no tai-Ka gaikō seisaku" 米國の對華外交政策 (United States policy toward China), 303–330. An essay on the principles of "Formulism", "Rationalism", and "Universalism" underlying American policy.

Reviewed by Onoe Masao 尾上正男 in **Kokusaihō gaikō zasshi**, 51.3 (July 1952), 318–320.

1.3.2 MOMOSE Hiromu 百瀬弘 and NUMATA Tomoo 沼田鞆雄, **Kindai Shina to Igirisu** 近代支那と英吉利 (Modern China and Great Britain), Keisetsu Shoin 螢雪書院, 1940, 390 pp.

Part one of this survey, "Political" (by Numata, down to p. 152), summarizes the well known East India Co.-to-Nationalist Revolution cycle in Anglo-Chinese relations. Part two, "Economic" (by Momose) surveys the growth of British trade and investment (to p. 242). The most interesting is part three, "Cultural" (by Momose), which breaks new ground in a flexible-minded study of the shock to traditional Chinese ways (especially the examination system) administered by Western learning, the activity of British Protestant missions both before and after the Boxer incident, and the eventual anti-British and anti-Christian movements of the 1920s (pp. 337–353). The author is well informed on the British literature and cultural institutions concerning China and goes far to present a balance sheet of the British record there.

1.3.3 YANO Jin'ichi 矢野仁一, **Kinsei Shina gaikōshi** 近世支那外交史 (History of modern Chinese foreign relations), Kōbundō, 1930 (3rd printing, 1938, 946 pp.+index 24 pp., 7 maps).

This large volume is divided into 72 chapters which summarize a series of episodes or historical phases that may be grouped roughly as follows: early contact with Portugal, Spain and Holland (ch. 1-14), British trade and diplomacy through the Opium War (15-26), the treaty conflicts and settlements down to the 1860s (27-47), missionary questions (48-54), Russian relations (55-63), foreign relations connected with peripheral states of Asia—Burma, Tibet, etc. (64-72). The narrative is carefully constructed but somewhat tedious and is often intermixed with lengthy moralistic appraisals. Brief source references inserted in the text indicate the use made of Chinese as well as Western materials; but publication of Chinese sources and further research since 1930 have now made much of this account out of date. See reviews by Nohara Shirō 野原四郎 in **Shigaku zasshi**, 41. 10 (Oct. [?] 1930), 1241–1244; by Nishiyama Yoshihisa 西山榮久 in **Tōa keizai kenkyū**, 14.3 (July 1930), 468–470; by Matsuura Kasaburō 松浦嘉三郎 in **Shinagaku**, 6. 1 (Jan. 1932), 162–163. Dr. Yano made his contribution not only by the extensive factual coverage of a wide range of subject matter connected with Chinese foreign relations but also by his great industry as a pioneer research worker, in a field and among materials which had for the most part lain neglected by Japanese scholars. He is now a professor emeritus of Kyōto University.

1.3.4 YANO Jin'ichi 矢野仁一, **Gendai Shina kenkyū** 現代支那研究 (Researches on modern China), Kōbundō, 1923, 351 pp.

A series of studies: "From the Open Door doctrine to the abolition of special interests"; on customs reform; "From the tripartite intervention to the Russian lease of Port Arthur and Dairen"; "Sino-Russian relations, past and present"; Sungari River navigation rights; the Mongolian problem; the Tibetan problem. Sources are occasionally indicated.

1.3.5 YANO Jin'ichi 矢野仁一, **Kindai Shina ron** 近代支那論 (Essays on modern China), Kōbundō, 1923 (8th printing, 1926, 373 pp.).

A reprint of articles and addresses in which Dr. Yano discusses a score of historical topics or problems of contemporary interest, many of them still matters of debate: whether China originally possessed Manchuria, Mongolia and Tibet; whether China is a country in the usual sense; what are her boundaries; her relations with tribal peoples; the fall of the dynasty; the Twenty-one Demands; international control for China; the fruits of the Washington Conference; Western studies (*hsi-hsueh* 西學) in China, and the like. While written for the general reader, the book is based on the author's factual research and his interpretations are often suggestive and stimulating, though not always persuasive.

1.3.6 ŌKUMA Tadashi 大熊眞, **Bakumatsuki Tōa gaikōshi** 幕末期東亞

外交史 (History of East Asian foreign relations in the late Tokugawa period), Kengensha 乾元社, 1944, 285+7 pp.

By a member of the Foreign Office research staff, the first four chapters of this volume have the particular merit of trying to treat the opening of China and of Japan together as a single phenomenon in East-West contact (pp. 1–125. The remainder is on Japan). However, the brevity and superficiality of the narrative which combines Chinese and Japanese events, and the use of textbook-type sources, severely limit the result achieved, while the 5 chapters on the opening of Japan are conventional and disregard the alleged need to view Japan and China together. See review by Banno Masataka 坂野正高 in **Tōyōbunka kenkyū,** 2 (Sept. 1946), 61–64.

1.3.7 SAITŌ Yoshie 齋藤良衞, **Kinsei Tōyō gaikōshi josetsu** 近世東洋外交史序說 (Introduction to modern Far Eastern diplomatic history), Ganshōdō, 1927, 509 pp.

This survey written in colloquial style by a scholarly Foreign Office official covers ground now well-trodden, from the opening of China to the Anglo-German agreement of 1900; but it has the particular merit that the author apparently had informal and unacknowledged access to Foreign Office documents as background for his work. The book was produced primarily as a cram book for applicants for diplomatic service examinations; as such it can be considered a well-balanced general survey for its day. No footnotes, bibliography or index.

1.3.8 TANAKA Suiichirō 田中萃一郎, **Tōhō kinsei shi** 東邦近世史 (Modern Far Eastern history), vol. 1, Tōhō Kyōkai 東邦協會, 1900, 459+14 pp.; vol. 2, Maruzen, 1902, 656+14 pp.

Though the author was keenly conscious of Japan's position as a late-comer in Far Eastern international rivalries and expressed an expansionist sentiment in the preface, this pioneer work in the field of international history in Japan was a factual and sober survey which made rather extensive use of the Chinese-Japanese-Western sources then available. The narrative treats India and South-east Asia as well as the Far East from 1498 down to 1900. This book has been considered a "classic" and, being also a readable general account, was republished in 3 vols. in the Iwanami bunko 岩波文庫 series, 1939–43. The author was at Keiō University.

1.3.9 TATSUMI Raijirō 巽來治郎, **Kyokutō kinji gaikōshi** 極東近時外交史 (History of contemporary Far Eastern foreign relations), Waseda University, 1910, 838+app. 112 pp.

Produced just before the Chinese Revolution, this volume provides a contemporary view of the *status quo ante* in power politics; it covers the Sino-Japanese War and Boxer crisis (to p. 387), and the Russo-Japanese War and its aftermath, and concludes with a section on the application of international law to the events of this troubled period (pp. 726–838), and an appendix of documents.

1.3.10 HASHIMOTO Masukichi 橋本增吉 and MAKINO Yoshitomo 牧野義智, **Shina no gaikō kankei** 支那の外交關係 (China's foreign relations), Gaikōjihōsha 外交時報社, 1920, 2 vols., 408 and 485 pp. (Vol. 1 by Hashimoto, vol. 2 by Makino).

15

Volume 1, to 1895, is a conventional account of foreign relations very similar to Dr. H. B. Morse's volumes of 1910–18. Volume 2 approaches the twentieth century by reference first to China's traditional world view, and then discusses the reform movement and revolution, the Twenty-one Demands, China in the World War and at Versailles, and concludes with a survey of China's current international problems (pp. 273–485). The last should be of interest in connection with the Japanese view about 1920.

1.3.11 KUBOTA Bunzō 窪田文三, **Shina gaikō tsūshi** 支那外交通史 (An historical survey of China's foreign relations), Sanseidō 三省堂, 1928, 506 pp.

A narrative by an ex-foreign service official, in 21 chapters from the Canton trade to the aftermath of World War I, stressing treaties and conflicts pretty much along the lines of Western blue-book history, without notes, bibliography or index; now superseded by later works. Some parts are apparently literal translations from H. B. Morse's **International Relations of the Chinese Empire.** The author was Japanese Consul General in Tientsin in 1913–14.

1.3.12 INASAKA Katashi 稲坂碕, **Kinsei Shina gaikōshi** 近世支那外交史 (History of China's modern foreign relations), Meiji University, 1929, 420 pp.

This conventional survey, Opium War to Nine-power Treaty, is mainly devoted to the period after 1894 and questions affecting Japan. It has no footnotes, bibliography or index, and little scholarly value.

1.3.13 NAKANO Hidemitsu 中野英光 (Major of Infantry and officer of the General Staff), contributor (*kikō* 寄稿), **Saikin Kyokutō gaikōshi** 最近極東外交史 (History of recent Far Eastern international relations), pub. by the Kaikōsha 偕行社 (the Army Club), 1931, 2 vols., 558 and 701 pp.

A factually competent but conventional and now superseded history of China's foreign relations, i.e. power politics centered around China. The first volume comes down to the post-Boxer period and the second runs from the Russo-Japanese War down to the period of the Nationalist Government. The narrative is concisely written and well organized; sources are not indicated.

1.3.14 INOMATA Tsunao 猪俣津南雄, **Kyokutō ni okeru teikokushugi** 極東に於ける帝國主義 (Imperialism in the Far East), Kaizōsha 改造社, 1932, 423 pp. (Keizaigaku zenshū, 24).

Drafted mainly in 1929, this is an orthodox but very explicit effort to put modern Chinese history in a Marxist-Leninist framework, applying all the terminology. Two chapters, apparently on the recent Japanese expansion on the mainland, were editorially excised during publication, and in their place the author's "Nippon no dokusen shihonshugi" 日本の獨占資本主義 (Monopolistic capitalism in Japan) was included.

1.3.15 AOYAGI Atsutsune 青柳篤恒, **Tōa gaikōshi ron** 東亞外交史論 (An historical discussion of East Asian international relations), Sekaidō

Shoten 世界堂書店, 1938 (3rd printing, 1942, 402 pp.). Prefaces by Hsü Shih-ch'ang 徐世昌 and Ōkuma Shigenobu 大隈重信 dated 1918.

Originaly written at the time of World War I and published at first under the title **Kyokutō gaikōshi gaikan 極東外交史概觀** (Survey of Far Eastern international relations) in 1938, this volume had its title changed in this 3rd printing, to suit the wartime market. It recounts the usual saga of China's tribulations in foreign affairs from 1840 to the end of Yuan Shih-k'ai without adding significantly to present knowledge on most topics. The latter portion, however, undoubtedly benefits from the fact that the author served as an adviser to Yuan Shih-k'ai under Dr. Ariga Nagao 有賀長雄. He taught Chinese language and problems for many years at Waseda University, from the late Meiji period, and looked after overseas Chinese studying in Japan. He also assisted Ōkuma Shigenobu as an interpreter or translator in his contact with Chinese notables.

1.3.16 SANO Kesami 佐野袈裟美, **Shina kindai hyakunen shi 支那近代百年史** (History of China during the last hundred years), Hakuyōsha, vol. 1, 1939, 519 pp.; vol. 2, 1940, 564 pp.

This survey of the century from the Opium War to the China Incident lays its stress on foreign relations rather than on domestic or institutional history, and might better be called a history of the great powers' encroachment on China— dollar diplomacy and "incidents". Sources, as noted in text and bibliography, appear to have been rather miscellaneous. While this work has been considered useful for Japaneses beginners, for research purposes it is a rather pedestrian compilation of facts and statistics, based uncritically on secondary writings and tinged with a schematically economic interpretation.

1.4 ON THE PERIODIZATION OF CHINESE HISTORY

Note: This section presents a number of books and articles of interest for their theoretical or general ideas or interpretations. In our ideological age, these often have implications for policy, or at least for political attitudes. According to these general views, many historians are grouped into camps or schools of interpretation. They formulate their research projects and search for facts accordingly. To do so unconsciously does not avoid the issue, and so most Japanese historians are well aware of these general theories even when they do not write about them. They form part of the present intellectual climate in Japan, which is of course uniquely Japanese. Broadly speaking, we suggest that two of the elements visible in this climate are (1) a greater awareness of the long sweep of Chinese history than is enjoyed in the West, with a deeper grasp of the political and cultural traits which accompany the intensive, irrigated agriculture of a peasant-based society; and (2) a wider acceptance of Marxist (not necessarily Stalinist or even communist) ideas concerning the socio-economic transition from early to modern times. As indicated in the Introduction above, our references to Marxism in this volume are usually made in the broad sense, to the general body of ideas stemming from 19th century writings of Marx and Engels which form a small but important part of Western intellectual history. How far the use of such ideas may have involved the user in further acceptance of Leninist ideas of imperialism and revolution, or in dogmatic submission to Stalinism and the current Communist Party line, is always a question, which we have not always been in a position, nor taken it upon ourselves, to decide. We have noted in various sections a number of party-line products, most of which are distinguished for the little that they have to offer an enquiring historian. These products of communist doctrine appear to be rated less highly in post-war Japan's free market-place of ideas than writings of independent researchers whose Marxism, if any, is part of their own individual mixture of ideas. This section and the one following, 1. 5, are of interest in connection with sec. 8, where somewhat similar questions of interpretation have arisen (see sec. 8. 2 in particular).

The first items below represent largely a general view of which Professor Naitō Torajirō of Kyōto Imperial University has been a chief protagonist. Subsequent items indicate the wide diversity of views presented in recent books and articles. The question of periodization (e. g., has Chinese society been "feudal"?) will no doubt remain important for its political as well as its historical implications.

1.4.1 NAITŌ Konan (Torajirō) 内藤湖南 (虎次郎), **Shina ron** 支那論 (Essays on China), Osaka: Sōgensha, 1938 (7th printing, 1943, 378 pp.).

As published posthumously in 1938 this volume combined Dr. Naitō's **Shina ron** of 1914 with his **Shin Shina ron** of 1924 (pp. 225-334) and a lecture of 1928 on "The cultural life of modern China" 近代支那の文化生活 (see Naitō's **Tōyō bunkashi kenkyū**, 1. 4. 2). The first essay of 1914 states his famous thesis that China until late T'ang had a government by aristocracy, the emperor being merely the topmost in status among the levels of aristocratic families; whereupon the demise of the latter made the late-T'ang and later emperors autocrats ruling through a full-fledged bureaucracy—this latter period is labelled by the author as "modern". While republican government will be new and different, certain historical problems will persist—relations with non-Chinese, the domestic administrative system, finance, political morality, and national policy. These problems are freely discussed against their historical background. Unlike his **Shinchō suibō ron** of 1912, 2. 1. 1, this survey seems hopelessly sceptical as to the future of the revolution. Dr. Naitō is bitterly critical of the reactionary trend of Yuan Shih-k'ai's regime, and as an alternative to its current orientation even sees some possibility of an international control of China.

The essay of 1924 strongly criticizes the ineptitude of Japan's China policy, and ominously predicts an almost inevitable and catastrophic Sino-Japanese clash and a possibility of Japanese-American collision. In analyzing China's sociopolitical structure Naitō doubts the genuine depth of her modern nationalism, and strikingly minimizes the probable consequences of the rising communist movement. He foresees that Japanese economic activities in China (especially the penetration of small-capital merchants into the interior) may work as a stimulus to the reconstruction of the Chinese socio-economic structure from within (leading eventually to total industrialization in the modern sense.)

These essays, condensing the insights of an influential historian on China's historical and contemporary problems, provide a key by which to discriminate among main trends in Japanese study of China. Naitō's misjudgement of the modern Chinese situation might be ascribed partly to an historically-minded classicist's contempt for republican chaos and partly to an inadequate analysis of the inner social structure of the villages and of the interrelationship between the local bosses and the mandarinate or warlords. Thus he propounds a thesis that state and society in China are utterly separate and indifferent to each other. See reviews by Tachibana Shiraki 橘樸 in pp. 360–408 of his **Shina shisō kenkyū**, 6.1.3; and by Nohara Shirō 野原四郎 in **Chūgoku hyōron 中國評論**, 1.4 (Nov.–Dec. 1946), 35–42. See also the bibliography of Dr. Naitō's writings (21 pp.) and appreciations of his career in **Shinagaku**, 7.3 (July 1934), 447–542.

1.4.2 NAITŌ Torajirō 內藤虎次郎, **Tōyō bunkashi kenkyū 東洋文化史研究** (Studies in the cultural history of East Asia), Kōbundō, 1936, 383 pp., illus.

Among these lectures and essays, three or four—on "The cultural life of modern China", Western culture vs. the culture of the native people, the Chinese view of China's future, and a criticism of the communist movement ("Go back to [the real] China!")—serve to apply Professor Naitō's ideas to the contemporary scene and are useful keys by which to understand his idea of periodization in Chinese history. See review by Naitō Boshin 內藤戊申 in **Tōyōshi kenkyū**, 1.5 (June 1936), 473–476.

1.4.3 OTAKE Fumio 小竹文夫, "Shinashi no jidai kubun, gendai Shina no igi" 支那史の時代區分──現代支那の意義 (The periodization of Chinese history, the meaning of the term modern China), **Shina kenkyū**, 44 (March 1937), 19–34.

A thoughtful early discussion of interpretations made by Naitō, Yano and others. Otake rejects Yano's view that China's pre-modern history might as well be classed as a single ancient period of unchanging institutions. He roughly follows Naitō's widely accepted interpretation of late-T'ang and Sung as the foundation period of "modern" (*kindai* 近代) Chinese society but inclines, for many stated reasons, to view the Opium War as the main turning point into contemporary times (*gendai* 現代), as distinguished from the "modern" age in Naitō's sense.

1.4.4 WADA Sei 和田清, "Shina no kokutai ni tsuite" 支那の國體について ("De la constitution nationale de la Chine"), **Orientalica**, 1 (1948), 97–113.

An eminent historian here compares the Chinese polity with that of Japan and the Western world to bring out its distinctive features; appraises the persistence of Chinese ideas of law, government, religion, etc. in the modern period; and takes a look at China's non-feudal bureaucratism and family-based principle of communal association as affecting the growth of nationalism. This brief survey is noteworthy as a summary of the views of an experienced, commmon-sense historian, concerning the controversial problems of Chinese political and social structure as well as the periodization of Chinese history.

1.4.5 MIYAZAKI Ichisada 宮崎市定, **Tōyōteki kinsei 東洋的近世** (The modern age of the Oriental type), Osaka: Kyōiku Taimusu Sha 教育タイムス社 (Educational Times Co.), 1950, 208 pp.

This small essay volume seeks to give the general reader a picture of the intercommunication between East and West, and the "modern" cultural, social, economic and political development of China, with attention also to the growth of nationalism. As a scholar deeply versed in China's past, Professor Miyazaki dwells especially on China's historical background, and this volume is noteworthy as a systematic and challenging presentation of the author's idea of periodization. Using a general three-stage scheme of world history, he sees in the ancient period, unity; in the medieval, disunity; and in the modern, reunification. He asserts that China entered the "modern" age in the Sung period, (which corresponds to the European modern age that began in the Renaissance in the 14th century) but remained in the stage of *ancien régime* until the late Ch'ing period. According to Prof. Miyazaki, Europe after the industrial revolution must be defined as in the "most recent" age, to be distinguished from the preceding "modern" age. Throughout this book the influence of Dr. Naitō Torajirō is considerably visible, though the latter rather proposed to establish an autonomous scheme of periodization for Chinese history, independent of and so not comparable to the European historical development (see p. 6 in Naitō's **Shina jōkoshi** 支那上古史 [The ancient history of China], Kōbundō Shobō, 1944, 344 pp.). Reviewed by Araki Toshikazu 荒木敏一 in **Shirin**, 34.4 (Aug. 1951), 405–408. The author is now professor of Oriental history at Kyōto University.

1.4.6 MIYAZAKI Ichisada 宮崎市定, **Ajiashi gaisetsu アジア史概説** (A general survey of Asian history), Jimbun Shorin 人文書林, vol. 1, seihen 正編, 1947,; vol. 2, zokuhen 續編, 1948, total 388 pp.

This popular short survey devotes the first volume to the origins and early contacts of India, China and peripheral peoples; and the second volume to the "modern" (i.e., from the Sung period) growth of nationalism in China, cultural influences from Islam and Europe, and Japan's historical position. These broad essays are perhaps mainly of interest for China's modern history as expressing one view of her historical periodization, which is more fully discussed in the author's **Tōyōteki kinsei** (1.4.5). A similarly useful statement is in Professor Miyazaki's wartime volume **Tōyō ni okeru soboku shugi no minzoku to bummei shugi no shakai 東洋に於ける素朴主義の民族と文明主義の社會** (Simple races and civilized societies in East Asia), Fuzambō, 1940 (2nd printing, 1942, 197 pp.), which is largely confined to the period down to the Ming dynasty, but sets forth a general interpretation of Chinese history. See review by Saguchi Tōru 佐口透 in **Tōyōshi kenkyū**, 11.1 (Sept. 1950), 66–69.

1.4.7 MIYAZAKI Ichisada 宮崎市定, "Sōdai igo no tochishoyū keitai" 宋代以後の土地所有形體 ("The demesne in China in the Sung Period and after"), **Tōyōshi kenkyū**, 12.2 (Dec. 1952), 97–130.

This article seeks to provide factual evidence for the author's idea of periodization in Chinese history. He sees the "modern" age as beginning in the Sung period and so is critical of the current interpretation that a medieval-feudal society began in the Sung period. Differing from Prof. Sudō Yoshiyuki, Dr. Miyazaki stresses the point that demesnes of large-scale landlords of the Sung period and after were composed of great numbers of dispersed and parcelled-out segments of land, unlike the concentrated land ownership of the preceding periods; that the peasants (*tien-hu* 佃戶) were not serfs or villeins but free men whose relations with their landlords were regulated through contracts; that there emerged a class of land-agents (*yeh-chu* 業主) between peasants and landlords; and that landlords rather tended to become *rentiers* and speculators. It will be seen that this view of the *tien-hu* system contradicts that accepted by Dr. Niida Noboru and other participants in this recent and continuing discussion. See also 7.15.8.

1.4.8 MATSUMOTO Yoshimi 松本善海, "Chūgoku shakaishi no aratanaru kadai, sono hashigaki" 中國社會史の新たなる課題，そのはしがき ("The New Problems in the Social History of China", an introductory note), **Shigaku zasshi,** 58.3 (Aug. 1949), 302–312.

In the context of the postwar reevaluation of history, a leading theoretician on Chinese history discusses the recent trends of research in that field. This brief but sophisticated note may be a useful clue to the scholarly atmosphere in post-war Japan. In this connection, note the article which touched off the post-war polemical reappraisal of the periodization of Chinese history, Maeda Naonori 前田直典, "Higashi Ajia ni okeru kodai no shūmatsu" 東アジアにおける古代の終末 (The end of the ancient period in East Asia), **Rekishi 歷史,** pub. by Shigakusha 史學社, 1.4 (April 1948), 19–31.

1.4.9 NIIDA Noboru 仁井田陞, "Chūgoku shakai no 'hōken' to *fyūdarizumu*" 中國社会の「封建」とフューダリズム ("The 'Fêngchien [封建]' system in China and Western feudalism"), **Tōyō bunka,** 5 (April 1951), 1–39.

This is a polemical approach to the controversial question of "periodization" in Chinese history. Starting from the recent interpretation, based on important factual research by Sudō Yoshiyuki 周藤吉之 (see below), that China became a medieval-feudal society with a kind of serfdom (佃戶制 *tien-hu* system) as its socio-economic substructure, Dr. Niida attempts to deal comprehensively with the problem of feudalism in China. After dwelling upon the socio-legal and historical analysis of the *tien-hu* system as existing from the Sung down to the Ch'ing period, he attacks the competing interpretation put forward by Prof. Kaizuka Shigeki 貝塚茂樹 and others which considers that the *feng-chien* 封建 system of the Chou period corresponded to European feudalism. On the contrary, argues Dr. Niida, the *feng-chien* system was merely the political superstructure of antique-slavery and in itself should be sharply distinguished from feudalism because of the lack of formally stated reciprocal obligations (*Treudienstvertrag*) in the strictly legal sense. China's political structure in later periods, after the Chou, is also distinguished from feudalism, and the perennial persistence of patriarchal relations is stressed. Finally, Confucianism of the Sung period represented by Chu Hsi 朱熹 is analyzed as the ideology of Chinese medieval society, a sort of Oriental natural law which served to stabilize the position of the new ruling class based on the rising *tien-hu* system. This vigorous and cogent article marks a further stage in an unfinished debate. Some points in it were later revised in Dr. Niida's **Chūgoku hōsei shi,** 3.6.1. The following six articles by Mr. Sudō, abovementioned, are noted for reference:

1) "Sō-Gen jidai no denko ni tsuite" 宋元時代の佃戶について (On the *tien-hu* in the Sung and Yuan periods), **Shigaku zasshi,** 44.10 (Oct. 1933), 1245–1279; 11 (Nov. 1933), 1435–1463, originally a graduation thesis.
2) "Sōdai no denkosei" 宋代の佃戶制 (The *tien-hu* system in the Sung period), **Rekishigaku kenkyū,** 143 (Jan. 1950), 20–40, a reappraisal of the thesis.
3) "Sōdai shōen no kanri ni tsuite—toku ni kanjin o chūshin to shite" 宋代莊園の管理について—特に幹人を中心として (Management of the manor by *kan-jen* 幹人 during the Sung), **Tōyō gakuhō,** 32.4 (April 1950), 1–32.

4) "Sōdai kanryōsei to daitochi shoyū" 宋代官僚制と大土地所有 (Bureaucracy and large-scale landownership in the Sung period), **Shakai kōseishi taikei** 社會構成史大系, 9.8.7, series 8, 1950, 152 pp.

5) "Sō-Kin jidai ni okeru shōen to denko no ichikōsatsu—tokuni Chōan fukin ni tsuite" 宋金時代に於ける莊園と佃戶の一考察―特に長安附近について (A study of the manors and tenants in the vicinity of Ch'ang-an in the Sung and Chin periods), **Tōhōgaku**, 2 (Aug. 1951), 51–63.

6) "Sōdai shōensei no hattatsu" 宋代莊園制の發達 ("The Development of the Manor System in Sung Dynasty"), **Tōyō bunka kenkyūjo kiyō,** 4 (March 1953), 3–81. (*See note on page 29.*)

1.4.10 IMAHORI Seiji 今堀誠二, "Chūgoku ni okeru hōkenteki shōkōgyō no kikō, seisan kankei yori mitaru Naimōko no toshi to nōson" 中國における封建的商工業の機構――生產關係より見たる內蒙古の都市ɀ農村 ("Feudal traits of industrial and commercial organizations in China", production-relations in Inner Mongolian cities and villages), **Tōyō bunka,** 3 (July 1950), 1–54.

This highly interpretative article is based on field work by the author in Inner Mongolia, and poses as its starting point the following two hypotheses in Ch. 1: (1) The "feudal" stage of Chinese society (in the Marxist sense) is characterized as "a feudal society to which a family-slavery adheres", which may be called "Asiatic feudalism"; and (2) in this "Asiatic feudalism", the rural production-relations determine the characteristics of the commerce and industry in the cities and towns. Against this theoretical backgronnd, Chapter 2 (20 pp.) deals with the process of growth in commerce and industry, stressing the predominance of merchant-usurer capital (*shōninteki shigotoba shoyūsei* 商人的仕事場所有制 [merchant-ownership of workshops]) and denying the current thesis that there have been "manufactures" in China since the Sung period. Chapter 3 (21 pp.) analyzes capital and labor in individual enterprises, stressing the decisive influence of the Asiatic system in contrast to the European medieval system, especially as concerns master-artisan-apprentice relations. The fourth and final chapter gives a sketchy survey of Asiatic cities. Dr. Imahori is professor of Oriental history in Hiroshima University and once assisted in Dr. Niida's field survey of gilds in Peking as an interpreter. The latter's influence is visible in this article (especially in the idea of family-slavery). See the critique by Prof. Muramatsu Yūji in his "Chūgoku ni okeru jiyū to shihonshugi", 1.4.12. The attempt to generalize concerning China on the basis of Chinese city growth in Inner Mongolia is only one of several aspects of this study which may stimulate criticism.

1.4.11 FUJII Hiroshi 藤井宏, "Chūgokushi ni okeru shin to kyū, 'Shokkōtai' no bunseki o meguru shomondai" 中國史に於ける新ɀ舊――「織工對」の分析をめぐる諸問題 ("Elements new and old in the history of China", some problems concerning the analysis of *Chih-kung-tui* 織工對), **Tōyō bunka,** 9 (June 1952), 21–40.

This is a good example of a polemical article on the problem of periodization of Chinese history. The author is fundamentally sympathetic with the Marxist theory of developmental stages and appreciative of the recent interpretation which has China's medieval-feudal stage beginning from the Sung period. However, he cautions against the tendency to detect "old" elements everywhere in the past while failing to appreciate what was relatively "new" against its historical background. From this point of view, he criticizes Dr. Niida Noboru's interpretation of the famous essay of Hsü I-k'uei 徐一夔 entitled "Chih-kung-tui" 織工對, which describes the life and mentality of silk-weaving manual laborers near Hangchow 杭州 in the early Ming period. The author is professor of Oriental history in Hokkaidō Daigaku.

1.4.12 MURAMATSU Yūji 村松祐次, "Chūgoku ni okeru jiyū to shihon-shugi" 中國における自由と資本主義 ("Freedom and capitalism in China"), **Tōyō bunka,** 6 (Sept. 1951), 27–50.

This stimulating theoretical attack on Dr. Imahori Seiji's interpretation of Chinese history (put forward in articles such as "Chūgoku ni okeru hōken-teki shōkōgyō no kikō", 1.4.10) is by a social historian who denies the scholarly soundness of applying the Marxist theory of social developmental stages to China. The author's own approach to the understanding of Chinese history is based on Max Weber's views. This article is noteworthy as challenging the recent idea of periodization according to which China proceeded into a stage of medieval-feudal society in the Sung period. The author is professor in Hitotsubashi Unversity.

1.4.13 HIRASE Minokichi 平瀬巳之吉, **Kanjin shihai to kokkateki tochi shoyū** 官人支配と國家的土地所有 (Mandarin rule and state landownership), **Shakai kōseishi taikei** (9.8.7), series 1, 1949, 108 pp.

The author, a specialist in economic history, proposes his own idea of a three-stage historical development of China: (1) a society of tribal communities (down to the foundation of the Ch'in 秦 empire, 221 B.C.), (2) a pre-modern society (*zenki shakai* 前期社會, down to the Opium War)—the period of "Mandarinismus" 官人支配, and (3) the modern society (since the Opium War). This polemical study is of interest for its criticism of alternative theories (concerning China's arrested development, protofeudalism, scholar government, Max Weber's idea of water control, and the like), and for its analysis of the nature of the officialdom and its powers.

1.4.14 MORITSUGU Isao 森次勳, "Shina nōgyō shakai no ichikōsatsu" 支那農業社會の一考察 (An investigation of Chinese rural society), **Mantetsu chōsa geppō,** 22.11 (Nov. 1942), 97–128.

A theoretical and historical analysis of the periodization of Chinese history which centers its discussion around the phenomenon of anti-feudal centralization of power, as tied in with the Chinese system of land use. Current writings on this topic are quoted.

1.4.15 SUGIMOTO Naojirō 杉本直治郎, "Tōyōshijō ni okeru jidai kubun no saikentō to jikyoku no ninshiki" 東洋史上に於ける時代區分の再檢討と時局の認識 (A reexamination of periodization in Oriental history and the understanding of the current situation), pp. 166–225 in Hiroshima Shigaku Kenkyūkai [Hiroshima Bunrika Daigaku nai] 廣島史學研究會 [廣島文理科大學內] (Hiroshima Society for Historical Research, with offices in the Hiroshima University of Letters and Sciences), **Sekai no**

gensei yori mitaru shiteki ronsō 世界の現勢より見たる史的論叢 (Collected historical writings from the viewpoint of the current world situation), Chūbunkan Shoten 中文館書店, 1939, 314 pp.

Raising questions as to whether periodization is after all necessary or feasible, and as to how to define "period" 時代 and "division" 區分, the author distinguishes tendencies toward a "Sinocentric Oriental history" (represented by Naka Michiyo 那珂通世, Kuwabara Jitsuzō 桑原隲藏, and others) and toward "Orientalistic Oriental history" 東洋本位主義の東洋史, both of which he deplores. This technical and painstaking article surveys extensively the various modes of periodization adopted by the text-books for middle schools and higher education, but concludes with a rather chauvinistic (1938) view of its own.

1.5   ON THE NATURE OF CHINESE SOCIETY

Note: Here we have brought together a variety of items in the realm of historical interpretation or theory, many of them Marxist to one or another degree, which relate to the same subject of Chinese social structure that is dealt with in sec. 8 below. Our division of materials between this section and sec. 8.1 is perhaps arbitrary, but our aim has been to exhibit here some of the main lines of doctrinaire interpretation, specifically concerning "Oriental society" and the "Asiatic mode of production", and to list in sec. 8.1 those studies which fall specifically within the academic field of sociology. (On the other hand, we have put items of all kinds, doctrinaire or not, under sec. 8.2, "Village studies", where the main point of argument has related to the "closed" or unclosed nature of the village community, etc.) In the present section, the first item (Shima) seems to stand alone in a field of international intellectual relations which deserves fuller exploration, as a key to the origin of some of the earth-shaking pronouncements which have been made by European ideologues and echoed by Japanese enthusiasts. In the remainder it will be evident that the Rekishigaku Kenkyūkai (see in sec. 9) has been a chief focus for broad and schematic interpretations of history. We include two items on Max Weber and several on K. A. Wittfogel, both of whom have had much influence in Japan. Item 1.5.6 by Hani Gorō was one of the most influential early Marxist pronouncements.

1.5.1 SHIMA Yasuhiko 島恭彦, **Tōyō shakai to Seiō shisō** 東洋社會と西歐思想 (Eastern society and Western European thought), Seikatsusha, 1941, 171 pp. Repub. by Sekai Hyōronsha, 1948, 268 pp., with an appendix of two chapters added, pp. 205–268.

This concise study quotes the political economists of England, France and Germany in order to appraise the early modern European interest in the nature of Oriental society. Particular attention is devoted to the early English economists (Richard Jones) and the English legal historians who studied India. The fourth and last chapter treats the question of capitalism in the Orient, particularly the views of Max Weber. The result is a noteworthy attempt at critical evaluation of the historical development of Western scholarly views on Oriental society, with special reference to the problem how the approach of each thinker was determined by his social and ideological background at home as well as by contemporary European activities in the East. The first chapter was published in a slightly different form in **Tōa keizai ronsō**, 1.3 (Sept. 1941), 619–635. The author is now professor in the Faculty of Economics, Kyōto Unversity.

1.5.2 OTAKE Fumio 小竹文夫, "Gendai Shina shakai ron" 現代支那社會論 (On the contemporary Chinese society), **Tōa keizai ronsō**, 1.1 (Feb. 1941), 26 pp.

This lucid essay attacks the application of the term "feudal" to China by comparing the Chinese situation with the feudal institutions of medieval Europe and Japan in six different respects. This essay was later published, slightly revised, in Professor Otake's **Kinsei Shina keizaishi kenkyū**, 7.1.3, 1–36. See review by Nishida Ta'ichirō 西田太一郎 in **Tōa jimbun gakuhō**, 2.1 (Mar. 1942), 143–148.

1.5.3 ASAMI Shirō 淺海士郎, comp., "Nippon ni okeru 'Ajiateki seisan yōshiki' ronsō bunken" 日本に於けるアジア的生産樣式論爭文獻 (Bibliography of articles in Japan concerning the "Asiatic mode of production"), **Rekishigaku kenkyū**, 3.1 (Nov. 1934), 86–89; 3.4 (Feb. 1935), 328–329.

This bibliography provides, under some 60 items, a list of polemical and other writings concerning the so-called Asiatic mode of production as applied to China and Japan, including Japanese translations of Russian, German and Chinese literature. Almost all of these are articles scattered in periodicals. See also the section on the Asiatic mode of production, written by this compiler, on pp. 306–314 of Rekishigaku Kenkyūkai, **Rekishigaku nempō, Shōwa jūnen ban** (9.1.5).

1.5.4 AIKAWA Haruki 相川春喜, **Rekishikagaku no hōhō ron** 歷史科學の方法論 (A treatise on the methods of historical science), Hakuyōsha, 1935, 352+index 10 pp.

While centered on Japan, this book gives a summary of the phases in the discussion of the "Asiatic mode of production" theory, and enthusiatically notes a variety of the arguments put forward concerning its various aspects.

1.5.5 Rekishigaku Kenkyūkai, ed., **Sekaishi no kihonhōsoku** 世界史の基本法則 (Basic laws of world history), Iwanami Shoten, 1949, 80 pp.

This special issue of the above society prints the reports and discussions at the annual meeting of the society for the year 1949, concerning social theory as applied to world history and Japan in particular. In predominantly Marxist terms it describes the contradictions in ancient and feudal societies and the crisis of capitalism, and thus states many of the assumptions underlying the Japanese Marxist approach to China.

1.5.6 HANI Gorō 羽仁五郎, **Tōyō ni okeru shihonshugi no keisei** 東洋に於ける資本主義の形成 (The formation of capitalism in the Orient), Kyōto: San'ichi Shobō 三一書房, 1948, 193 pp.

A reprint of articles appearing under the same title in **Shigaku zasshi** 43.2,3,6,8 (1932), this essay has chapters on India, Japan and China (pp. 67–152). The last, entitled "Chinese society and the imperialist powers", explores and expounds the Marxist theory of Oriental Society as applied to China. This was one of the leading discussions of the then controversial problem of "the Asiatic mode of production". The author's view is that the Asiatic mode of production is the peculiar form which slavery and then serfdom took in China, based on the universal Marxist three-stage scheme of development of world history, and that the feudal stage in China (Asiatic feudalism) incorporated remnants of older ages (i. e., the tribal system). Against this background, he then discusses the problem of modernization in China during the stage of imperialism.

1.5.7 AKIZAWA Shūji 秋澤修二, **Shina shakai kōsei** 支那社會構成 (China's social structure), Hakuyōsha, 1939, 407 pp.

This far from modest work states in no uncertain terms the basic features of China's "Asiatic" type of society, with a panoramic appraisal and application of the Marxist assumptions—the Asiatic mode of production (which is rejected by the author), slavery and feudalism, closed villages and despotism, arrested development, etc. (On contemporary China, however, the author's views are labelled a "draft" 草案). The result is useful as a full-blown statement of dogma or hypothesis which appears in many other works in piecemeal fashion.

1.5.8 MORITANI Katsumi 森谷克巳, **Tōyōteki shakai no rekishi to shisō** 東洋的社會の歷史ご思想 (History and thought of the Oriental type of society), Jitsugyō no Nipponsha, 1948, 241 pp.

The author was at Keijō Imperial University in Korea for 19 years, 1927–1945. Becoming entranced with the Marxist approach, he published in 1934 a book entitled **Shina shakai keizai shi** 支那社會經濟史 (Shōkasha 章華社, 427 pp.), a broad pioneer survey based on the materialistic interpretation (reviewed by Sano Riichi 佐野利一 in **Rekishigaku kenkyū**, 3.5 [Mar. 1935], 485–488; by Aoki Tomitarō 青木富太郎 in **Shigaku zasshi**, 46.6 [June 1935], 802–805). In 1937 he published a volume entitled **Ajiateki seisan yōshiki ron** アジア的生産様式論 (Essays on the Asiatic mode of production) which was widely read (5th printing, Ikuseisha 育生社, 1941, 334 pp.) and put all of Chinese and Korean history into a broad Marxist framework. This later volume of 1948, though less sweeping, retains the same approach in a somewhat more sophisticated guise. Sections deal with Chinese and Korean economic history, in general terms concerning institutional origins, and with Chinese socio-economic thought of the classical period, as seen rather wishfully at second- or third-hand.

1.5.9 TŌMA Seita 藤間生大, "Seijiteki shakai seiritsu ni tsuite no joron, 'Ajiateki seisan yōshikiron' no gutaika no tame ni" 政治的社會成立についての序論——「アジア的生産様式論」の具體化のために (A preliminary discussion of the establishment of political society, toward making concrete the "Theory of the Asiatic mode of production"), **Rekishigaku kenkyū,** 133 (May 1948), 2–10; 134 (July 1948), 19–37.

This application of Marx to anthropology and Asia, by a specialist in Japanese ancient history, deals mainly with antiquity in Japan but may be taken as an interesting example of recent discussion in a field of Marxist theory which is also applicable to China. This is one of those writings by Japanese scholars called forth by the publication in 1947 of the Japanese translation of Marx's posthumously published draft on the various forms of society preceding the capitalist mode of production.

1.5.10 AOYAMA Hideo 青山秀夫, "Wēbā no Shina shakaikan josetsu" ウェーバーのシナ社會觀序説 (An introduction to Weber's view of Chinese society), **Tōkō**, 4 (April 1948), 2–21; 6 (Oct. 1948), 39–60.

Subtitled "Max Weber and Naitō Konan", this study discusses Weber's views on bureaucracy, feudalism, and particularly the family-based type of bureaucracy, as well as the stationariness or arrested development of Chinese society, and in conclusion takes up the points of similarity and dissimilarity with Naitō's ideas. The author is a careful scholar, though somewhat pedestrian, versed in modern economics, and also a leading expositor of Max Weber's social theory. See his book entitled **Makkusu Wēbā no shakai riron マックス・ウェ ーバーの社會理論** (The social theory of Max Weber), Iwanami Shoten, 1950, 293+index 32 pp., which includes this article (pp. 151–218). He is now professor in Kyōto University.

1.5.11 KAIZUKA Shigeki 貝塚茂樹, "Wēbā no Jukyōkan" ウェーバー の儒教觀 ("Weber's view of Confucianism"), **Tōyō bunka**, 7 (Nov. 1951), 1–26.

This is a rather descriptive explanation of Max Weber's view (in his **Gesam-melte Aufsätze zur Religionssoziologie)** of the patrimonial bureaucracy in China and of Confucianism in its original form. By a leading specialist in Chinese ancient history, this study stresses the rationalistic tendencies of Confucius' thought. The author is director of the Jimbunkagaku Kenkyūjo, Kyōto University.

1.5.12 BABA Jirō 馬場次郎, "Wittofōgeru hakushi no Shina kenkyū kinkyō ウイットフォーゲル博士の支那研究近況 ("Recent Studies on China by Doctor Wittvogel"), **Rekishigaku kenkyū**, 8.6, whole number 54 (June 1938), 663–684.

A lengthy account and commentary on Dr. Wittfogel's project for the study of Chinese social history, which translates in full his report published in **Pacific Affairs** (March 1938). See also the account by Nezu Masashi 禰津正 in **Shinagaku**, 8.1 (Oct. 1935), 133–141, on the occasion of Dr. Wittfogel's visit to Japan. A section of his **Wirtschaft und Gesellschaft Chinas**, on capital in China as an "Asiatic" society, was translated in **Mantetsu chōsa geppō**, 13.6 (June 1933), 162–231. The whole work was later translated by Hirano Yoshitarō as **Shina no keizai to shakai**, pub. by Chūō Kōronsha, 1934 (7th printing, 1940, 2 vols., 507+30 pp. and 410+23 pp.). Yokokawa Jirō 横川次郎 translated 5 articles as **Shina keizaishi kenkyū** (pub. by Sōbunkaku 叢文閣, 1935, 261 pp.). Moritani Katsumi 森谷克己 and Hirano Yoshitarō translated 4 articles as **Tōyōteki shakai no riron** (pub. by Nippon Hyōronsha, 1939, 322+bibliog. and index 54 pp.), and a further translation of 5 articles was put out by Hirano and Usami Seijirō 宇佐美誠次郎 as **Shina shakai no kagakuteki kenkyū**, Iwanami Shoten, 1939, 192+11 pp. (Iwanami shinsho, 35).

1.5.13 SATŌ Haruo 佐藤晴生, "Wittofōgeru 'Tōyōteki shakai no riron' ni tsuite" ウィットフォーゲル「東洋的社會の理論」に就いて (On the Wittfogel "Theory of Oriental Society"), **Mantetsu chōsa geppō,** 19.11 (Nov. 1939), 1–17.

Subtitled "a raising of certain doubts", this article quotes Russian, German and other authors in discussing such questions as whether a slave system existed in ancient China or whether the Chou was an "early feudal" period.

1.5.14 NODA Hirohiko 野田大彦, "Shina keizai no teitaisei no ichi sokumenkan" 支那經濟の停滯性の一側面觀 (A view of the stationariness of the Chinese economy), **Tōa jimbun gakuhō,** 1.4 (Feb. 1942), 1176–1194.

A critical appraisal of a tenet of the theory of "Oriental Society" (as propounded by Dr. K. A. Wittfogel) that the stationariness of its social development was caused by the concentration of political power, which was in turn based on the need for "inter-local" and large-scale irrigation and water control. The writer quotes classical and modern authors.

1.5.15 TAMURA Jitsuzō 田村實造, "Amerika ni okeru Tōyō shigaku kenkyū no ichidōkō, Wittofōgeru 'Chūgoku seifuku ōchō riron' sono ta" アメリカにおける東洋史學研究の一動向——ウィットフォゲル「中國征服王朝理論」その他 ["A Tendency of Asiatic studies in U.S.A.— On the Theory of 'Dynasty of Conquest' by Dr. Karl A. Wittfogel"], **Shirin,** 33.1 (Jan. 1950), 78–88.

A review and brief appraisal of recent publications by **Dr.** Wittfogel and Feng Chia-sheng 馮家昇, principally their volume **History of Chinese Society, Liao,** 1949.

1.5.16 FUJIWARA Sadamu 藤原定, "Shindai ni okeru hōken shisō to hōkensei no zanson" 清代に於ける封建思想と封建制の殘存 (Remnants of feudal thought and the feudal system in the Ch'ing period), **Mantetsu chōsa geppō,** 20.4 (April 1940), 1–61.

This has interest mainly as an ideological work-out in which the author finds feudalism adhering to the thought of Huang Tsung-hsi 黃宗羲, Ku Yen-wu 顧炎武 and other worthies of the early Ch'ing, and feudalism similarly rampant among the bureaucracy.

1.5.17 TAMURA Jitsuzō 田村實造, "Ajia shakai no kōshinsei to yūboku minzoku to no rekishiteki kankei" アジア社會の後進性と遊牧民族との歷史的關係 ("Effects of the nomadic invasions of China upon the so-called retardiness of Asiatic society"), pp. 1–8 in **Yūboku minzoku no shakai to bunka** 遊牧民族の社會と文化 ("Studies in Nomadic Society and Culture"), ed. by the Society for Eurasian Studies ユーラシア學會, Kyōto: Shizenshi Gakukai 自然史學會, 1952, 219 pp.

A leading specialist on the northern peoples suggests that by allying themselves with conservative ruling elements (bureaucracy, landlords and gentry), the nomadic invaders of China hampered her normal development. The author is professor of Oriental history in Kyōto University.

1.5.18 SUZUKI Shun 鈴木俊, "Shina ni okeru shakaishi kenkyū no gaikan" 支那に於ける社會史研究の概觀 ("Disputes among Chinese Historians on Social History of China"), **Rekishigaku kenkyū,** 3.2 (Dec. 1934), 167–181; 3.3 (Jan. 1935), 268–280.

This is a translation of ch. 13 of Wu Ch'i-yuan 伍啓元, **Chung-kuo hsin-wen-hua yün-tung kai-kuan** 中國新文化運動概觀, published in Shanghai in 1934, which surveys the leading theories of Chinese social development put forward by historians and others in modern China. It is noted here because comments were added by the translator, and his product contributed to Japanese discussion of the subject.

1.5.19 DEGUCHI Yūzō 出口勇藏, "Hausuhōfā no Tōa bunka seisaku" ハウスホーファーの東亞文化政策 (The East Asia cultural policy of Haushofer), **Tōa keizai ronsō,** 1.3 (Sept. 1941), 25 pp.

This wartime article quotes the thoughts of Dr. Karl Haushofer on Geopolitik in Asia, which may well be of historical interest. Many more such items can of course be found in literature of the Tripartite Pact period.

Note to item 1.4.9 (added while in press): Nos. 2, 3, 5 and 6 of the above-cited articles have been revised and included in Sudō's new book: **Chūgoku tochiseidoshi kenkyū** 中國土地制度史研究 (A study of the history of land systems in China), Tōkyō Daigaku Shuppankai 東京大學出版會, 1954, 726+index 13 pp.

2. LATE CH'ING POLITICAL HISTORY: DOMESTIC REBELLION AND FOREIGN AGGRESSION 內亂外患 TO 1900

Note: Believing that modern China's foreign relations and domestic politics are practically inseparable, as objects of historical study, we have included in this section both the "Western impact" and China's "response" to it. Both of these topics, for example, are dealt with in the surveys of Ch'ing history grouped in sec. 2.1. Details of the treaty system are reserved for sec. 3.7. Sec. 2.6 on the Reform Movement is of course part and parcel of China's modern intellectual history taken up in part 6 below.

2.1 SURVEYS OF CH'ING HISTORY

Note: Here also the influence of Dr. Naitō is very great. The writings of Inaba Iwakichi were often derived from Naitō's personal suggestions.

2.1.1 NAITŌ Torajirō 内藤虎次郎, **Shinchō suibō ron** 清朝衰亡論 (On the fall of the Ch'ing dynasty), Kōdōkan 弘道館, 1912, 123 pp.

This printed version of lectures given by Dr. Naitō in late 1911 was later reprinted as part of the volume entitled **Shinchōshi tsūron** (Kōbundō, 1944), 2.1.2. The lectures trace the developments in the military, fiscal-economic, and intellectual spheres under the Ch'ing from beginning to end. In the final section, Dr. Naitō put forth an optimistic prediction as to the future of the revolutionary movement.

2.1.2 NAITŌ Torajirō 内藤虎次郎, **Shinchōshi tsūron** 清朝史通論 (A survey of the history of the Ch'ing dynasty), Kōbundō, 1944, 423 pp.

This title is used to cover two works. The first, called **Shinchōshi tsūron,** is a stenographic reproduction of Dr. Naitō's lectures of 1915 at Kyōto Imperial University. They concern various aspects of the position of the Ch'ing emperor, Manchu relations with peripheral states, the Jesuit influence, classical, historical, and literary studies and fine arts. They form in general a brilliant survey in a colorfully colloquial style. The second work (pp. 325-423) is **Shinchō suibō ron** 清朝衰亡論 (On the fall of the Ch'ing dynasty) originally published in 1912 (2.1.1). See the reviews by Kanda Nobuo 神田信夫, in **Shigaku zasshi,** 55.12 (Dec. 1945), 1353-1357; and by Banno Masataka 坂野正高 in **Kokka gakkai zasshi,** 60.1 (Jan. 1946), 52-59.

2.1.3 Sambō Hombu 參謀本部 (The General Staff, Japanese Army), ed., **Shina seijishi kōryō** 支那政治史綱領 (Essentials of Chinese political history), Waseda University, 1918, 318 pp.

Written by Inaba Iwakichi 稲葉岩吉 for the Army, this volume divides its summary of Chinese political institutions and practices into periods of feudal government (to 221 B.C.), aristocratic government, and autocracy (since late T'ang, pp. 177 ff.). This analysis of the social, economic and political institutions of the autocratic Chinese empire of the Sung-Yuan-Ming-Ch'ing periods, following the interpretation of Dr. Naitō Torajirō, presents a general view of China which deserves study as having probably had a positive influence upon Japanese military thinking and continental expansion policies. The last section appraises the Revolution of 1911. Despite its early date, this work touches many subjects and problems of recent scholarly interest, e.g., the advisers of Chinese officials (*mu-yu* 幕友), the scholar-official Feng Kuei-fen 馮桂芬, influence of the family system on political order, the degree of local self-government. Although rather unknown even in Japan, this book is highly appreciated by specialists as one of the best general surveys of Chinese history.

2.1.4 INABA Iwakichi 稲葉岩吉, **Shinchō zenshi** 清朝全史 (History of the Ch'ing dynasty), Waseda University, 1914, 2 vols., 804 and 744+56 pp., maps, illus.

This history begins with the Nüchen 女眞 people under the Ming, and the rise of the Manchu state down to its seizure of China (to p. 376). It continues as a history of the dynasty, its rule in China and relations elsewhere—rather than as a history of China. Volume 2 treats mainly the nineteenth century, including literary and intellectual trends, to the revolution. While much of this work has presumably been superseded or revised by later research, the point of view is well worth noting both as an aspect of Japan's continental relations and as an alternative to customary Western or Chinese views, e.g. the chapter on the "T'ung-chih restoration" 同治中興. This book was translated into Chinese and had a wide circulation in China, where studies of the same kind had not yet been made. Liang Ch'i-ch'ao is said to have had a high regard for it.

In his studies of the Manchu dynasty, Dr. Inaba appears to have been stronger on its origins and early organization than on its period of decline. His general ideas for the most part followed the lead of his teacher, Naitō Torajirō, but his rather journalistic competence as a writer and his broad perspective made his writings popular in Japan as well as in China. While his immediate influence on younger scholars was not great (he was not a professor engaged in teaching), Inaba probably had some influence on the thinking of the Japanese military.

2.1.5 INABA Iwakichi 稲葉岩吉, **Kinsei Shina jikkō** 近世支那十講 (Ten lectures on modern China), Kaneo Bun'endō 金尾文淵堂, 1916, 496 pp., illus.

These miscellaneous lectures by the leading Japanese specialist of his day on the Manchu regime include a number of interesting approaches which should repay attention today. Major titles include: "Modern Chinese thought and the future of the revolution", "Chinese classical studies 經學 and government", "The K'ang-hsi Emperor" (and Arai Hakuseki's 新井白石 view of him), "The three main reasons for the fall of the Ch'ing Dynasty", "The T'ung-chih

Restoration and the Meiji Restoration" (pp. 169–179), "The Taipings", "Mongol-Chinese relations", "Present state of Chinese learning", "Prohibited books of the Ch'ing period".

2.1.6 INABA Kunzan 稲葉君山, **Kindai Shinashi** 近代支那史 (History of modern China), Osakayagō Shoten, 1920, 528 pp., illus.

This condensed and popularized volume, derived from the author's **Shinchō zenshi** (2.1.4), recounts the rise and glory of the Manchu regime down to the Ch'ien-lung period (to p. 293) and then deals with the process of internal revolt and revolution, abetted by foreign contact and the Opium War, down to the end of the Taiping Rebellion. Though this ground has been much traveled by others, all of Inaba's writings seem worth noting for their grasp of the dynastic point of view.

2.1.7 INABA Kunzan 稲葉君山, **Saikin Shinashi kōwa, kinsei no bu** 最新支那史講話近世之部 (Lectures on recent Chinese history, section on the modern period), Nippon Hyōronsha, 1920, 415 pp., map.

With a preface by Major General Ugaki Issei 宇垣一成, principal of the Military Staff College 陸軍大學, these lectures originally delivered there deal with a variety of topics in Ch'ing dynastic history down to the end of the Taiping Rebellion. Though arranged somewhat differently, the contents are very close to Inaba's **Kindai Shinashi** (2.1.6), which in turn reflects the conclusions he had worked out through his study of the Manchu dynasty in **Shinchō zenshi** (1914), 2.1.4, and other writings. Inaba's **Shina kinseishi kōwa**, (2.1.8), is an enlarged version of this book, continuing the narrative down to the revolution of 1911.

2.1.8 INABA Iwakichi 稲葉岩吉, **Shina kinseishi kōwa** 支那近世史講話 (Lectures on modern Chinese history), Nippon Hyōronsha, 1938, 437 pp., maps., illus.

Based on lectures at the Military Staff College (Rikugun Daigaku 陸軍大學), these 11 chapters deal with a variety of subjects, but they all fall within Inaba's special competence in the history of the Ch'ing down to the 1911 revolution. Included are 44 pp. on the life and times of the Empress Dowager, and there are interesting illustrations. This volume for the general reader can be considered a revised and condensed edition of the author's **Shinchō zenshi**, (2.1.4). A separate essay included as an appendix (pp. 381–405) is entitled "The current state of research in Manchurian history" and gives an interesting appraisal of Naitō Torajirō's efforts in this field. See review by Yokota Seizō 横田整三, in **Shigaku zasshi**, 49.7 (July 1938), 954–956.

2.1.9 INABA Iwakichi 稲葉岩吉, **Zōtei Manshū hattatsushi** 増訂滿洲發達史 (History of the development of Manchuria, revised edition), Nippon Hyōronsha, 1935 (6th printing 1940, 584 pp., original edition, 1915.), illus. Preface to the original edition by Naitō Torajirō, 8 pp.

A systematic narrative of historical developments in the region later known as "Manchuria", up through the Ming regime, the rise of the Manchus and their closing of their homeland to Chinese migration, followed by the emergence of international dangers (Russian in particular), the opening of the area to

Chinese migration, and modern relations with Korea and Mongolia. Aside from its apparent factual value, this volume is of interest as a Manchuria-centered study of what has generally been regarded by Chinese and Western historians as a sub-area of China. The volume lacks footnotes, bibliography or index, but references are roughly given in the text. This was a by-product of the author's participation in the famous research in Manchu-Korean historical-geography, sponsored originally by the South Manchurian Railway Co. and headed by Dr. Shiratori Kurakichi 白鳥庫吉 (see reference articles in Rekishi-gaku Kenkyūkai, comp., **Manshūshi kenkyū** (4.5.1), pp. 1-8, 263-292.).

2.1.10 INABA Iwakichi 稻葉岩吉, **Manshūkokushi tsūron** 滿洲國史通論 (An historical survey of the Manchu state), Nippon Hyōronsha, 1935, 393 pp., illus.

Drawing on his wide background knowledge of the earlier history of the Manchurian area and particularly on his lectures given at Kenkoku Daigaku 建國大學 in Ch'ang-ch'un, Dr. Inaba constructs a survey account of the historical experience of the people (or succession of peoples) of that area—their relations with China, Korea, the Hsiung-nu and other states, and their continual contacts with Chinese culture, Buddhism, etc. The Ch'i-tan 契丹, Nüchen 女眞 and Mongols and the eventual rise of the Manchus all are included in this framework. The last section (from p. 359) deals with Russian encroachment, leading to the Russo-Japanese war. Unlike his **Zōtei Manshū hattatsushi,** (2.1.9), in this work for the general reader the author attempts to be interpretative rather than factual. While this volume would seem to some Chinese historians to create an historical continuity out of successive developments which actually lacked it, this in itself is an interesting historical fact.

2.1.11 YANO Jin'ichi 矢野仁一, **Kindai Shina shi** 近代支那史 (Modern Chinese history), Kōbundō, 561 pp., preface dated 1926.

A systematic account of the Ch'ing dynasty's fortunes in establishing and maintaining its domination over the people of China and Inner Asia. After describing the establishment of Ch'ing relations with the Mongols, Jungars, Moslems, Tibetans, and other tributaries, the author then recounts the White Lotus, Taiping, Nien-fei, Panthay, Northwest and other rebellions of the 19th century. The resultant volume is a factual study of the waxing and waning of Manchu power over subject peoples, with the addition of provocative comments based on Dr. Yano's own particular stationary theory of Chinese development (chapter 1). The last four chapters concerning the opening of China and the foreign pressures leading to the breakdown of the dynasty form an interesting and suggestive survey partly because of this peculiar interpretation.

2.1.12 YANO Jin'ichi 矢野仁一, **Shinchō matsushi kenkyū** 清朝末史研究 (Late Ch'ing historical researches), Ōsaka: Yamato Shoin 大和書院, 1944, 342 pp.

Historical essays on aspects of the 1898 Reform Movement (53 pp.), court politics (7 pp.), Boxer Movement (66 pp.), Republican Revolution (42 pp.), Ch'ing relations with Russia over Manchuria, beginning with the boundary settlements of 1858-60 (90 pp.), and Ch'ing relations with Christian missions from Ricci on down (82 pp.). These essays are largely in narrative form, without footnotes; but sources are mentioned in the text.

33

2.1.13 WADA Sei 和田清, editorial supervisor, **Shindai no Ajia** 清代の アジヤ (Asia in the Ch'ing period), vol. 6 in the series **Tōyō bunkashi taikei** 東洋文化史大系 (A general cultural history of the Orient), Sei-bundō Shinkōsha 誠文堂 新光社, 1938, 397 pp., photos and illus., maps.

This glossily printed and profusely illustrated popular survey, compiled by Dr. Wada and other competent specialists outlines both the history of China and the encroachments of the powers in India and all the Far East. Emphasis is laid upon cultural as well as institutional developments, down to 1912. The unique distinction of this series, however, in view of its relative age, lies in the many rare and interesting illustrations which accompany the text page by page. Note the companion volume on the Republic (5.1.8).

2.1.14 SANO Manabu 佐野學, **Shinchō shakai shi** 清朝社會史 (Social history of the Ch'ing dynasty), Bunkyūdō 文求堂: Part 1, **Kokka to shakai** 國家と社會 (State and society) — vol. 1, **Kokka,** 1947, 145 pp.; vol. 2, **Kazoku** 家族 (Family), 1948, 118 pp.; Part 2, **Shakai kaikyū** 社會 階級 (Social classes)—vol. 1, **Jinushi narabini kanryō to risho** 地主並び に官僚と吏胥 (Landlords along with bureaucrats and clerks), 1947, 112 pp.; vol. 2, **Nōmin** 農民 (Farming population), 1948, 180 pp.; vol. 3, **Gō-shin, dorei to semmin, shōnin** 豪紳・奴隷と賤民・商人 (Powerful gentry, slaves and lowly people, merchants), 1947, 114 pp.; Part 3, **Nōmin bōdō** 農民暴動 (Peasant rebellions) — vol. 1, **Shindai minran no honshitsu narabini hatten, Byakurenkyō no ran** 清代民亂の本質並に發展・白蓮教 の亂 (The nature of popular rebellions in the Ch'ing period and their development, the White Lotus Society rebellion), 1947, 103 pp.; vol. 2, **Kaikō, Nen, Kempi** 海寇・捻・拳匪 (Pirates, Nien-fei, Boxers), 1947, 132 pp.; vol. 3, **Taihei Tengoku kakumei** 太平天國革命 (The Taiping revolution), 1947, 117 pp.

This ambitious work, as indicated in the above titles, is based mainly on Chinese primary sources (although **Ch'ing shih-lu** is not used) and represents an effort at socio-historical analysis of the development of the modern Chinese agrarian-based revolution. The small volumes lack footnotes and bibliographical apparatus, though sources are roughly indicated in the text, and they paint a broad canvas. The first volumes deal comprehensively with the Ch'ing political structure, the family system, the system of land-holding, and the official-dom, and then proceed to concentrate increasingly on elements in the process of agrarian revolt. Some attention is paid to China as an "Asiatic" type of society. The author is one of those founders of the Japanese Communist Party who left the party later. The manuscripts of this work were written in prison, the source materials having been provided by Suzue Gen'ichi 鈴江言一 and others. See the critical appraisal of the author's view on the Chinese social and political structure by Matsumoto Yoshimi 松本善海 in his two articles: "Kyū-Chūgoku shakai no tokushitsuron e no hansei," 8.2.8., and "Kyū-Chūgoku kokka no tokushitsuron e no hansei," 8.2.8. See also the review of the three volumes on the peasant rebellions by Suzuki Chūsei 鈴木中正 in **Chūgoku kenkyū,** 2 (Nov. 1947), 82–85, and the review of the volume on the state by Satoi Hiko-shichirō 里井彦七郎 in **Tōyōshi kenkyū,** 10.5 (May 1949), 69–74. Mr. Sano was professor in Waseda University until his death a few years ago.

2.1.15 KANDA Nobuo 神田信夫, "Manshū minzoku no suibō" 滿洲民族の衰亡 (The decline of the Manchu race), in **Kindai Chūgoku kenkyū,** 1948, 9.8.6, pp. 271–296.

A factual survey tracing factors such as poverty, inadequate stipends, luxurious living, and increase of numbers which adversely affected the Manchu rank-and-file as the dynasty continued, together with late-Ch'ing efforts to save the situation. Prof. Kanda is now at Meiji University, Tōkyō.

2.1.16 KANAI Yukitada 金井之忠, and SASSA Hisashi 佐佐久, **Shimmatsu shichijūnen shi** 清末七十年史 (A seventy-year history of the late Ch'ing period), Kōbundō, 1942, 139 pp. (Kyōyō bunko, 109).

This small vest-pocket survey summarizes the Ch'ing decline, the Opium War and the Taiping Rebellion and then, rather unexpectedly, presents an interesting view of the T'ung-chih Restoration and the beginning of the Reform Movement (pp. 77–115), with some further remarks on bureaucratic degeneration and revolutionary stirrings.

2.1.17 KITAMURA Hirotada 北村敬直, "Shindai no jidaiteki ichi, Chūgoku kindaishi e no tembō" 清代の時代的位置──中國近代史への展望 (The historical position of the Ch'ing era, a perspective on Chinese modern history), **Shisō** 思想, pub. by Iwanami Shoten, 292 (Oct. 1948), 625–635.

A brief disquisition posited on the thesis that the premodern and stationary society of China was already in the process of breaking up from within at the time of the Opium War. The author sees the key factor in the spread of a money economy, which had been accelerated through the separation of political and economic centers by Emperor Yung-lo 永樂 of the Ming, and also through the large-scale import of foreign silver and its economic repercussions. See also the same author's "Chūgoku no jinushi to Nippon no jinushi" 中國の地主と日本の地主 (Landlords in China and in Japan), **Rekishi hyōron** 歷史評論, (ed. by the Association of Democratic Scholars 民主主義科學者協會, pub. by Koishikawa Shobō 小石川書房), 4.2 (Feb. 1950), 19–25; and "Shindai no shōhinshijō ni tsuite" 清代の商品市場について (On the [inter-local] commercial market in the Ch'ing period), **Keizaigaku zasshi** 經濟學雜誌 (ed. by the Association for Economic Research at the Commercial College of Ōsaka 大阪商科大學經濟研究會 and pub. by Nippon Hyōron Shinsha 新社), 28. 3-4 (April 1953), 1—22. Prof. Kitamura is at the Faculty of Economics of the Ōsaka Metropolitan University 大阪市立大學.

2.2 THE CANTON TRADE

Note: These monographic studies, of a period before Japan had entered the Chinese scene, are of uneven value; but the studies by Etō Shinkichi on the country trade (2.2.5), by Momose Hiromu on foreign silver (2.2.18), and by Inaba Iwakichi (2.2.6) and others on the Canton officials and factories all seem worthy of note.

2.2.1 YANO Jin'ichi 矢野仁一, **Shina kindai gaikoku kankei kenkyū**

支那近代外國關係研究 (A study of modern Chinese foreign relations), Kōbundō, 1928, 569 pp.

A solid factual account of Portuguese-Chinese relations in the Ming and Ch'ing periods from ca. 1511 (the Malaccan appeal to China) down to 1749 (the pact concerning Macao). Portuguese relations are studied through both Western and Chinese sources checked against each other. Dr. Yano also deals with the Ming tribute system in some detail (ch. 3 and 5).

2.2.2 UCHIDA Naosaku 內田直作, "Jūrokuseikidai ni okeru Kōshū fukin no kaikō no shozai ni tsuite" 十六世紀代に於ける廣州附近の海港の所在に就いて (On the locations of sea-ports near Canton in the 16th century), **Shina kenkyū,** 41 (June 1936), 123–139, map; "Okuda Otojirō shi 'Tonmon'ō' jitchi chōsa no shōkai" 奧田乙次郎氏「屯門澳」實地調査の紹介 (A report and comment on Mr. Okuda Otojirō's field investigation of Tamao [T'un-men-ao]), ibid., 42 (Nov. 1936), 181–192.

The first of these studies seeks to ascertain the location of places in or near the Canton delta which figured in the early Portuguese contact with China—Tamao (T'un-men-ao, identified as modern Castle Peak Bay opposite Hongkong), Nan-t'ou 南頭 and Lampacao 浪白窖. The second article reports confirmation of the identification of Tamao by a member of the Japanese consulate general at Hongkong.

2.2.3 FUJITA Masanori 藤田正典, "Jūshichi-hachi seiki ni okeru Ei-Shi tsūshō kankei" 十七・八世紀に於ける英支通商關係 (Anglo-Chinese trade relations in the seventeenth and eighteenth centuries), **Tōa ronsō,** 1 (July 1939), 19–122.

This long article surveys the course of East India Co. relations at Canton down through the failure of the Macartney mission in considerable detail. While its source materials are chiefly Western (e.g. Morse, **Chronicles**) and it appears to cover much of the same ground as E. H. Pritchard's **Crucial Years of Early Anglo-Chinese Relations, 1750–1800** (1936), it has the merit of being an independent appraisal, solidly based, and was one of the authoritative Japanese studies of its day. See also Nozoe Shigekatsu 野副重勝, "Gokō zen ni okeru Shina gaikoku bōekishi no shitsuteki kentō" 五港前に於ける支那外國貿易史の質的檢討 (A qualitative study of China's foreign trade before the opening of the five ports), **Shina 支那,** 23.2 (Feb. 1932), 37–54; 3 (March 1932), 40–56; 7 (July, 1932) (not seen by us); 10 (Oct. 1932) (not seen by us); 11 (Nov. 1932), 54–62; 24.2 (Feb. 1933), 38–46.

2.2.4 UEDA Toshio 植田捷雄, "Eikoku Higashi Indo Kaisha no tai-Shi katsudō" 英國東印度會社の對支活動 ("Actions of the East India Company to China"), **Kokka gakkai zasshi 國家學會雜誌,** 56.7 (July 1942), 863–883; 8 (Aug. 1942), 985–1006; 10 (Oct. 1942), 1265–1280.

A survey of British and so mainly of the Company's relations with China down to the early 19th century, based largely on Western sources but also using pertinent Chinese and Japanese materials.

2.2.5 ETŌ Shinkichi 衛藤瀋吉, "Ahen Sensō izen ni okeru Eikoku shōnin no seikaku" 阿片戰爭以前における英國商人の性格 ("Activities of the British 'Country Traders' in China before the Opium War"), **Tōyō bunka kenkyūjo kiyō,** 3 (June 1952), 5–80.

Traces the rise of the "country trade" from India to the east, its involvement in opium and growth at Canton, and the factors causing the end of the East India Company monopoly in 1834. A solid survey making extensive use of the parliamentary papers and other Western sources and evaluating the role of the private traders, independently of M. Greenberg's recent **British Trade and the Opening of China.** This is a part of a completed manuscript; the remaining parts (down to the aftermath of the Opium War) are also to be published. Prof. Etō is now at the Tōkyō Kōgyō Daigaku 東京工業大學.

2.2.6 INABA Iwakichi 稻葉岩吉, "Shindai no Kanton bōeki" 清代の廣東貿易 (The Canton trade in the Ch'ing period), **Tōa keizai kenkyū,** 4.2 (April 1920), 239–274; 4.3 (July 1920), 359–380; 4.4 (Oct. 1920), 591–612; 5.1 (Jan. 1921), 103–113.

This long monograph was published before the appearance of Dr. H. B. Morse's **Chronicles of the East India Company** and has been neglected by later specialists. Dr. Inaba goes into the origin and status of the Hoppo, the effort to concentrate the foreign trade at Canton, various aspects of the tariff and other levies, the Cohong, restrictions on foreigners and prohibitions concerning various aspects of the trade, and the hong debts. He uses the **Yüeh hai-kuan chih** 粵海關誌 and draws on his background knowledge of the Ch'ing bureaucracy, in a way that should still be of interest, with reference to the Hoppo's relations to the court and to his colleagues at Canton. See also Matsumoto Tadao 松本忠雄, "Kanton no Kōshō oyobi ikan" 廣東の行商及夷館 (Hong merchants and factories at Canton), **Shina** 支那, 22.12 (Dec. 1931), 26–35; 23.1 (Jan. 1932), 52–67; 23.4 (April 1932), 63–73; 23.5 (May 1932), 51–56.

2.2.7 TANAKA Suiichirō 田中萃一郎, "Jūsankō" 十三行 (The thirteen hongs [Cohong]), pp. 61–74 in **Tanaka Suiichirō shigaku rombunshū 史學論文集**, Maruzen, 1932, 685 pp.

A brief account of the location and names of the Canton hong merchants, and derivation of the title of Hoppo, first published in **Mita gakkai zasshi 三田學會雜誌**, 12.4 (April 1918), and since largely superseded.

2.2.8 MUTŌ Chōzō 武藤長藏, "Kanton Jūsankō zusetsu" 廣東十三行圖說 ("Explication des gravures de 13 Factoreries à Canton"), **Tōa keizai kenkyū,** 15.1–2 (April 1931), 297–365, 6 plates.

An extensive and probably useful, though uninspired, working over of the evidence, mainly Western, concerning the names and locations of the old Canton factories, with long quotations of sources and some photographs.

2.2.9 UCHIDA Naosaku 內田直作, "Kanton Jūsankō to Jūsan Yōkō no meishō ni tsuite" 廣東十三行と十三洋行の名稱に付いて (On the designations of the Thirteen Factories [*shih-san-hang*] and the Thirteen Hong Merchants [*shih-san yang-hang*] at Canton), **Shina kenkyū,** 38 (Nov. 1935), 123–142.

The writer documents the confusion sometimes achieved by historians between these two, because each was referred to as "thirteen", but fails to find an integral causal connection between them. The later book by Liang Chia-pin 梁嘉彬 (**Kuang-tung shih-san-hang k'ao 廣東十三行考**) has probably made this article out of date.

2.2.10 SASAKI Masaya 佐々木正哉, "Etsu-kaikan no rōki" 粤海關の陋規 ("The Illegal Tariff of the Custom House at Canton during the Ch'ing Dynasty"), **Tōyō gakuhō,** 34.1–4 (March 1952), 132–161.

While this interesting (though not always lucid) study of the extra customs charges, customary gifts, and surcharges which accumulated at Canton during the eighteenth century naturally relies heavily on H. B. Morse's **Chronicles of the East India Company,** it makes a particular contribution by tracing these fiscal developments in Chinese records, mainly of the Yung-cheng period. This includes the unscrambling of some of the Chinese terminology. The author is now at Ōsaki High School 大崎高等學校 in Tōkyō.

2.2.11 SHIMODA Reisuke 下田禮佐, "Kanton bōeki no kenkyū" カントン貿易の研究 (A study of the Canton trade), **Shirin,** 15.1 (Jan. 1930), 42–53; 2 (April 1930), 184–200.

This brief article continues the author's earlier study of 16th century European trade in China and covers the subject of customs charges, regulations and manner of foreign life and trade in pre-treaty Canton on the basis of H. B. Morse's **Chronicles of the East India Company** and some Ch'ing regulations (e. g. **Ta-Ch'ing hui-tien).**

2.2.12 SHIMODA Reisuke 下田禮佐, "Ahen Sensō made, Ahen Sensō izen ni okeru ahen bōeki" 阿片戰爭まで——阿片戰爭以前に於ける阿片貿易 (Up to the Opium War, the opium trade before the Opium War), **Chiri ronsō 地理論叢** (Kyōto), 8 (1936), 93–120.

A scholarly but rather superficial account of the opium trade as a British evil; traces the growth of the traffic and Chinese efforts to suppress it, using Western and some Chinese sources.

2.2.13 SUZUE Gen'ichi 鈴江言一, "Sensō-zen no ahen kaikoku jijō narabini Higashi Indo Kaisha kankei ni tsuite" 戰爭前の阿片開國事情竝東印度會社關係に就いて (On the opening of China through the opium trade before the Opium War and its relation to the East India Company), **Mantetsu chōsa geppō,** 16.8 (Aug. 1936). 59–97.

An "exposure" of British imperialism which uses Western sources to trace the early growth of the Indo-Chinese opium traffic and the increasing financial dependence of the E. I. Co. in India upon it, up to 1840. This article makes pioneer use of a few Western primary sources not formerly used in Japan. The writer also studies, mainly through the **Sheng-hsün 聖訓,** the problem of what Chinese strata used to smoke opium, and tries to analyze the causal connection between opium smoking and the Chinese social and political structure.

2.2.14 NAKAE Kenzō 中江健三, "Ka-kei nenkan no Eikoku no Makao senryō ni tsuite" 嘉慶年間の英國の澳門占領について (On the British occupation of Macao in the Chia-ch'ing period), **Shien 史淵**, 19 (Dec. 1938), 247–264.

This brief study uses Dr. H. B. Morse's **Chronicles of the East India Co.** and the published Chinese documents to trace the late 18th century development of the British interest in Macao, the subsequent Anglo-French friction near Canton, and the British plan of 1802 for seizure of Macao.

2.2.15 NAKAE Kenzō 中江健三, "Ka-kei nenkan no Makao senryō ni tsuite" 嘉慶年間の澳門占領について ("Über die englische Eroberung Makaos in Kakei-Zeitalter"), **Rekishigaku kenkyū,** 9.10, whole number 71 (Nov. 1939), 1090–1111; 10.2, no. 74 (Feb. 1940), 123–149; 10.3, no. 75 (Mar. 1940), 225–239.

A research article which uses the available sources (mainly British but with the addition of Japanese, Chinese and Portuguese materials) to work out in detail the story behind the British seizure of Macao in 1808, as well as the events of the incident itself.

2.2.16 SONODA Yōjirō 園田庸次郎, "Eikan Arusesuto oyobi Rira no tōkō" 英艦アルセスト及びリラの東航 (The oriental voyage of the British frigate Alceste and brig Lyra), **Seikyū gakusō,** 18 (Nov. 1934), 123–157.

A carefully documented study of the British geographical exploration of the Korean coast and local contact in 1816, at the time of the Amherst embassy to China, using some Chinese and Korean as well as many English sources.

2.2.17 TANAKA Suiichirō 田中萃一郎, "Bokugin kō" 墨銀考 (A study of Mexican silver dollars); also "Bokugin kō hoi" 補遺 (Supplement to...), pp. 41–51 and 52–60 in **Tanaka Suiichirō shigaku rombunshū 史學論文集,** Maruzen, 1932, 685 pp.

Early articles on Mexican silver dollars in the Far East, published in **Mita gakkai zasshi 三田學會雜誌**, 9.11 (Nov. 1915) and 10.3 (Mar. 1916), respectively, and superseded by studies of Momose Hiromu, 2.2.18 and others.

2.2.18 MOMOSE Hiromu 百瀬弘, "Shindai ni okeru Supein doru no ryūtsū" 清代に於ける西班牙弗の流通 (Circulation of the Spanish dollar in the Ch'ing period), **Shakai keizai shigaku,** 6.2 (May 1936), 129–153; 6.3 (June 1936), 290–312; 6.4 (July 1936), 429–451, illus., tables.

This basic study uses both Chinese and Western materials (e.g. Blair and Robertson, **The Philippine Islands**) to survey the evidence on the use of Western silver coins in early modern China. This continues the author's study of the flow of foreign treasure and coin into China; see his article "Mindai no ginsan to gaikokugin ni tsuite" 明代の銀產と外國銀に就いて (On the production of silver in the Ming period and silver from foreign countries),

Seikyū gakusō 青丘學叢, 19 (Feb. 1935), 90–147; and also his article "Mindai ni okeru Shina no gaikoku bōeki" 明代に於ける支那の外國貿易 (The foreign trade of China in the Ming period), **Tōa,** 8.7 (July 1935), 95–110.

2.2.19 MASUI Tsuneo 増井經夫, "Ginkei happi to Yōgin bensei" 銀經發秘と洋銀辨正 ("The Yin-ching-fa-pi and the Yang-yin-pien-cheng, Two Chinese Books on Spanish Dollar"), pp. 625–639 in **Wada hakushi kanreki kinen Tōyōshi ronsō,** 9.8.12.

These two handbooks, published in 1826 and 1854, give information on the handling and detection of counterfeits among foreign ("Spanish") dollars, used increasingly in China's growing money economy during the 19th century.

2.2.20 KATŌ Shigeshi 加藤繁, "Dōkō Kampō chū Shina nite chūzō serraretaru yōshiki ginka ni tsuite" 道光咸豊中支那にて鑄造せられたる洋式銀貨に就いて ("On some Chinese silver coins of European style minted during the Tao-kuang and Hsien-feng era, A.D. 1821–1868"), **Tōhō gakuhō, Tōkyō,** 2 (Dec. 1931), 284–292, plate; continued in ibid., 3 (Dec. 1932), 351–355.

Brief scholarly comment on 6 pieces of silver coinage of European style, one of which was officially minted in Taiwan, three in Chang-chou 漳州, and two by Shanghai merchants, in the period when foreign silver dollars were beginning to circulate more widely in connection with China's maritime trade. Also printed in **Shina keizaishi kōshō** (9.8.8), 2.450–462.

2.2.21 SASAMOTO Shigemi 笹本重巳, "Kanton no tekka ni tsuite" 廣東の鐵鍋について ("Market for the Canton-Made Iron Pan after XV Century"), **Tōyōshi kenkyū,** 12.2 (Dec. 1952), 131–44.

An interesting study of the records regarding the export of iron goods and of this article in particular, down into the Ch'ing period.

2.3 THE OPENING OF CHINA 1840–1860

Note: The chief developments to be noted in this section are the increased use during the last few years of the published Chinese documents on foreign relations (chiefly the **Ch'ou-pan i-wu shih-mo** series), in which the lead has been taken by Ueda Toshio, and the gradual permeation of an approach based on political science.

2.3.1 MATSUDA Tomoo 松田智雄, **Igirisu shihon to Tōyō** イギリス資本と東洋 (English capital and the Orient), Nippon Hyōronsha, 1950, 290 pp.

Subtitled "The pre-modern nature and the modern nature of Oriental trade", this illuminating and interpretative monograph deals with the century-long development of British commercial activity, from the latter days of the East India Co. monopoly and its policy developments in Malaya under Raffles to the commercial opening of China by the free traders at Hongkong and the gradual concentration, if not monopoly, of capital and enterprise there under such firms as Jardine, Matheson and Company. The author, a professor at St. Paul's University in Tōkyō, has specialized in German social and economic history and approaches this subject well versed in the writings of Marx and Max Weber, as well as Protestantism. While he relies heavily throughout on secondary works, often of marginal value, he builds up an impressive picture, based on the field investigation sponsored by the Tōa Kenkyūjo, of capital accumulation at Hongkong and the proliferation of (e.g., Jardine's) subsidiary interests, down into the 20th century. On the author's somewhat idealistic interpretation, see the critical review by Etō Shinkichi 衞藤瀋吉, in **Tōyō bunka,** 4 (Nov. 1950) 115–125. See also reviews by Yanai Katsumi 楊井克巳 in **Shakai keizai shigaku,** 社會經濟史學, 16.3 (Sept. 1950), 148–151; and by Banno Masa-taka 坂野正高 in **Rekishigaku kenkyū,** 147 (Sept. 1950), 48–51.

2.3.2 UEDA Toshio 植田捷雄, **Tōyō gaikōshi gaisetsu, Chūgoku kaikoku hen** 東洋外交史概說——中國開國編 (A survey of Far Eastern diplomatic history, the opening of China), Nikkō Shoin, 1948, 266 pp.

This survey for the general reader traces China's early contacts with the Western powers and then presents an account of the process of negotiation and conflict by which China was finally opened up to full Western contact in the period 1834–1860. This story is illumined by Dr. Ueda's comments on the many points of international law which bulk so large in this formative period. The last chapter gives an interesting analysis of Sinocentrism. Chapters down through the first treaty settlements are based on the author's own researches, but later chapters on the Arrow War and second treaty settlements are apparently based on Japanese and Western secondary writings only. Succeeding volumes, originally planned, have not been published.

2.3.3 UEDA Toshio 植田捷雄, "Ahen Senso ron" 阿片戰爭論 ("A treat-ise on the Opium War"), **Kokusaihō gaikō zasshi,** 42.1 (Jan. 1943), 22–47; 42.2 (Feb. 1943), 135–158; 42.3 (Mar. 1943), 237–270.

This study makes wide use of British and also some Chinese documenta-tion to analyze the aims and procedures of British policy, and the Chinese attitudes towards the opium question, together with some attention to the effect of the war on other powers, including Japan. This is a critical appraisal of controversial points rather than an original fact-finding research.

2.3.4 UEDA Toshio 植田捷雄, "Ahen Senso to Shimmatsu kammin no shosō" 阿片戰爭と清末官民の諸相 ("The actual Attitude of the Chinese Mandarins and common people towards the Opium War"), **Koku-saihō gaikō zasshi,** 50.3 (July 1951), 235–271.

An interesting description of the relations and attitudes among Chinese officialdom, the local populace and the British invaders. Dr. Ueda makes wide use of contemporary Chinese and especially British sources to indicate the cleavages within Chinese society.

2.3.5 UEDA Toshio 植田捷雄, "Nankin Jōyaku no kenkyū" 南京條約 の研究 ("On the Treaty of Nanking"), **Kokusaihō gaikō zasshi,** 45. 3–4 (Mar. 1, 1946), 93–123; 45. 5–6 (May 1, 1946), 154–175; supplement in 46.3 (Feb. 1, 1947), 123–155.

A comprehensive analysis of the motivating circumstances, procedures and actual development of the Anglo-Chinese negotiations, which makes extensive use of the Chinese IWSM documents as well as contemporary British accounts.

2.3.6 UEDA Toshio 植田捷雄, "Shina no kaikoku to kokusaihō" 支那の開國と國際法 (The opening of China and international law), **Tōyō bunka kenkyū,** 1 (Oct. 1944), 31-48.

An examination, from Sino-Western sources, of the discordant legal assumptions on the two sides of the treaty settlement of 1842-44.

2.3.7 MIYAZAKI Sonoji 宮崎其二, "Ahen Sensō no keizaiteki igi" 阿片戰爭の經濟的意義 (" Economic Significance of the Opium War "), **Shakai keizai shigaku,** 2.2 (May 1932), 150-180.

This is a pioneer interpretative (though somewhat naive) survey, based on Western works of H.B. Morse and others and on Chinese (including IWSM) sources and now outdated. It deals with (1) the emergence of the opium question, in the course of British economic expansion into the Chinese empire, (2) the increased British influence after the War, and (3) the consequent making of China into a "semi-colony".

2.3.8 MATSUMOTO Morinaga 松本盛長, "Ahen Sensō to Taiwan no goku" 鴉片戰爭と臺灣の獄 (The Opium War and the Taiwan case), pp. 442-563 in **Shigakuka kenkyū nempō** 史學科研究年報 (Annual research report of the department of history), ed. and pub. by the Taihoku Teikoku Daigaku Bunseigakubu 臺北帝國大學文政學部 (Faculty of literature and politics of the Taihoku Imperial University), vol. 4 (1937), 583 pp.

A long and thorough delving into the notorious case of the seizure and execution of British personnel from the vessels *Nerbudda* and *Ann* during the Opium War by the over-eager Formosan authorities, together with their false reports, the Anglo-Chinese negotiations, and their removal. The author uses the IWSM and other Chinese sources; he sympathizes with the Taiwan authorities and is critical of I-liang, who denounced them, and Ch'i-ying and Mu-chang-a, who supported the denunciation.

2.3.9 HANABUSA Nagamichi 英修道, "1842 nen Nankin Jōyaku ni tsuite" 一八四二年南京條約について (On the Nanking Treaty of 1842), **Hōgaku kenkyū** 法學研究 (" Journal of Law, Politics and Sociology "), pub. by the society for the same with offices in the Faculty of Law, Keiō Gijuku Daigaku 慶應義塾大學 (" Keio University "), Mita, Minato-ku, Tōkyō, 22.1 (Jan. 1949), 1-30.

Dr. Hanabusa in this article gives Japanese translations in extenso of the English and Chinese texts of the treaty, which forms a useful way of indicating the disparities between them. He also comments on the supplementary agreement on 8 items, made at Nanking in the form of an exchange of notes, which began the negotiations for the Supplementary Treaty of October 1843 and which is quoted in full in Liang T'ing-nan 梁廷枏, **I-fen chi-wen 夷氛記聞** (printed in 1937). The author is a professor at Keiō University.

2.3.10 TOYAMA Gunji 外山軍治, "Shanhai dōdai Go Ken-shō" 上海道臺吳健彭 (The Shanghai Taotai, Wu Chien-chang), **Gakkai 學海**, 1.7 (Dec. 1944), 45–54.

A short but colorful note on the career of the notorious ex-hong merchant "Samqua" at Shanghai in the 1850s, drawing on the Chinese (IWSM) documents. The author stresses Wu's opportunistic adaptability to treaty port ways (e.g. his use of pidgin English) and his eventual impeachment for collaboration with foreign merchants. Prof. Toyama is now at the Ōsaka Gaikokugo Daigaku 大阪外國語大學.

2.3.11 YANO Jin'ichi 矢野仁一, "Kō Shaku-ji oyobi Rin Soku-jo no ahen sōgi ni tsuite" 黄爵滋及び林則徐の鴉片奏議に就いて (On the memorials discussing the opium question submitted by Huang Chueh-tzu and Lin Tse-hsü), pp. 757–771 in **Takase hakushi kanreki kinen Shinagaku ronsō** 高瀬博士還曆記念支那學論叢 (Collected essays in Chinese studies celebrating the sixtieth birthday of Dr. Takase), Kyōto: Kōbundō, 1928.

A brief article quoting arguments presented by these two protagonists during the opium controversy of the 1830s.

2.3.12 YANO Jin'ichi 矢野仁一, **Ahen Sensō to Honkon** アヘン戦爭と香港 (The Opium War and Hongkong), Kōbundō, 1939, 316 pp., map, chronological table.

A reworking of the story of Anglo-Chinese relations from the early Canton period up through the Treaty of Nanking, which is roughly the same thing as the pertinent section of Dr. Yano's **Kinsei shina gaikō shi** (1.3.3) published in 1930. Many footnotes for the general reader are added, mainly on technical terms of Ch'ing institutions. Chang Hsi's diary, **Fu-i jih-chi**, which was not available at the time of writing the former volume, is used, but the published Chinese documents (IWSM) are not mentioned.

2.3.13 YANO Jin'ichi 矢野仁一, **Arō Sensō to Emmeien** アロー戦爭と圓明園 (The Arrow War and Yuan-ming-yuan), Kōbundō, 1939, 309 pp,. 2 maps.

A republication of sections of the same author's **Kinsei Shina gaikō shi** (1930), 1.3.3, recounting Anglo-Chinese relations from the time of the Canton city question and Arrow War through the various treaty settlements of 1858–60–76, with attention also to Burma and Central Asian relations. Many footnotes for the general reader are added, mainly on technical terms relating to Ch'ing institutions.

2.3.14 MIYAZAKI Ichisada 宮崎市定, "Shinagawa shiryō yori mitaru Ei-Futsu rengōgun no Pekin shinnyū jiken, tokuni shusenron to waheiron" 支那側史料より見たる英佛聯合軍の北京侵入事件──特に主戰論と和平論 (The incident of the intrusion into Peking of the Anglo-French allied force as viewed from the Chinese sources, with special reference to the counsels of war and peace), **Tōa kenkyū shohō**, 24 (Oct. 1943), 852–884.

Condensed from a study carried out for the Tōa Kenkyūjo, this article in a colloquial style represents an appraisal of a relatively recent event by a scholar versed in earlier periods of Chinese history: the background of the ill-fated negotiations of Kuei-liang 桂良 and Ch'i-ying 耆英 at Tientsin in 1858; the pourparlers of Kuei-liang at Shanghai, the hostilities of 1859, and the ensuing policy debate over resistance or conciliation when the Anglo-French forces retaliated in 1860. This is an illuminating analysis of the behind-the-scenes politics of the Ch'ing regime, mainly based on an imaginative use of the IWSM documents.

2.3.15 BANNO Masataka 坂野正高, "Ahen sensō go ni okeru saikeikoku taigū no mondai" 阿片戰爭後における最惠國待遇の問題 ("The problem of the most favoured nation treatment after the Opium War"), **Tōyō bunka kenkyū 東洋文化研究**, 6 (Oct. 1947), pp. 19–41.

This article is carefully based on the Ch'ing documents (IWSM) and analyzes the way of thought of the mandarins such as Ch'i-ying in charge of foreign contact, who had to deal with this problem.

2.3.16 BANNO Masataka 坂野正高, "Gaikō kōshō ni okeru Shimmatsu kanjin no kōdō yōshiki — 1854 nen no jōyaku kaisei kōshō o chūshin to suru ichikōsatsu" 外交交涉に於ける清末官人の行動樣式――一八五四年の條約改正交涉を中心とする一考察 ("Behaviours of Mandarins as diplomats late in the Ch'ing Dynasty — with special reference to Treaty revision negotiations in 1854"), **Kokusaihō gaikō zasshi,** 48.4 (Oct. 1949), 502–540; 48.6 (Dec. 1949), 703–737.

This is a case study mainly based on the Chinese and American documents (IWSM and the McLane correspondence), checked against each other, of a typical diplomatic negotiation in the 1850s. The article describes how the mandarins on the spot, in charge of foreign contact, behaved toward foreign officials as well as toward Peking, and thus attempts to clarify the actual working of the Ch'ing machinery for foreign affairs in this period.

2.3.17 BANNO Masataka 坂野正高, "'Sōrigamon' setsuritsu no haikei" 「總理衙門」設立の背景 ("Determining factors in the instituting of the Tsungli Yamen"), **Kokusaihō gaikō zasshi,** 51.4 (Aug. 1952), 360–402; 5 (Oct. 1952), 506–541; 52.3 (June 1953), 89–109.

Making extensive use of Western and Chinese documents, this article reappraises the factors which led to the setting up of a new agency to handle China's foreign relations. Special attention is given to the shift in the balance of power in the Chinese political scene at the time of the Arrow War, under decisive pressures from abroad as well as from below.

2.3.18 OKUDAIRA Takehiko 奥平武彦, "Kurimiya sensō to Kyokutō" クリミヤ戰爭と極東 ("Crimean War and the Far East"), **Kokusaihō gaikō zasshi,** 35.1 (Jan. 1936), 42–68; 35.4 (April 1936), 313–343.

A study in diplomatic history based largely on British despatches in the Public Record Office in London and on the published Western literature. Sections deal with the pre-war activities of the powers in the Far East, the

British (Stirling) treaty with Japan of 1854, naval operations at Petropavlovsk and De Castries Bay and of the Anglo-French fleet in 1855, and the effect of the war in the Far East.

2.3.19 TAMAKI Hajime 玉城肇, "Bei-Shi bōekishi no ichikōsatsu, Ten-shin Jōyaku teiketsu made" 米・支貿易史の一考察――天津條約締結まで (An inquiry into the history of Sino-American trade, up to the conclusion of the Treaty of Tientsin), **Tōagaku,** 3 (Dec. 1940), 57–114.

Though based on the usual Western sources, this rather well-informed survey of Sino-American trade, from the Canton period down to the 1860s, is one of the few Japanese studies in this field.

2.4 THE TAIPING REBELLION

Note: As in China, the Taiping Rebellion is one of the liveliest subjects of current Japanese research. Scholars like Ichiko Chūzō provide a valuable and critical, though non-schematic, alternative to the doctrinaire approach. The first item below, by Suzuki Chūsei, also opens up new ground.

2.4.1 SUZUKI Chūsei 鈴木中正, **Shinchō chūkishi kenkyū** 清朝中期史研究 (A study of mid-Ch'ing history), Toyohashi: Aichi Daigaku Kokusai Mondai Kenkyūjo 愛知大學國際問題研究所 (Aichi University Research Institute on International Problems), 1952, 225 pp., map.

This recent, original and important study of the White Lotus Society rebellion of 1796–1805 uses the 411 volumes of the official history and a wealth of similar Chinese sources to explore its socio-economic background and the long and costly process by which it was put down. Addressing himself to the problem of anti-dynastic revolt, the author first presents an interesting analysis of factors of population increase, profligacy, corruption, etc., which ushered in the Ch'ing dynasty's period of decline, followed by chapters on the movement of population into the Szechwan-Shensi-Hupei border region and their circumstances, the nature of the White Lotus Society, its uprising, the early campaigns of the imperial commanders, and the vigorous policies by which suppression was finally achieved. The latter included the building of village walls and training of militia, in ways which foreshadowed military developments in China during the next century and a half. See the review by Sasamoto Shigemi 笹本重巳 in **Tōyōshi kenkyū,** 12.1 (Sept. 1952), 80–83. Prof. Suzuki is now at Aichi University in Toyohashi.

2.4.2 MIYAZAKI Ichisada 宮崎市定, "Chūgoku kinsei no nōmin bōdō—tokuni Tō Mo-shichi no ran ni tsuite" 中國近世の農民暴動，特に鄧茂七の亂について ("Peasant Riots in Chinese Modern History, especially on the Event of Teng Mao-ch'i"), **Tōyōshi kenkyū,** 10.1 (Dec. 1947), 1–13.

A case study of a Fukien peasant revolt in the middle of the fifteenth century (1448–49). The author is sceptical about the current tendency to consider all popular uprisings since the Sung as "peasant" revolts, but this case is presented as an exceptional example of a genuine "peasant" revolt, which had no connection with the literati, urban outlaws or heterodox secret societies.

2.4.3 ICHIKO Chūzō 市古宙三, "Chūgoku ni okeru Taihei Tengokushi no kenkyū" 中國における太平天國史の研究 ("On the Studies of the Taiping Rebellion in China"), **Shigaku zasshi,** 60.10 (Oct. 1951), 937–946.

A valuable critical review, by a leading Japanese specialist on 19th century China, which lists and briefly appraises some 69 different books, articles or volumes of documents concerning the Rebellion. It is followed by a "Memorandum of the Taiping Rebellion's Sources in Library of Congress, U.S.A." (ibid. 947–8) by Professor Yamamoto Tatsurō 山本達郎. Prof. Ichiko is now at Ochanomizu Daigaku お茶の水大學 in Tōkyō.

2.4.4 ICHIKO Chūzō 市古宙三, "Bakumatsu Nipponjin no Taihei Tengoku ni kansuru chishiki" 幕末日本人の太平天國に關する知識 (Japanese awareness of the Taiping Rebellion in the late-Tokugawa period), pp. 451–495 in **Kaikoku hyakunen kinen Meiji bunkashi ronshū** 開國百年記念 明治文化史論集 (Collected essays on Meiji cultural history, commemorating the centennial of the opening of the country), Kengensha, 乾元社 1952, 844 pp.

Professor Ichiko here traces the reports of the Rebellion reaching Japan, beginning with news brought to Nagasaki by Chinese merchant vessels in 1852 and later, as recorded in **Bakumatsu gaikoku kankei bunsho** 幕末外國 關係文書 and similar collections. He concludes that the news of the Rebellion, although carefully noted, was probably less of a stimulus to policy changes in Japan than news of the Opium and Arrow Wars. This article forms a critique of Prof. Masui's views on the same topic (Masui, **Taihei Tengoku,** 2.4.11).

2.4.5 ICHIKO Chūzō 市古宙三, "Taihei Tengoku shōsho no kaisei" 太 平天國詔書の改正 ("Revision of the Imperial Declaration of the T'ai-p'ing Dynasty"), **Tōyō gakuhō,** 33.2 (Oct. 1950), 183–199.

An important analysis of the complex causes for the successive revisions of these key dynastic documents (in the period from March 1853 to June 1854), which provides indications of Taiping domestic developments.

2.4.6 ICHIKO Chūzō 市古宙三, "Shu Kyū-tō kō" 朱九濤考 ("On the life of Chu Chiu-t'ao, a T'ai-p'ing leader"), **Tōhōgaku,** 3 (Jan. 1952), 23–31.

This brief article brings an impressive body of bibliographical and historical knowledge to bear on the identification of the uncertain role played by an early Taiping figure.

2.4.7 ICHIKO Chūzō 市古宙三, "Haijō kō, Taihei Tengoku seido kanken" 拜上考，太平天國制度管見 ("The God-worshippers; a study of the Taiping System"), **Ochanomizu daigaku jimbunkagaku kiyo** お茶の水大學人文科 學紀要 (Humanistic science bulletin of Ochanomizu University), 1 (Mar. 1952), 31–45.

A study of Taiping "Christian socialism" as a crude mixture of puritanism and communism, in which the select "God-worshippers" served and were served on a quite different level from the class of "contributors" who also labored for the state.

2.4.8 ICHIKO Chūzō 市古宙三, "Taihei Tengoku no sambishi" 太平天國の讃美詩 ("The Doxology of the T'aipings"), pp. 85–102 in **Wada hakushi kanreki kinen Tōyōshi ronsō, 9.8.19.**

An analysis of the semi-Christian daily hymn of the Taipings, comparing it with genuine Christian doctrine and noting its political usefulness, as well as its revisions which reflected the shift in power relations among the top-level leaders.

2.4.9 FUJIWARA Sadamu 藤原定, "Shimmatsu ni okeru jinkō kajō no shogenshō to Taihei Tengoku undō" 清末における人口過剰の諸現象と太平天國運動 (Phenomena of overpopulation in the late Ch'ing and the Taiping movement), **Mantetsu chōsa geppō,** 19.7 (July 1939), 1–71.

This much-quoted pioneer article marshals a variety of evidence from the early 19th century under the headings of vagrants 流民, migrants 移民, the coolie trade, rises in the price of land, etc., ideas of equalization 均田 or limitation of landholding 限田, and bandits—as indices of population pressure in the pre-Taiping society of the Ch'ing. After beginning with a survey of population estimates, the author uses the **Shih-ch'ao sheng-hsün 十朝聖訓** and **Tung-hua lu 東華錄,** among many other sources, to outline these phenomena.

2.4.10 ŌSHIMA Susumu 大島晉, "Taihei Tengoku undō no shoyōsō" 太平天國運動の諸様相 (Various aspects of the Taiping movement), **Rekishi hyōron 歷史評論** (ed. by the Association of Democratic Scholars 民主主義科學者協會, pub. by Kawade Shobō), 36 (May 1952), 1–11.

A painstaking interpretative article, based among other things on local gazetteers and some new sources. The author pays special attention to conditions in the upper basin of the Yangtze, and also points to elements of leadership in the movement which distingushed it from earlier peasant revolts. See review by Ichiko Chūzō 市古宙三 in **Shigaku zasshi,** 62.5 (May 1953), 466–467.

2.4.11 MASUI Tsuneo 増井經夫, **Taihei Tengoku 太平天國** (The Taiping Rebellion), Iwanami Shoten, 1951, 172 pp., illus. (Iwanami shinsho, 70).

This popularly-written but rather unscholarly volume, in addition to surveying the rise and fall of the Rebellion, the question of its modernity and the careers of Li Hsiu-ch'eng 李秀成 and Liu Yung-fu 劉永福, also has a noteworthy chapter on Japanese contemporary knowledge of it (pp. 1–32) and a useful bibliography of 103 Japanese writings concerning the Taipings. See review by Satoi Hikoshichirō 里井彦七郎 in **Chūgoku kenkyū,** 16 (Sept. 1952), 90–93. Prof. Masui is now at Kanazawa 金澤 University.

2.4.12 KAWABATA Genji 河鰭源治, "Tenchō dempo seido no seiritsu" 天朝田畝制度の成立 ("Social and economic background of the compilation of the **T'ien-ch'ao t'ien-mou chih-tu"**), **Tōyō gakuhō,** 33.2 (Oct. 1950), 200–222.

A scholarly analysis of this key document of the Taiping land reform program, which raises doubts as to how far the reality of Taiping action ever conformed to it and tries to assess its real political significance. Although the author does not at all deny the revolutionary character of the Taiping movement as a whole, he considers its land reform program to have been little more than a tool of a traditional dynastic despotism. The author is now at Odawara Jōtō High School 小田原城東高等學校 in Kanagawa Prefecture.

2.4.13 HATANO Yoshihiro 波多野善大, "Taihei Tengoku ni kansuru nisan no mondai ni tsuite" 太平天國に關する二三の問題について (Concerning two or three problems related to the Taiping Rebellion), **Rekishigaku kenkyū,** 150 (Mar. 1951), 32–42.

This article criticizes the conventional Marxist interpretation (usually made against the background of fundamental class "contradictions" in the Chinese social structure, etc.) that the Rebellion was caused by the intrusion of British cotton goods and the resulting depression of Chinese rural industry. The author urges the importance of the conflict between the Hakka 客家 and the natives 本地人 in the initial stage of the Rebellion and of the increased unemployment caused by the transfer of foreign trade to Shanghai. As for the social background of the Nanking stage of the movement, he hypothesizes the devastating impact on the Sungkiang 松江 cotton industry of the import of cotton goods through Shanghai and the rural impoverishment resulting from the contagious spread of opium-smoking accelerated by the opening of Shanghai. See critique by Satoi Hikoshichirō 里井彦七郎, in **Rekishigaku kenkyū,** 153 (Sept. 1951), 46–49. Professor Hatano is now at Nagoya University.

2.4.14 HATANO Yoshihiro 波多野善大, "Taihei Tengoku" 太平天國 (The Taiping Rebellion), **Rekishigaku kenkyū,** 142 (Nov. 1949), 53–57.

Though written as a guide for beginners, this brief note by a competent specialist in late-Ch'ing history is a useful survey of the main historical approaches to the Taiping movement and the controversial points involved in its interpretation.

2.4.15 SUZUKI Tadashi 鈴木正, "Seki Tatsu-kai no shisō ni tsuite" 石達開の思想に就いて (On the thought of Shih Ta-k'ai), **Shikan,** 19 (Dec. 1939), 98–104.

A brief comment based on the diary, **Shih Ta-k'ai jih-chi** 日記, including the attitude of this Taiping leader toward Christianity.

2.4.16 MIYAGAWA Hisashi 宮川尚志, "Shoki Taihei Tengoku no shūkyōsei" 初期太平天國の宗教性 ("On the Religious Character in the T'ai-p'ing Rebellion in its early period"), **Jimbun kagaku,** 1.3 (Dec. 1946), 212–229.

A brief essay describing the adaptive transition of the Taiping religion from a radical and popular stage in its initial period to a conservative compromise after the occupation of Nanking. This article is apparently based on Toriyama Kiichi 鳥山喜一, "Taihei Tengoku ran no honshitsu" 太平天國亂の本質 (The real character of the Taiping Rebellion), pp. 3–125 in Keijō Teikoku Daigaku Bungakkai 京城帝國大學文學會 (The association for literary studies, Imperial University of Keijō [Seoul]), ed., **Tōhō bunkashi sōkō** 東方文化史叢考 (Collected researches in Eastern cultural history), 1935, Ōsakayagō Shoten, 532 pp. (Vol. 1 of the series). This latter article is noteworthy as a substantial pioneer work which stressed the personality of Hung Hsiu-ch'üan and his relationship to Christianity and made extensive use of the Western bibliography on the subject.

2.4.17 UEDA Toshio 植田捷雄, "Taiheiran to gaikoku" 太平亂と外國 ("The Taiping Rebellion and the Foreign Powers"), **Kokka gakkai zasshi,** 62.9 (Sept. 1948), 464–494; 12 (Dec. 1948), 669–687; 63.1–3 (Mar. 1949), 31–95.

This article summarizes the general circumstances and course of events in the development of Western relations—successively friendly, neutral and hostile—toward the Taipings, relying mainly on standard Western sources. The author also discusses the efforts of Tseng Kuo-fan and others towards the modernization of China, as stimulated by the Rebellion. A condensed translation in English has appeared in **Japan Science Review.**

2.4.18 AKINAGA Hajime 秋永肇, "Taihei Tengoku gaikōshiron" 太平 天國外交史論 (On the history of the foreign relations of the Taiping Rebellion), **Seigakuka kenkyū nempō** 政學科研究年報 (Annual research report of the political science department), pub. for the Faculty of Letters and Political Science of the Taihoku Imperial University, Taiwan, by Noda Shobō, 野田書房, 7 (1941) 171–220.

Covering the years 1853–1864, this study uses the usual and limited Western sources to trace the course of foreign neutrality and intervention. Nothing new seems to have been found.

2.4.19 SUZUE Gen'ichi 鈴江言一, "Taihei Tengoku to gaikoku kankei" 太平天國と外國關係 (The Taiping Rebellion and its foreign relations), **Mantetsu chōsa geppō,** 14.10 (Oct. 1934), 99–116.

A brief, documented, and doctrinaire survey, dealing with (1) Hung Hsiu-ch'üan and Christianity, (2) the neutrality of the foreign powers, (3) the opium question as affecting the British policy of aid to the Ch'ing (arguing that the Taipings' intransigence against opium drove the British to the imperial side), and (4) the initial industrialization (for making modern arms, etc..) motivated by the rebellion. Reviewed by Nohara Shirō 野原四郎 in **Rekishigaku kenkyū,** 3. 4, whole number 16 (Feb. 1935), 384–386.

2.4.20 TOYAMA Gunji 外山軍治, "Taiheiran ni okeru Shinchō no gaikoku ni taisuru enjo yōsei" 太平亂に於ける清朝の外國に對する援助 要請 ("Aid given by Foreign Powers to the Chinese Empire in the Taiping Rebellion"), **Shirin,** 31.3–4 (Dec. 1947), 32–45.

Although somewhat more detailed than the corresponding chapter in the author's book of the same year, 2.4.21, this article does not add substantially to it.

2. 4. 21 TOYAMA Gunji 外山軍治, **Taihei Tengoku to Shanhai** 太平天國と上海 (The Taiping Rebellion and Shanghai), Kyōto: Kōtō Shoin 高桐書院, 1947, 169 pp., illus.

This small volume tells the colorful story of events at Shanghai in the 1850s and '60s against the background of the great rebellion. The Small Sword Society 小刀會 uprising of 1853, the writer Wang T'ao 王韜, Ward and Gordon and foreign arms, certain Japanese diarists who recorded the Shanghai scene, and various aspects of Ch'ing and Taiping policy are touched upon. The narrative is chiefly based on the Chinese documents (IWSM). Some basic Western primary sources available in Japan were apparently not used, and so misrepresentations in the memorials of Ch'ing officials are uncritically accepted. Index 11 pp. See also his article "Ō Tō to Chōhatsuzoku" 王韜と長髮賊 (Wang T'ao and the long-haired rebels), **Gakkai 學海**, 2. 8.

2. 4. 22 TOYAMA Gunji 外山軍治, "Shanhai no shinshō Yō Bō" 上海の紳商楊坊 (The wealthy Shanghai merchant Yang Fang), **Tōyōshi kenkyū** (new series), 1.4 (Nov. 25, 1945), 17–34.

A study, mainly based on Chinese sources, of the famous "Yang Tze-tang", so-called "paymaster" of the Ever-victorious Army under F. T. Ward. Yang became an official and worked closely with Governor Hsueh Huan and other leaders in foreign contact at Shanghai.

2. 4. 23 ŌTANI ("Ohtani") Kōtarō 大谷孝太郎, "Taihei Tengoku no ichizuke" 太平天國の位置付け ("The Character of Taiping Rebellions"), **Hikone ronsō 彦根論叢** ("The Hikone Ronso"), 8 (Mar. 1952), 33–50; 10 (July 1952), 26–45. (Edited by The Institute for Economic Research, Shiga 滋賀 University, Hikone.)

The first of these articles, subtitled "from simplicity to classicism" 素朴的から古典的へ ambitiously seeks to characterize the essential nature of the Rebellion but in effect discusses mainly the generalizations of other writers like Lo Er-kang 羅爾綱 and Ojima Sukema 小島祐馬. The second article, subtitled "socially scientific" 社會科學的, enthusiastically quotes Marxist interpreters, with equally generalized and insubstantial results.

2. 4. 24 NASHIMOTO Yūhei 梨本祐平, **Taihei Tengoku kakumei 太平天國革命** (The Taiping rebellion), Chūō Kōronsha, 1942, 440 pp.

An account of the rebellion from first to last, based on a conventional Marxist framework of assumptions (e. g. that imperialism caused it), with frequent quotation of secondary works (not always acknowledged) and some source materials. The narrative gives a rather full description of the activities and interrelations of the various Taiping leaders.

2. 4. 25 Sō Sō-ju 曹宗儒 (Ts'ao Tsung-ju), "T'ai-p'ing-chün tsai k'o Wu-ch'ang chih shih-liao" 太平軍再克武昌之史料 (Some materials concerning the reoccupation of Wu-ch'ang by the Taiping army), **Tōyōshi kenkyū**, 4.3 (Jan.—Feb., 1939), 238–42.

This brief item prints the Chinese texts of 8 manuscripts (found in possession of a bookdealer in Peking in 1935, when Mr. Ts'ao was working at the Palace Museum there), which were written as drafts of letters or reports by imperial officials at Wu-ch'ang in 1855. Although written in Chinese this article does not appear to have been published elsewhere and is therefore noted here.

2.4.26 Ō Sei-ei (Wang Shih-ying) 王世英, "Taihei Tengoku ni okeru keizai oyobi bunka no kenkyū" 太平天國に於ける經濟及び文化の研究 (A study of economics and culture in the Taiping Rebellion), **Tōa keizai kenkyū,** 27.3 (Aug. 1943), 234–255; 27.4 (Nov. 1943), 317–345.

An amateurish arrangement of materials concerning Taiping policies on land equalization, common property, conscription, sharing of produce, trade and customs, diplomacy, religion, bureaucracy, equality of men and women, calendar, justice, etc. A good deal of documentation is quoted but uncritically and without source reference. The author, a Chinese, wrote this in Japanese, with some assistance.

2.4.27 TANAKA Suiichirō 田中萃一郎, "Taihei Tengoku no kakumeiteki igi" 太平天國の革命的意義 (The significance of the Taiping Rebellion as a revolution), **Shigaku zasshi,** 23.7 (July 1912), 766–786.

Notes of a lecture delivered to the History Society. This is a pioneer scholarly appraisal of the Taiping documents found in the British Museum. The speaker remarks especially upon the importance of the Taiping doctrine and their calendar as well as their Sinocentric and traditionalistic anti-foreignism. Reprinted on pp. 21–40 of **Tanaka Suiichirō shigaku rombunshū** 史學論文集, Maruzen, 1932, 685 pp. For a similar notice of the British Museum documents, see Naitō Torajirō's article in **Shirin,** 10.3 (1925), 412–417.

2.4.28 (Chūgoku Kenkyūjo) Chōsabu (中國研究所) 調査部 Research Section ("Institute of China Research"), "Taihei Tengoku hyakushūnen no gakkai tembō" 太平天國百周年の學界展望 ("On the One Hundredth Anniversary of Taiping Rebellion"), **Chūgoku kenkyū,** 14 (June 1951), 40–48.

In addition to translating the appropriate **Jen-min jih-pao** 人民日報 editorial, the editors describe 9 major historical works of recent years, including some published under the communist regime to commemorate the Rebellion.

2.5 THE T'UNG-CHIH RESTORATION

Note: This generally neglected aftermath of the Taiping Rebellion has been touched upon also in works listed in the following section on the Reform Movement. Note also 7.12.5 (Miyazaki).

2.5.1 MOMOSE Hiromu 百瀬弘, "Fū Kei-fun to sono chojutsu ni tsuite" 馮桂芬と其の著述について (On Feng Kuei-fen and his writings), **Tōa ronsō,** 2 (Jan. 1940), 95–122.

A well-informed summary of the career and major works of one of the intellectual leaders of the T'ung-chih restoration period, who served as secretary to Li Hung-chang and others and helped to formulate the doctrine of "Self-strengthening". A Chinese condensation of this article appeared in China in the wartime journal **Chung-ho yueh-k'an**, 3.3 (1942), 53–66. The original gives extensive references to Feng's writings and the few other sources available. The author graduated at Tōkyō University in Chinese history, studied in the United States and during the war worked under the Tōa Kenkyūjo; he is now a professor at Kōbe.

2.5.2 FUMOTO Yasutaka 麓保孝, "Shindai Juka shichōjō ni okeru Wa Bun-tan-kō no chii" 清代儒家思潮上に於ける倭文端公の地位 ("Wo Jen († 1871), his man and his influence to the cultural world of China"), **Tōyō gakuhō**, 32.1 (Oct. ["Sept." sic] 1948), 92–101.

A brief analysis of the scholarly life and work of the famous anti-foreign Mongol official, who was a leading philosopher of his time and was posthumously titled Wo Wen-tuan-kung, seen mainly through the comments of his contemporaries, e.g. his correspondence with Tseng Kuo-fan 曾國藩.

2.5.3 SHIMMI Hiroshi 新美寛, "Sō Koku-han no shukan" 曾國藩の手翰 (Letters of Tseng Kuo-fan), **Tōyōshi kenkyū,** 8.2 (March-April, 1943), 115–121.

Printing of four unpublished letters of Tseng Kuo-fan, written in 1855, which were found at An-ch'ing in 1938; with some comment on them.

2.5.4 SAKURAI Nobuyoshi 櫻井信義, **Sō Koku-han** 曾國藩 (Tseng Kuo-fan), Koga Shoten 古賀書店, 1943, 330 pp.

The author, a minor official in Manchoukuo interested in studying Chinese leaders, writes about Tseng from all angles but from a traditionalistic viewpoint, without contributing anything new from the point of view of research.

2.5.5 NOMURA Masamitsu 野村政光, "Tenshin Kyōan ni tsuite" 天津教案に就いて ("On the Anti-missionary-riot in Tientsin"), **Shirin,** 20.1 (Jan. 1953), 67–99.

A detailed but somewhat superficial monographic study of the antecedents and course of the "Tientsin massacre" of 1870, using the Chinese (IWSM) documents.

2.5.6 TABOHASHI Kiyoshi 田保橋潔, "Shin Dōchichō gaikoku kōshi no kinken" 清同治朝外國公使の覲見 (The audiences of the foreign ministers in the T'ung-chih period of the Ch'ing), **Seikyū gakusō,** 6 (Nov. 1931), 1–31.

A brief account of the first audience granted to the Japanese ambassador and the foreign ministers, using mainly Western sources and written before the availability of the Chinese (IWSM) documents.

2.6 THE REFORM MOVEMENT

Note: While this topic forms an integral part of China's modern intellectual history, and is also treated under sec. 6.1 below, it is taken up here because of its transcendant political importance. Arrangement is roughly chronological by subject matter. The recent studies of Itano Chōhachi, Onogawa Hidemi, and Ichiko Chūzō, are all important contributions in a rapidly developing field. On K'ang Yu-wei's **Ta-t'ung shu,** note also the long chapter by Kimura Eiichi, in 6.1.11, and also item 1.1.18.

2.6.1 OTAKE Fumio 小竹文夫, "Mimmatsu irai Seisho kō" 明末以來西書考 (A study of Western books published in Chinese since the late-Ming period), **Shina kenkyū,** 43 (Jan. 1937), 35–71.

A useful bibliographic listing of Western writings which appeared in Chinese during the late-Ming and early-Ch'ing period (mainly the seventeenth century, listed on pp. 38–46), and during the late-Ch'ing period (pp. 46–66),—a large body of materials awaiting further study. Now superseded by more recent lists such as those of P. Henri Bernard in **Monumenta Serica,** 10 (1945), 1–57, 309–388.

2.6.2 KIKUCHI Takaharu 菊地貴晴, "Kōgakkai to hempō undō (jo), Kōgakkai no setsuritsu ni tsuite" 廣學會ご變法運動(序)―廣學會の設立について ("Reform movement and the Society for the Diffusion of Christian and General Knowledge among the Chinese"), pp. 305–317 in **Tōyōshigaku ronshū** 東洋史學論集 ("Studies in the Oriental History"), ed. by "Department of The Oriental History in the Tokyo University of Education" 東京敎育大學東洋史學研究室, Shimizu Shoin 清水書院, 1953, 340 pp.

This brief preliminary study uses Chinese and foreign sources to describe the founding of the Society, its sponsors and their aims in the Chinese scene of 1887. The author proposes to publish another article concerning the influence of the Society upon the reform movement, and its connections with the foreign powers.

2.6.3 ITANO Chōhachi 板野長八, "Kō Yū-i no daidō shisō" 康有爲の大同思想 (K'ang Yu-wei's concept of the *Ta-t'ung* or "Great Harmony"), in **Kindai Chūgoku kenkyū,** 1948, (9.8.6), pp. 167–204.

An important analysis of K'ang's fundamental idea, by one of the most competent specialists in the history of Chinese thought. Prof. Itano analyzes penetratingly the original concept of *ta-t'ung* in the **Li-chi** 禮記 against its social background, and exposes its despotic, traditionalistic and Sinocentric characteristics, which he sees as cloaked under the integrating idea of "harmony". The author considers the later development of traditional Chinese thought as a cyclical movement of this *ta-t'ung* concept and so K'ang's way of thought is explained as a self-conscious but typical response to the Western impact on the part of this traditional thinking. Itano regards K'ang's theory as having nothing modern about it at all, and he detects a persistence of the *ta-t'ung* concept in Sun Yat-sen as well as in later Nationalist thinkers. This interpretation may be compared with the different view of Onogawa Hidemi, 2.6.5–7. Prof. Itano is now at Hokkaidō University 北海道大學.

2.6.4 ITANO Chōhachi 板野長八, "Ryō Kei-chō no daidō shisō" 梁啓超の大同思想 ("The Idea of Ta-t'ung of Liang Ch'i-ch'ao"), pp. 69–84 in **Wada hakushi kanreki kinen Tōyōshi ronsō,** 9.8.12.

A study of Liang's development of his own version of the "Great Harmony" conception, in which the author stresses the fundamentally Confucian mold of Liang's thought, especially in his conservative later years. Other articles by this author are printed in **Gendai Chūgoku jiten,** 9.5.3

2.6.5 ONOGAWA Hidemi 小野川秀美, "Shimmatsu yōmuha no undō" 清末洋務派の運動 ("On the Movement of Europeanizing at the End of the Ch'ing Era"), **Tōyōshi kenkyū,** 10.6 (Feb. 1950), 429–466.

The author distinguishes the two phases of the reform movement directed at introduction of Western technology 技藝, and reform of institutions 變法, respectively, and traces how the early interest in arms, railroads, mining and other technological developments gave way by the 1890s to the interest in political and institutional change. He quotes the views of Feng Kuei-fen 馮桂芬, Hsueh Fu-ch'eng 薛福成, Wang T'ao 王韜, Liu Ming-ch'uan 劉銘傳, Tseng Chi-tse 曾紀澤, and similar writers on these questions in a useful survey of Chinese essay materials. Prof. Onogawa is now at the Jimbunkagaku Kenkyūjo, Kyōto University.

2.6.6 ONOGAWA Hidemi 小野川秀美, "Shimmatsu hempōron no seiritsu" 清末變法論の成立 ("The formation of the Reformatory Thought at the End of the Ts'ing Dynasty"), **Tōhō gakuhō, Kyōto,** 20 (Mar. 1951), 153–184.

This study sees late-Ch'ing political thought as divided into the movements for Westernization (*yōmu* 洋務), institutional reform (*hempō*), and revolution (*kakumei* 革命), both the former being reformist rather than revolutionary. The transition from the first to the second having been precipitated by the Sino-Japanese War, the author traces the rise of the demand for parliamentary government (as put forward by T'ang Chen 湯震 and others such as Cheng Kuan-ying 鄭觀應), which in turn broke the ground for K'ang Yu-wei. A clear and illuminating study.

2.6.7 ONOGAWA Hidemi 小野川秀美, "Shimmatsu no shisō to shinkaron" 清末の思想と進化論 ("Political Thoughts and the Evolution Theory at the End of the T'sing (清) Dynasty"), **Tōhō gakuhō, Kyōto,** 21 (Mar. 1952), 1–36.

Studies the influence of Yen Fu's translation of Huxley's **Evolution and Ethics,** which was finished in 1896, and published in the **Kuo-wen-pao** 國聞報 in 1897 and as a separate book in 1898. The main stress is laid upon the development of Liang Ch'i-ch'ao's thought, under the decisive influence of the evolution theory—from the theory of *ta-t'ung* 大同 received from his teacher, K'ang Yu-wei, to an almost revolutionary radicalism, and then back to a moderate progressivism.

2.6.8 Iснiко Chūzō 市古宙三, "Hokyō to hempō" 保教と變法 (The preservation of the Confucian teaching and the reform of the laws), pp. 113–138 in Niida Noboru, ed., **Kindai Chūgoku no shakai to keizai, 9.8.7.**

This brief article by a leading student of the subject discusses succinctly the relationship in K'ang Yu-wei's reform program between the movement to preserve Confucianism and the movement to reform political institutions. This is studied in the context of his shift of emphasis from reformist politics before the first revolution to traditionalistic improvement through edification in the republican period.

2.6.9 Iснiко Chūzō 市古宙三, "Ryō Kei-chō no hempō undō" 梁啓超の變法運動 ("The canvassing of Reform movement of Liang Chi-chao"), **Kokushigaku** 國史學 ("Kokushigaku, The Journal of Japanese History"), pub. by the Kokushigakkai ("The Society of Japanese Historical Research"), Kokugakuin 國學院 University, Tōkyō, 54 (Jan. 1951), 71–83.

A cogent and percipient summary of Liang's views on reform before the 1898 episode and an examination of how they developed thereafter and diverged from those of K'ang Yu-wei, based on a careful perusal of Liang's writings.

2.6.10 Ojima Sukema 小島祐馬, "Kuyōka no risō to suru daidō no shakai" 公羊家の理想とする大同の社會 (The society of the *Ta-t'ung* [or Great Harmony, "Utopia"] as an ideal of the *Kung-yang* school), **Keizai ronsō,** 8.6 (June 1919), 730–742.

A scholarly study of the classical origins, in the **Li-chi** 禮記 and *Kung-yang* commentary on the **Ch'un-ch'iu** 春秋, of the utopian conception and its terminology, later used and developed in K'ang Yu-wei's famous **Ta-t'ung shu** 書.

2.6.11 Ojima Sukema 小島祐馬, "Tan Shi-dō no 'Jin-gaku'" 譚嗣同の「仁學」 (The **Jen-hsüeh** of T'an Ssu-t'ung), **Geimon,** 8.11 (Nov. 1917), 983–1014.

A competent study of one of the more difficult works of the Reform movement, one of the few such studies available. Note Professor Ojima's recent survey of Chinese revolutionary thought (6.1.1.), as viewed 30 years later.

2.6.12 Yoshikawa Katsuji 吉川勝治, "Tan Shi-dō ni okeru jingaku no kōzō" 譚嗣同に於ける仁學の構造 ("The Structure of 'Jen-hsüo' in 'T'an-szǔ-t'ung'"), **Ritsumeikan bungaku** 立命館文學 ("The Ritsumeikan Literature Review"), ed. by "The Research Institute of Cultural Sciences, Ritsumeikan University", Kyōto, 84 (May 1952), 46–59.

Defining the major influences on the martyred philosopher T'an Ssu-t'ung (1866–1898) as (1) the K'ang-Liang school and the *ta-t'ung* 大同 concept, (2) the Buddhist ideas of the Nanking scholar Yang Jen-shan 楊仁山 (Wen-hui 文會), (3) Western learning, and (4) Wang Ch'uan-shan 王船山, this pioneer study explores T'an's relation to the revolution and to the Confucian tradition in general, in addition to studying his **Jen-hsüeh.**

2.6.13 UNO Seiichi 宇野精一, "Shurai Ryū Kin gisakusetsu ni tsuite" 周禮劉歆僞作說について ("On the theory that Liu Hsin forged the 'Rites of Chou'"), **Tōa ronsō,** 5 (Oct. 1941), 237–273.

This historical survey of the forgery theory links the references to it of ancient and pre-modern scholars with those of K'ang Yu-wei, Liang Ch'i-ch'ao, Hu Shih, Kuo Mo-jo and others in recent times. This and other controversial problems relating to the **Rites of Chou** are fully disscused in the same author's **Chūgoku kotengaku no tenkai** 中國古典學の展開 (Development of the study of the classics in China), Hokuryūkan 北隆館, 1949, 381+38 pp.

2.6.14 TADOKORO Gikō 田所義行, **Chūgoku ni okeru sekai kokka shisō** 中國に於ける世界國家思想 (Thought concerning a world-nation in China), Shōhei Kōshi 昌平公司, 1947, 228 pp.

By a Tōkyō graduate now a professor at Tōkyō Womens' College 東京女子大學, this is a rather idealistic exposition and commentary on K'ang Yu-wei's utopian commonwealth (ta-t'ung 大同) rather than a research study contributing new evidence. Sources are not indicated. The author seems to have sympathetically and rather pathetically identified himself with K'ang Yu-wei's anachronistic aspirations. Reviewed by Saitō Akio 齋藤秋男 in **Chūgoku kenkyū,** 4 (June, 1948), 71–77.

2.6.15 YAMAGUCHI Ichirō 山口一郎, "Kō Yū-i no shōgai to shisō no hatten" 康有爲の生涯と思想の發展 (Development of K'ang Yu-wei's life and thought), **Chūgoku bungaku,** 62 (June 1940), 97–110.

A quick overall appraisal, by periods and with bibliography, by a specialist in intellectual history.

2.6.16 SATŌ Saburō 佐藤三郎, "Nisshin Sensō no Chūgoku ni oyoboshita eikyō ni tsuite" 日清戰爭の中國に及ぼした影響について ("On the influences of the Sino-Japanese War upon China"), **Yamagata daigaku kiyō (jimbun kagaku)** 山形大學紀要 (人文科學) ("Bulletin of the Yamagata University, [Cultural Science]"), 1 (Mar. 1950), 65–80.

Subtitled as centering on Chinese opinion concerning the affairs of the day, this study examines the newspaper materials collected by Ch'en Yao-ch'ing 陳耀卿 and published as **Shih-shih hsin-pien** 時事新編 (6 chüan, preface 1895). The conclusion is that Chinese opinion at first was overconfident, underestimating the Japanese, but with the progress of disaster became critical of the Ch'ing officials and army, and began to view Western civilization as a means by which to strengthen China.

2.6.17 INADA Masatsugu 稻田正次, "Bojutsu seihen ni tsuite" 戊戌政變について (On the coup d'état of 1898), in **Kindai Chūgoku kenkyū,** 1948, (9.8.6), pp. 207–242.

A well-informed, factual summary narrative of the 1898 affair—part of an unpublished larger study. The author stresses the idea that Itō Hirobumi's 伊藤博文 visit to Peking accelerated the coming of the catastrophe in September (pp. 230–237). The writer is a specialist in constitutional law who served in wartime as a professor in Peking and is now at Tōkyō Kyōiku Daigaku.

2.6.18 KANO Naoki 狩野直喜, "Shina kinsei no kokusui shugi" 支那近世の國粹主義 (The doctrine of preserving the national characteristics in modern China), **Geimon,** 2.10 (Oct. 1911), 1655–1666; 3.1 (Jan. 1912), 89–96.

Dr. Kano notes evidences of the growth of a traditionalistic nationalism or effort to preserve Chinese culture (in response to the inroads of Western culture), as seen in the program of Chang Chih-tung 張之洞 and others in education, and in the new legal codes, respectively.

2.6.19 KUMANO Shōhei 熊野正平, "Bankin Shina ni okeru gikoha narabini seitōha no shisō ni tsuite" 輓近支那に於ける疑古派並びに正統派の思想に就いて (On the ideas of the school sceptical about classical learning, and of the orthodox school, in the recent period in China), **Shina kenkyū,** 22 (May 1930), 537–559.

The author takes K'ang Yu-wei 康有爲 and T'an Ssu-t'ung 譚嗣同 as protagonists of the sceptics and Chang Ping-lin 章炳麟 (T'ai-yen 太炎) of the traditionalists, the former involving the *chin-wen* 今文 and *kung-yang* 公羊 schools, the latter connected with the *ku-wen* 古文 school and *k'ao-chü* 考據 tradition, etc. The author then proceeds to describe, with quotations, the views of these three scholars.

2.6.20 MOTOMURA Shōichi 本村正一, "Ban-Shin yōmu undō shiron" 晚清洋務運動史論 (An historical discussion of the Westernization movement of the late Ch'ing), **Shien 史淵**, 29 (Mar. 1943), 1–29.

An interpretative but somewhat elementary account of the efforts of Li Hung-chang to inaugurate Western armament and industrialization and the eventual response of Chang Chih-tung and others to the impact of the defeat by Japan, etc. The author quotes a number of memorials by title but with inadequate references.

2.7 THE BOXER REBELLION

Note: The current interpretation of this famous incident, in China, lends special interest to the independent-minded and scholarly studies listed below, which set a high standard well above the results of Western research on this topic.

2.7.1 ICHIKO Chūzō 市古宙三, "Giwaken no seikaku" 義和拳の性格 (The nature of the Boxer movement), in **Kindai Chūgoku kenkyū,** 1948, (9.8.6), pp. 245–267.

A model of historical condensation, this authoritative essay sums up the record of the Boxers' origin, affiliations and social-political status, concluding that they were essentially a traditional rebel-bandit type of organization and the degree of intentional Ch'ing dynasty support of them has been exaggerated.

2.7.2 ICHIKO Chūzō 市古宙三, "Giwaken zakkō" 義和拳雜考 (Notes on the Boxers), **Tōa ronsō,** 6 (April 1948), 90–97.

A scholarly but brief article on the origin and status of the Boxers, which concentrates on a critical appraisal of the views of G. N. Steiger in his **China and the Occident, 1927.**

2.7.3 MURAMATSU Yūji 村松祐次, "Giwakenran, 1900 nen, no seijiteki haikei" 義和拳亂――一九〇〇年――の政治的背景 ("Political Background of the Boxer Rebellion [1900]"), **Hitotsubashi ronsō,** 26.5 (Nov. 1951), 529–554.

A careful review of the perennial question whether the Boxers represented a popular anti-foreign secret society movement or a court-inspired militia organization. Using the **Ch'ing shih-lu** and other documentation, Prof. Muramatsu, like Prof. Ichiko (2.7.1), comes to a somewhat mixed conclusion, depending on the time to which the question is applied. His main approach to this knotty problem is to analyze the Ch'ing effort at systematic organization of militia throughout the country, which began in 1898.

2.7.4 MURAMATSU Yūji 村松祐次, "Giwadanran no shakai keizai teki haikei" 義和團亂の社會經濟的背景 ("Socio-economic Back Grounds of the Boxer Movement"), **Hitotsubashi ronsō,** 28.4 (Oct. 1952), 413–435.

A companion piece to the author's article on the Boxers' political background, 2.7.3, this article reexamines the views developed by previous writers (including Prof. Ichiko Chūzō 市古宙三 among the most recent, and with whom the present author finds some disagreement). The result is a highly sophisticated survey of this historical problem, which points up the decisive effect of the bad harvest in 1897 and the resulting famine and increased vagrancy in North China. The author proposes to publish further articles on the Boxers.

2.7.5 SOMURA Yasunobu 曾村保信, "Giwadan Jiken to Shinchō seifu" 義和團事件と清朝政府 ("The Boxer Outbreak in 1900 and the Chinese Government"), **Kokusaihō gaikō zasshi,** 51.2 (June 1952), 137–170.

This is an intelligent reappraisal of the important question how the Boxers were related to the dynasty. The author ascribes the vacillating and obscure attitude expressed in the edicts issued during the earlier stages of the movement to the delicate balance between the competing pro- and anti-Boxer factions at court. The former faction was led by die-hard Manchu nobles and the latter by enlightened and nationalistic Chinese officials in and out of Peking. The pro-Boxer Manchu faction prevailed in the end, identifying themselves with the Boxer movement probably because they wished to divert it from attacking the dynasty and at the same time felt themselves so much isolated and deserted by the Chinese that they were desperate to find allies.

2.7.6 YAMAMOTO Sumiko 山本澄子, "Giwaken no seikaku ni tsuite" 義和拳の性格に就いて ("Social character of Boxers' Rebellion"), **Shikan,** 33 (April 1950), 45–61.

Descriptive discussion of the various causes suggested as having contributed to the 1900 outbreak—anti-foreignism, dynastic encouragement, anti-official and anti-Manchu feeling, religious enthusiasm, the Boxers' possible status as a self-defense corps, and the like.

2.7.7 YANO Jin'ichi 矢野仁一, "Giwakempiran no shinsō ni tsuite" 義和拳匪亂の眞相に就いて (On the facts of the Boxer Rebellion), **Shigaku zasshi,** 38.9 (Sept. 1927), 815–840.

A lecture surveying the subject broadly as of 1927. See also his **Shinchō matsushi kenkyū** (2.1.12).

2.7.8 TABOHASHI Kiyoshi 田保橋潔, "Giwakempiran to Nichi-Ro" 義和拳匪亂と日露 (The Boxer Rebellion and Japan and Russia), in **Tōzai kōshō shiron,** 9.8.11, pp. 1051–1106.

This scholarly and critical article is a solid addition to the Boxer literature. Dr. Tabohashi makes extensive use both of Westerners' memoirs and of the **Ch'ing shih-lu** 實錄 and other Chinese and Japanese materials. In the first half, the author analyses the political process down to the Sino-foreign hostilities and stresses the point that the coming of the catastrophe must to a considerable extent be ascribed to the unnecessarily intransigent pressures imposed upon the Chinese court and provincial officials by the English, French and American ministers. In the latter half, the author traces the shift in the attitudes of the Japanese and Russian authorities, who were at first passive and almost indifferent, then gradually became active, and finally despatched large expeditionary forces.

2.7.9 INŌ Tentarō 稲生典太郎, "Giwadan Jihen to Nippon no shuppei gaikō" 義和團事變と日本の出兵外交 (The Boxer Incident and the negotiations concerning the sending of Japanese troops), pp. 497–562 in **Kaikoku hyakunen kinen Meiji bunkashi ronshū** 開國百年記念明治文化史論集 (Collected essays on Meiji cultural history, commemorating the centennial of the opening of the country), Kengensha 乾元社, 1952, 844 pp.

A systematic and detailed study of the Japanese response to the Boxer crisis, including the preparations of the General Staff, Foreign Office negotiations with Britain and others, mobilization of the Fifth Division and its despatch, etc., based mainly on Foreign Office publications, memoirs and other Japanese sources.

2.7.10 EGUCHI Bokurō 江口朴郎, "Giwadan Jiken no igi ni tsuite" 義和團事件の意義について (On the significance of the Boxer Rebellion), **Rekishigaku kenkyū,** 150 (Mar. 1951), 74–83.

A brief survey by a specialist in European history, of the events of 1900—origin, attitude of the court, scope, suppression by the powers—which adds little to previous studies but attempts to appraise the incident in the context of world politics and of the problem of Asian nationalism. Cogently criticized by Ichiko Chūzō 市古宙三 in **Rekishigaku kenkyū,** 153 (Sept. 1951), 49–52.

2.7.11 HATTORI Unokichi 服部宇之吉, **Pekin rōjō nikki** 北京籠城日記

(Diary of the siege of Peking), Kaimeidō 開明堂, 1926 (reprinted 1939, 164+39 pp., map, illus.).

Professor Hattori's diary (pp. 1–126) is accompanied by his reflections on the events of 1900, written in 1923 (pp. 127–164). An extract from the diary of the author's wife, Hattori Shigeko 繁子, who lived at the time of the siege in Ōsaki 大崎 near Tōkyō, is printed as an appendix (39 pp.).

2.7.12 SUGAWARA Sagae 菅原佐賀衛, Major General, Japanese Army, **Hoku-Shin Jihen shiyō** 北清事變史要 (Essentials of the North China incident), Kaikōsha 偕行社, 1926, 131 pp., map, illus.

By a general officer who had written other popularized accounts of Japanese military operations, this narrative relies on the Army General Staff's documented account, **Hoku-Shin jihen senshi** 北清事變戰史 (not seen by us), and acknowledges the assistance of other officers in its compilation. The narrative concisely summarizes the tactical operations of the international relief column, with sketch maps, the relief of the Peking legations, and the subsequent "pacification" operations against Boxers elsewhere.

2.7.13 INŌ Tentarō 稻生典太郎, "Giwadan Jihen to rengōgun sōshikikan sennin mondai" 義和團事變と連合軍指揮官選任問題 ("The consideration of the appointment of the Head commander of U. N. forces in Giwadan-incident"), **Kokushigaku** 國史學, pub. by Kokushigakkai 國史學會 ("The Society of Japanese Historical Research") with offices in Kokugakuin 國學院 University, Tōkyō, 55 (July 1951), 29–40.

This fact-finding article tries to throw light on the controversial point whether Marshal von Waldersee was appointed to be commander of the Allied forces at the time of the Boxer incident through the urging of the Tsar or by the Kaiser's own initiative, and suggests that the appointment may have been made as the result of a Russo-German bargain. Though not clearly stated, this article seems to be based on the Japanese Foreign Office archives.

2.7.14 TAMURA Kōsaku 田村幸策, "Iwayuru Ei-Doku Yōsukō kyōtei no saikentō" 所謂英獨揚子江協定の再檢討 (A reexamination of the so-called Anglo-German Yangtze Agreement), **Tōagaku**, 4 (April 1942), 1–37.

The author uses the British and German war-origin documents to analyze this agreement of October 1900 as an aspect of great power diplomacy in the aftermath of the Boxer incident.

3. POLITICAL INSTITUTIONS

Note: This section concerns institutions rather than events. Strictly speaking, it should begin with sec. 1.3 above, "surveys of government", since most of the items here listed are supplementary to **Shinkoku gyōseihō** (1.2.1) and similar general works. On the whole, modern political science studies of Ch'ing institutions are remarkable for their rarity. The reader should perhaps be reminded that Western political science avoids regional specialization in order to be universal. As a result it has thus far proved incapable of embracing and understanding that half of man's political experience which has occurred east of Suez, and has remained parochial. The disastrous result is evident in our modern divided world, in which the Asian passivity toward authoritarian government remains incredible to the Western mind. In the sections which follow, particularly 3.3 on "Alien rule over China", Japanese scholars have broken new ground which deserves our exploration. At the start, it would be salutary to acknowledge the fruitful stimulus which their studies have received from Marxism and particularly from K. A. Wittfogel, as well as the contribution they have made entirely without benefit of Marxist stimuli (see, e.g., sec. 3.5). In sec. 3.7 it will be noted that we include the Western treaty system under the political institutions of the Ch'ing dynasty. The far-reaching implications of this simple and obvious fact have not yet been seriously studied, East or West.

3.1 THE MONARCHY AND BUREAUCRACY

Note: The recrudescence of a centralized authoritarian power in China, in an intensified form, should stimulate Western research into China's traditional political institutions, beginning with the twin foci of the ruler and his officials. For this, Japanese studies beginning with those of Inaba Iwakichi noted above (sec. 2.1) and continuing with those of Miyazaki Ichisada and others in this section, have much to offer. Other items, on the yamen clerks and minor officials, are in sec. 3.4 below. Note 6.7.1 on the examination system.

3.1.1 MIYAZAKI Ichisada 宮崎市定, **Yō-sei-tei, Chūgoku no dokusai kunshu** 雍正帝，中國の獨裁君主 (The Yung-cheng Emperor—China's autocratic ruler), Iwanami Shoten, 1950, 168+6 pp., 2 plates, map (Iwanami shinsho, 29).

A popularly-written study of the emperor's career and incidents in it (e.g. the Rites Controversy), stressing the despotic nature of the imperial institution in this period. Based on an intimate knowledge of the subject, though sources are omitted.

3.1.2 MIYAZAKI Ichisada 宮崎市定, "Shinchō ni okeru kokugo mondai no ichimen" 清朝に於ける國語問題の一面 (A view of the problem of national languages in the Ch'ing dynasty), **Tōhōshi ronsō** 東方史論叢, 1 (July 1947), 1-56.

This interesting article goes into the use made of Manchu and Chinese in the Ch'ing administration, as an aspect of the general problem of the rule of foreign invaders in China. (For Prof. Miyazaki's part in the Tōa Kenkyūjo study of this larger topic during the war, see its publication, 3.3.1.) Using obvious sources like the **Tung-hua lu** 東華錄, he here puts together a story never before worked out, as to the use of languages and translation between them in the central organs of the Ch'ing regime such as the Grand Secretariat (*Nei-ko* 內閣), Grand Council (*Chün-chi-ch'u* 軍機處) and the various other yamen at Peking, including the inevitable decline of the use of Manchu. This article is noteworthy also as providing an analysis of the emergence and vicissitudes of the Ch'ing Grand Secretariat, closely intertwined with the shift in high-level powerholding from the Manchu magnates to the emperor, on the one hand, and from the Manchus to the Manchu-Chinese bureaucracy, on the other, leading to a consolidation of the despotic power of the emperor.

3.1.3 Tōyō Bunka Kenkyūkaigi 東洋文化研究會議 (Eastern cultural research conference), **Tōyō no ie to kanryō** 東洋の家と官僚 (Family and bureaucracy in the Orient), Seikatsusha, 1948, 296 pp.

In this symposium the family is studied in India, China, Islam and Japan, and bureaucracy in Europe-America, China and Japan. The China sections are: Niida Noboru 仁井田陞 on the Chinese family, pp. 37–85; Ono Shinobu 小野忍, on literature critical of bureaucracy in China, pp. 199–219; and Wada Sei on Chinese bureaucracy, pp. 232–243. Each presentation is followed by a transcript of discussion in which a highly competent panel participated. This conference was sponsored by the Tōyō Bunko, Tōyō Bunka Kenkyūjo, Chūgoku Bungaku Kenkyūkai 中國文學研究會 (Chinese literature research association), and Chūgoku Kenkyūjo in order to promote the joint study of Asian problems, especially to stimulate collaboration among different sections of the social sciences. Niida's report, subtitled "Discipline of rural family labor in China", is a socio-legal analysis based on field survey reports and was reproduced, enlarged and revised, as one chapter in his **Chūgoku no nōson kazoku**, 8.4.1, see pp. 145–217. Ono's report comments on **Ju-lin wai-shih** 儒林外史 and **Lao-ts'an yu-chi** 老殘遊記, analysing the approach of the authors as traditional intellectuals. Wada's report is a handy historical survey. The transcripts of discussion provide an interesting panorama of the main scholarly approaches in Japan to the problems of China.

3.1.4 KANDA Nobuo 神田信夫, "Shinsho no gisei daijin ni tsuite" 清初の議政大臣について ("The I-cheng Ta-ch'en in the Earlier Ch'ing Dynasty"), pp. 171–189 in **Wada hakushi kanreki kinen Tōyōshi ronsō**, 9.8.12.

This study of this early and informal Ch'ing administrative institution provides interesting background information on the Grand Council (Chün-chi-ch'u), inasmuch as the latter succeeded it.

3.1.5 TAKIGAWA Masajirō 瀧川政次郎, "Shindai bumbukan fukusei kō" 清代文武官服制考 (A study of the system of uniforms for civil and military officials in the Ch'ing period), **Shigaku zasshi,** 53.1 (Jan. 1942), 1–65, illus.

Based on the author's collection of old costumes, on testimony secured from certain ex-mandarins and on the official Ch'ing regulations, this article discusses (1) the historical origins of the Ch'ing system of official costume, (2) the various types of dress, differentiated according to social status, official rank, official position, circumstances of the occasion, and the season and the weather, (3) various penal regulations concerning infringements of the system of uniforms, (4) granting of uniform or ornament as a reward of merit, and (5) contrarily, the suspension of such privileges as a disciplinary measure. The rest of the article (pp. 28–62) then describes the various articles of official dress, such as official caps, peacock feathers, mandarin squares and necklaces, and finally touches on the general characteristics of the system of costume as a combination of Manchu and Chinese traditions.

3.1.6 ASAI Torao 淺井虎夫, "Sōtoku jumbu ken gyoshi kō" 總督巡撫兼御史考 (A study of governors general and governors with joint appointments as censors), **Shigaku zasshi,** 15.7 (July 1904), 783–801.

A brief study evidently based mainly on the **Ta-Ch'ing hui-tien,** but referring to the background of the earlier periods.

3.1.7 HOSOI Shōji 細井昌治, "Shinsho no shori, shakaishiteki ichi-kōsatsu" 清初の胥吏——社會史的一考察 (The official clerks of the early Ch'ing, a socio-historical investigation), **Shakai keizai shigaku,** 14.6 (Sept. 1944), 1–23.

This article, widely based on the Ch'ing documentation concerning administration, considers the legal status of the yamen underlings, their social position and relationship to the official funds. It is one of the few studies of this important class and gives useful bibliographical leads for further work. The author expresses indebtedness to Wada Sei, Katō Shigeshi, Niida Noboru and Masui Tsuneo.

3.1.8 HATANO Yoshihiro 波多野善大, "Chūgoku kanryō no shōgyō-kōrikashi-teki seikaku" 中國官僚の商業高利貸的性格 ("The Chinese Bureaucrat as Merchant-Usurer under the Ch'ing Dynasty"), **Tōyōshi kenkyū,** 11.3 (Oct. 1951), 233–252.

Subtitled as "centering upon two examples of the Liang-huai 兩淮 salt merchants of the late Ch'ing," this article studies the official-landlord-merchant-usurer connection, as illustrated by activities of the Ch'iao 喬 and Lu 盧 families. This illuminating and suggestive article is based on the records of lawsuits, and so is considerably detailed.

3.2 MILITARY SYSTEM: THE MANCHU BANNERS

Note: The early development of the banner system in Manchuria has been studied in Japan with great intensity and thoroughness. While outside our period, this topic may be pursued through the bibliography of 32 studies and 64 sources described on pp. 467–495 of Sudō Yoshiyuki's magistral monograph (3.2.2). The work of Abe Takeo (3.2.1) is also authoritative. On the growth-process of the Peiyang army see 8.3.6.

3.2.1 ABE Takeo 安部健夫, "Hakki Manshū *niru* no kenkyū" 八旗滿洲ニルの研究 ("On the Manchu Niru System of the Eight Banners"), **Tōa jimbun gakuhō,** 1.4 (Feb. 1942), 799–875; continued in ibid., 2.2 (July 1942), 174–238; and **Tōhō gakuhō, Kyōto,** 20 (Mar. 1951), 1–134, maps, tables.

> While this long, authoritative and detailed study deals mainly with the origin and early development of the Banner system in its political, military, economic and other aspects, it should be noted here as of major background importance for an understanding of the system in its later decline. Prof. Abe is at the Jimbunkagaku Kenkyūjo, Kyōto University.

3.2.2 SUDō Yoshiyuki 周藤吉之, **Shindai Manshū tochi seisaku no kenkyū, tokuni kichi seisaku o chūshin toshite** 清代滿洲土地政策の研究 ——特に旗地政策を中心として (A study of Manchu land policy under the Ch'ing dynasty, with special attention to the policy concerning banner lands), Kawade Shobō, 1944, 495+index 13 pp.

> This thorough monograph, by a specialist best known for his study of Sung land relations, begins with the development of the banner land system before 1644 and then describes its subsequent administration in each of the three areas of Fengtien, Kirin, and Heilungkiang. Special note is taken of the problem produced by Chinese migration into Manchuria and its prohibition. Prof. Sudō is now at the Tōyō Bunka Kenkyūjo, Tōkyō University.

3.2.3 SUDō Yoshiyuki 周藤吉之, "Shinchō ni okeru Manshū chūbō no tokushusei ni kansuru ichikōsatsu" 清朝に於ける滿洲駐防の特殊性に關する一考察 ("A study on the Characteristics of the Bannerman Garrisons in Manchuria during the Ch'ing dynasty"), **Tōhō gakuhō, Tōkyō,** 11.1 (Mar. 1940), 176–203.

> A useful summary of the official figures for the period ca. 1644–1850, with 20 pages of data tabulated from the major sources to show the garrisons which were, at least, recorded for the area of Manchuria. Statistical comparisons are given with the garrisons in China proper.

3.2.4 SUDō Yoshiyuki 周藤吉之, "Shinchō chūki ni okeru kichi no kosaku kankei" 清朝中期に於ける旗地の小作關係 ("The Relation of the Tenant to the Bannerman Land Lord in the Middle of the Ch'ing Dynasty"), **Tōhō gakuhō, Tōkyō,** 12.1 (May 1941), 1–25.

Using a collection of *Hu-pu* 戶部 ("Board of Revenue") documents, Prof. Sudō analyzes the land use system on a bannerman's estate in the 18th and early 19th centuries, pointing out the customary iniquities of the estate managers in their relations with the domestic slaves (*chia-jen* 家人) and tenant-farmers (*tien-hu* 佃戶).

3.2.5 KITAYAMA Yasuo 北山康夫, "Shindai no chūbō hakki ni tsuite" 清代の駐防八旗について ("On the garrisons of the Eight Banners of the Ch'ing Period"), pp. 489–504 in **Haneda hakushi shōju kinen Tōyōshi ronsō**, 9.8.3.

A brief account of the setting up of the provincial garrisons of Manchu banner forces after the founding of the Ch'ing regime.

3.2.6 YONEKURA Jirō 米倉二郎, "Shindai Hoku-Man no tonkon, sōjōho tonden no yosatsu hōkoku o chūshin to shite" 清代北滿の屯墾――雙城堡屯田の豫察報告を中心として (The military colonization in North Manchuria during the Ch'ing period, with special reference to the preliminary report on the military colonists' lands in Shuang-ch'eng-pao 雙城堡), **Tōa jimbun gakuhō**, 1.3 (Dec. 1941), 698–713, maps, illus.

This is a brief field survey report, written against the historical background of the vicissitudes of the military colonization system in North Manchuria.

3.3 ALIEN RULE OVER CHINA

Note: These studies, especially item 3.3.1, explore a basic aspect of Chinese politics which has had unusual interest for Japan but is now of vital interest to the West as well. They relate, in short, to that age-old and still current problem, how to keep the Chinese populace under the control of a central authority, especially when that authority is held by an alien élite or by an élite whose highest loyalty is to an alien power.

3.3.1 Tōa Kenkyūjo 東亞研究所, **Iminzoku no Shina tōchi gaisetsu** 異民族の支那統治概說 (Outline of the rule of alien peoples over China), Tōa Kenkyūjo, 1943, 312+chronology 17 pp.; **Iminzoku no Shina tōchishi** (History of ...), 1944, 424 pp. (reprint of preceding in smaller page-size with addition of pp. 34–72 on "Liao" 遼).

This wartime collaborative survey of the "dynasties of conquest" was produced by a specially-formed group of 16 Kyōto and Tōkyō scholars who were mobilized to provide historical knowledge as an aid to policy in the period 1938–41. They wrote 19 essays, from which the final volume was drawn. (For authorship of sections, see preface). The dynasties of Northern Wei, Chin, Yuan and Ch'ing are each analyzed as to origin, official and military systems, economic and religious institutions, and policies toward China and Inner Asia, as a basis for comparative study. The sections on the Ch'ing (pp. 159–312, 1943 ed.), are by Momose Hiromu 百瀬弘, Miyazaki Ichisada 宮崎市定, Tamura

Jitsuzō 田村實造, Abe Takeo 安部健夫, Sudō Yoshiyuki 周藤吉之, and similar specialists.   In addition to extensive treatment of the official, military, and fiscal systems, there is an analysis of Manchu measures in the field of thought and ideology and in matters of social policy, as well as toward the Mongols, moslems, and Koreans.   In so brief a space no attempt is made to indicate sources.   While the academic caution of the authors may have been occasionally affected by the circumstances of compilation in wartime, this volume still stands almost alone as an over-all study of the phenomena of foreign conquest of China.

3.3.2 MOMOSE Hiromu 百瀬弘,, "Shinchō no iminzoku tōchi ni okeru zaisei keizai seisaku" 清朝の異民族統治に於ける財政經濟政策 (The fiscal and economic policies used during the Ch'ing period in the control of an alien race over China), **Tōa kenkyū shohō,** 20 (Feb. 1943), 1–116, many tables.

This excellent analysis is one result of the joint study of the alien control of China, sponsored by the Tōa Kenkyūjo (see Tōa Kenkyūjo, **Iminzoku no Shina tōchi gaisetsu** also published in 1943, 3.3.1, to which the author was a contributor.   He appears to interpret his title here as referring to Ch'ing control over alien, i.e. Chinese and other, races—but this is a purely grammatical difference).   He analyzes the Manchus' fiscal inheritance from the Ming, their early fiscal changes and increases of both expenditure and income (noting also remissions of taxes), and then takes up in detail the various types of Ch'ing expenditure and the major sources of income, comparing one period with others in a number of tables and seeking totals of income and outgo. This study of central government finance, useful in itself, is illumined by the author's attention to the underlying question of the Manchu administrators' methods of adopting Chinese forms, maintaining official salaries, encouraging cultivation and the like, as a basis of continued political control.   The result is an extremely illuminating contribution.

3.3.3 MIYAGAWA Hisayuki 宮川尚志, "Shin no sembu kyūshin kōsaku" 清の宣撫救賑工作 (The Ch'ing policies of non-forcible control and relief works), **Tōa kenkyū shohō,** 19 (Dec. 1942), 966–1064.

Based on a report compiled in connection with the Tōa Kenkyūjo project on alien control of China, 3.3.1, this study brings together evidence mainly from the earlier period of the Ch'ing dynasty on the imperial regime's methods of control supplementary to military and forcible means, beginning with an historical examination of *hsüan-fu* 宣撫 ("pacification") policies in action and the traditional ideas behind them.   Early imperial policies, treatment of Ming remnants and Chinese who surrendered, equal treatment of Chinese and Manchus, the wearing of the queue, treatment of rebels, relief of disaster and indigence, etc., are then pursued as examples of this policy.   The author is now at Okayama University.

3.3.4 ARITAKA Iwao 有高巖, "Gen Shin nichō no tai-Kan seisaku sōi no yurai" 元清二朝の對漢政策相違の由來 ("An Inquiry into the Diplomatic Attitudes towards the Chinese People adopted by the Gen and Shin Dynasties"), **Shichō,** 4.1 (Mar. 1934), 16–32.

This comparison of Mongol (Yuan) and Manchu (Ch'ing) policy is a brief but well informed essay which notes the two peoples' differences regarding their knowledge of agricultural life, reliance upon military force, and relationship to the rest of Asia, and the contrast in their success in overcoming Chinese opposition.

3.3.5 KUWABARA Jitsuzō 桑原隲藏, "Shinajin bempatsu no rekishi" 支那人辮髪の歴史 (A history of the queue among the Chinese), **Geimon,** 4.2 (Feb. 1913), 199–212.

This brief account traces the queue back into the Chin 金 period and among the Mongols, analyzes its socio-political use by the Ch'ing rulers, and notes early Western references to its symbolic value in the 19th century, and its widespread survival after the first revolution. Reprinted in the author's collected writings, 9.8.10.

3.4 LOCAL GOVERNMENT

Note: This basic aspect of political institutions, relatively neglected in the West, has been greatly advanced by the volume edited by Dr. Wada and particularly by the work of Matsumoto Yoshimi. Note that the study of Kano Naoki (3.4.7) touches on the relationships among lower strata of the bureaucracy which are also analyzed by Hosoi Shōji (3.1.7) and others in sec. 3.1 above. Sec. 8.2 contains further materials on this subject (see especially 8.2.9). See also 1.1.17.

3.4.1 WADA Sei 和田清, ed., for Chūkaminkoku Hōsei Kenkyūkai 中華民國法制研究會 (Research society on the legal institutions of the Chinese Republic), **Shina chihō jichi hattatsu shi** 支那地方自治發達史 (History of the development of local self-government in China), Chūō Daigaku 中央大學, 1939, 278 pp.

Projected at the request of a research-promoting group headed by Matsumoto Jōji 松本烝治, this volume has chapters on Sung, Yuan, Ming, Ch'ing (pp. 128–182), and the Republic (pp. 183–220), and a documentary appendix on the *pao-chia* 保甲, militia (*t'uan-lien* 團練) and other local institutions. The Ch'ing section by Matsumoto Yoshimi 松本善海, like the other sections, treats the local administrative system from the point of view of the ruling dynasty's problem of political control over the countryside, and goes on to deal with the tax system, development of the *pao-chia*, the use of militia against rebels, and the like. This is one of the most penetrating studies of this important subject in succession to the work of Inaba Iwakichi, 2.1.3, although limited mainly to the official, rather than the popular, aspects of it. See review by Ichiko Chūzō 市古宙三 in **Shigaku zasshi,** 51.4 (Apr. 1940), 552–555. See also 8.2.6 (Shimizu).

3.4.2 MATSUMOTO Yoshimi 松本善海, "Shindai ni okeru sōkōseido no sōritsu" 清代における總甲制度の創立 ("The *Tsung-chia* System as Formulated by the Ch'ing Government"), **Tōhō gakuhō, Tōkyō,** 13.1 (May 1942), 109–142.

This is a substantial study continuing the author's theoretical discussion of the same topic of local control in Wada Sei, ed., **Shina chihō jichi hattatsushi,** 3.4.1. It throws a scholarly light on the circumstances and the consequences of the setting up of machinery for maintaining order in local districts in the early Ch'ing period. Special attention is devoted to the relationship of this system to the problem of preventing the escape of Chinese farming slaves from the bannermen's estates in Chihli province. Prof. Matsumoto is now at the Tōyō Bunka Kenkyūjo, Tōkyō University.

3.4.3 MATSUMOTO Yoshimi 松本善海, "Chūgoku ni okeru chihō jichi seido kindaika no katei" 中國における地方自治制度近代化の過程 (The process of modernization of the institutions for local self-government in China), pp. 61–81 in Niida Noboru, ed., **Kindai Chūgoku no shakai to keizai,** 5.7.1.

A brief historical survey of Sun Yat-sen's local self-government ideas and program in 1924 and the centralization and other administrative changes attempted under the KMT law of 1940, the latter resulting in a reactionary revival of the *pao-chia* 保甲 system as a tool for repressing "democratization from below".

3.4.4 MATSUMOTO Yoshimi 松本善海, "Kyōchin seido" 鄉鎮制度 (The village system), in **Kindai Chūgoku kenkyū,** 1948, 9.8.6., pp. 27–51.

A comparison of the basic local government of the late-Ch'ing period with developments in the first years of the Republic—part of a projected larger study. Although rather sketchy, this article gives many suggestive interpretations.

3.4.5 YAMADA Hideji 山田秀二, "Min-Shin jidai no sonraku jichi ni tsuite" 明清時代の村落自治について ("Self-government in Village Communities of Ming-Ching China"), **Rekishigaku kenkyū,** 2.3 (July 1934), 214–230; 2.5 (Sept. 1934), 15–22; 2.6 (Oct. 1934), 2–30.

This long posthumous article studies (1) the structure of the organs of collective responsibility (*li-chia* 里甲, *pao-chia* 保甲, etc.) which formed the institutional basis of "local self-government" under the Ming and Ch'ing, respectively; and (2) the customary forms of cooperative activity (*hsieh-t'ung* 協同) regarding irrigation, famine or natural calamity, roads, etc. By adding a discussion of social factors like patriarchal authority and the cult of ancestors, the author rounds out a rather old-fashioned description of local government.

3.4.6 YAMAMOTO Yoshizō 山本義三, "Kyū-Manshū ni okeru kyōson tōchi no keitai" 舊滿洲に於ける鄉村統治の形態 (Forms of village control in old Manchuria), **Mantetsu chōsa geppō,** 21.11 (Nov. 1941), 1–56.

A study of the organs of local control maintained by the Ch'ing regime in official villages and those of bannermen and commoners, with special attention to the various measures and devices used in the last—such as selection of leading persons for semi-official positions or duties, etc. This articles carries forward the general approach of the volume edited by Dr. Wada, 3.4.1, and also of Shimizu Morimitu's **Shina shakai no kōzō,** 8.1.2. It also adds data on 7 major sources (archives of local governmental agencies preserved at the Mukden Library.).

3.4.7 KANO Naoki 狩野直喜, "Shinchō chihō seido" 清朝地方制度 (The local institutions of the Ch'ing dynasty), pp. 133–176 in his **Dokusho san'yo** 讀書纂餘, Kōbundō, 1947, 289 pp.

Originally published in Nov. 1907 in **Eizan kōenshū** 叡山講演集 (Collected lectures given at Hieizan 比叡山 [the site of Enryakuji 延曆寺, a famous temple near Kyōto]), when the Empress Dowager was still in power and had pronounced in favor of a constitution, this somewhat rambling lecture discusses the various posts in the provincial hierarchy and their interrelation (particularly between governor and governor general), mentions the intendants of circuit (taotais), and then gets down to details (from p. 148) on the functions of the hsien yamen and the literati and their relations. This last seems well informed and is one of the few illuminating accounts of the subject. Professor Kano had been a participant in compiling **Shinkoku gyōseihō**, 1.2.1, and was for many years a leading sinologue at Kyōto University.

3.4.8 NAKAMURA Jihei 中村治兵衞, "Shindai ni okeru chihō seiji no ichikōsatsu" 清代における地方政治の一考察 ("Eine Betrachtung über die Landesverwaltung in Chin-Dynastce [*sic*]"), **Rekishigaku kenkyū,** 11.2, whole number 86 (Feb. 1941), 86–94.

In view of the Nationalist Government's "new hsien system" 新縣制 announced to begin in 1940, the author outlines briefly the system below the hsien level, with comments on it. His article is based on local gazetteers.

3.4.9 KOBAYAKAWA Kingo 小早川欣吾, "Shinjidai ni okeru chihōjichi dantai no hai no keishiki ni tsuite, tokuni hokō seido o chūshin toshite" 清時代に於ける地方自治團體の牌の形式について――特に保甲制度を中心として (On the forms of tablets [*p'ai* 牌] used in the local self-governing bodies during the Ch'ing period, with special reference to the *pao-chia* 保甲 system), **Tōa jimbun gakuhō,** 1.2 (Sept. 1941), 370–422.

A rather formalistic and meticulously detailed and illustrated study of the various forms of tablets required to be hung at every door to indicate certain prescribed data about persons dwelling within, which thus served as tools of governmental control through the *pao-chia* system. Half a dozen types are described.

3.4.10 KOBAYAKAWA Kingo 小早川欣吾, "Shinjidai ni okeru hokōsatsu no keishiki to sono hensei ni tsuite" 清時代に於ける保甲冊の形式と其の編制について (On the forms of *pao-chia* registers and their compilation in the Ch'ing period), **Tōa jimbun gakuhō,** 3.1 (Mar. 1943), 73–143.

The author uses dynastic compilations, gazetteers, and other sources to describe various kinds of *pao-chia* registers, provides several samples of their form, and briefly mentions the procedure by which they were compiled. These registers being of key importance in the local control of population, the system of registration was applied to various types of persons, e.g. priests, and occupations, e.g. the fishing industry, in a great variety of forms. This is a basic textual study. The author was a specialist in Japanese legal history of the Meiji period.

3.4.11 IMAHORI Seiji 今堀誠二, **Peipin shimin no jichi kōsei** 北平市民の自治構成 (The structures for self-government among the populace of the city of Peiping), Bunkyūdō 文求堂, 1947, 195 pp., map, illus. Preface by Niida Noboru.

A systematic survey (by items and categories) of popular institutions, based on careful first-hand investigation as well as textual research, which describes the communal arrangements for public utilities and public safety (fire, water, roads, watchmen, etc.), religious, philanthropic and eleemosynary activities, and the like, all organized on the basis of neighborhood associations. See the reviews by Dr. Niida, in **Tōyō bunka kenkyū,** 9 (Sept. 1948), 53–55; by Ubukata Naokichi 幼方直吉 in **Chūgoku kenkyū,** 6 (Jan. 1949), 61–64.

3.4.12 FUJIWARA Sadamu 藤原定, "Shindai ni okeru mimpon shisō to shomin no chii" 清代に於ける民本思想と庶民の地位 (The concept of the populace as the foundation of the state in the Ch'ing period and the status of the common people), **Mantetsu chōsa geppō,** 20.11 (Nov. 1940), 1–68.

This socio-historical discussion takes up the traditional position of the urban populace in a semi-feudal bureaucratic despotism, the modern city and the restraints imposed upon it by the local governmental system, forms of popular association in the context of the closed village economy, traditional forms of protest, and against all this background the Ch'ing philosophers' assertion of the importance of the populace. Huang Tsung-hsi, Ku Yen-wu and others are quoted in a context which seems to have been derived partly from Marx and Max Weber.

3.4.13 HIRANO Yoshitarō 平野義太郎, "Chūgoku ni okeru kensei" 中國に於ける縣政 (Hsien government in China), in **Kindai Chūgoku kenkyū,** 1948, 9.8.6, pp. 2–23.

A quick sketch, outlining modern developments, with a few pages also on the regions under communist control.

3.4.14 WATANABE Sōtarō 渡邊宗太郎, "Jichi rippō no dōkō" 自治立法の動向 (Tendencies in the legal establishment of self-government), **Tōa jimbun gakuhō,** 2.3 (Dec. 1942), 299–316.

Distinguishing between self-government spontaneously developed from gradual association and self-government imposed by leadership (i.e. totalitarianism), this article briefly discusses the possibility of a choice between these two approaches under Wang Ching-wei's regime. The author is a legal scientist specialized in public law.

3.5 TRIBUTARY CONTROL OF INNER ASIA

Note: This section carries on from sec. 3.3 above and in turn is continued in sec. 4.8 below, the political institutions involved in "alien rule over China" being related to those used in a Chinese dynasty's "control of Inner Asia," and to the loss of it in the modern period. Owen Lattimore's **Inner Asian Frontiers of China** (1940) is one of the few Western studies which pursue the same topic.

3.5.1   Tōa Kenkyūjo 東亞研究所, ed., **Iminzoku no Shina tōchi kenkyū, Shinchō no henkyō tōchi seisaku** 異民族の支那統治研究——清朝の邊疆統治政策 (Studies on the rule of foreign peoples over China, the Ch'ing dynasty policy for control of border areas), Shibundō 至文堂, 1944, 398 pp.

A product of scholarly research directed to problems of wartime importance, this study grew out of the general project which also produced the survey volume **Iminzoku Shina tōchishi gaisetsu**, 3.3.1.   The chief sections of the present volume deal with Ch'ing policy concerning (1) Mongolia (pp. 15–97) by Tamura Jitsuzō 田村實造, (2) the Moslems (pp. 101–213) by Haneda Akira 羽田明, and (3) Korea (pp. 217–255) by Hatada Takashi 旗田巍.   The remainder of the volume, by Toyama Gunji 外山軍治, deals with the policies of the Chin 金 Dynasty toward the Sung and the Khitan.   Though written as topical surveys rather than monographic research, under a general project allegedly sponsored as a guide to Japan's wartime racial policies in the Co-prosperity Sphere of Greater East Asia, these writings by trained specialists in each field nevertheless have considerable scholarly value.

3.5.2   YANO Jin'ichi 矢野仁一, "Biruma no Shina ni taisuru chōkō kankei ni tsuite" 緬甸の支那に對する朝貢關係に就いて (On the tributary relationship of Burma to China), **Tōyō gakuhō**, 17.1 (April 1928), 1–39.

A factual account, based on Chinese sources, of the vicissitudes of Chinese control over Burma through the tribute system, mainly in the Ming and Ch'ing periods to the 19th century.   The *t'u-ssu* 土司 system by which native tribal chieftains were invested with Chinese official status, and the divide-and-rule policy which accompanied it, are noted in detail.

3.5.3   YANO Jin'ichi 矢野仁一, "Shina no doshi ni tsuite" 支那の土司に就いて ("On T'u-Ssu, the Half-Independent Native Principalities on the Borderland of China"), **Shinagaku**, 3.3 (Dec. 1922), 161–183.

A brief factual and institutional sketch of the tribal chieftains on the Inner Asian frontier, with a list, drawn from the **Hui-tien** 會典, gazetteers and Western writers.   Later included in Dr. Yano's **Kindai Shina shi**, 2.1.11.

3.5.4   YANO Jin'ichi 矢野仁一, "Kindai Chibetto shi kenkyū" 近代西藏史研究 (An historical study of modern Tibet), 76 pp., in **Tōyōshi kōza** 東洋史講座 (Lectures on Oriental history), vol. 14, Yūzankaku 雄山閣, 1930.

In this survey article Dr. Yano deals with the names applied to the Tibetan area, its extent, the course of Sino-Tibetan relations down to 1792, and their institutional basis, with a quick look at later British contact.

3.5.5 SUZUKI Chūsei 鈴木中正, "Shin to Guruka oyobi Eiryō-Indo tono kankei" 清とグルカ及び英領印度との關係 ("Political relations between Gurkha, British India and China under the Ch'ing 清"), **Tōyō gakuhō,** 32.1 (Oct. ["Sept." sic] 1948), 38–61.

Subtitled "At the time of the British-Gurkha war in 1814–1816", this factual study is a part of the author's general research on Sino-British relations in Tibet and Nepal from the late 18th down to the mid-19th century. While English and French works provide most of the background, the Chinese documents tell much of the actual story.

3.5.6 SUZUKI Chūsei ("Nakamasa") 鈴木中正, "Ahen sensō tōji ni okeru Shinchō to Nepāru oyobi Ei-In to no kankei" 阿片戰爭當時に於ける清朝とネパール及び英印との關係 ("China's Relationships with Nepaul and British India during the Opium War"), (**Aichi Daigaku**) **Bungaku ronsō** (愛知大學)文學論叢 ("The Journal of the Literary Association of Aichi University"), 3 (Nov. 1950), 49–79.

Tracing the course of Chinese relations with the Gurkas, Professor Suzuki analyzes the impact of the British upon this aspect of Ch'ing tributary relations.

3.5.7 SUZUKI Chūsei 鈴木中正, "Taihei Tengoku tōji ni okeru Shinchō to Chibetto oyobi Nepāru tono kankei" 太平天國當時に於ける清朝とチベット及びネパールとの關係 ("Expansion and Withdrawal of Chinese Power in Tibet and Nepal during the Ch'ing"), pp. 345–358 in **Wada hakushi kanreki kinen Tōyōshi ronsō,** 9.8.12.

An account of the three-cornered relations of the Manchu power with Tibet and the Gurkhas, mainly in the 1840s and 1850s; see the author's **Shinchō chūkishi kenkyū,** 2.4.1, and his articles noted just above. In these studies Professor Suzuki breaks new ground in looking at Ch'ing tributary relations in the nineteenth century from the side of Peking.

3.5.8 SAGUCHI Tōru 佐口透, "Shinkyō Uiguru shakai no nōgyō mondai, 1760~1820 nen," 新疆ウィグル社會の農業問題, 一七六〇～一八二〇年 ("Agrarian problems of Hsin Chiang Uighurs, mainly during the period 1760–1820"), **Shigaku zasshi,** 59.12 (Dec. 1950), 1094–1122.

Mainly based on Chinese documents (**Shih-lu,** and local gazetteers), this article attempts to grasp the social structure of the Uigur population of Sinkiang who came under Ch'ing control in the late 18th century. It deals carefully with their land-system, agricultural technology and class relations. The author is a leading specialist in the socio-historical study of Chinese Turkestan, now assistant professor of oriental history in Toyama University 富山大學.

3.5.9 SAGUCHI Tōru 佐口透, "Kazaffu to Shinteikoku tono kemba bōeki" カザッフと清帝國との絹馬貿易 ("The Silk-Horse Trade between the Ch'ing and the Kazakhs"), pp. 9–48 in **Yūboku minzoku no shakai to bunka**

遊牧民族の社會と文化 ("Studies in Nomadic Society and Culture"), ed. by the Society for Eurasian Studies ユーラシア學會, Kyōto: Shizenshi Gakkai 自然史學會, 1952, 219 pp.

Mainly based on a careful perusal of the **Ch'ing shih-lu,** this article analyzes the relations between the Ch'ing government and the Kazakhs, who were a Ch'ing tributary from the mid-18th century until annexed by Russia after 1840, through the barter of Chinese silk for Kazakh horses. The author pays attention to the seasons of trade, the quantities and prices of goods, the method of barter, etc., and also to the role played by the Kazakhs as a channel for Sino-Russian trade.

3.5.10 SHIMADA Jōhei 島田襄平, "Shindai Kaikyō no jintōzei" 清代囘疆の人頭稅 ("Poll tax in Kashgaria under Manchurian Dominion"), **Shigaku zasshi,** 61.11 (Nov. 1952), 995–1010, one table.

This is a brief but heavily annotated analysis of the Ch'ing tax system applied to Kashgaria, in its historical development down to the 19th century. Ten pages of footnotes provide extensive bibliography.

3.5.11 ANZAI Kuraji 安齋庫治, "Mōkyō shakai kōzō no kōshinsei" 蒙疆社會構造の後進性 (The backwardness of the Mongolian social structure), **Chūgoku hyōron** 中國評論, pub. by Nippon Hyōronsha for the Chūgoku Kenkyūjo (predecessor to **Chūgoku kenkyū**), 2.1 (Jan. 1947), 21–33.

With a heavy use of Marxist terminology, this article describes the Jassak 札薩克 system in Mongol society of the Ch'ing period and the controls superimposed upon it, together with the chief features of socio-economic structure and types of landlord-tenant relations in the ensuing period, noting the links between this region and the domestic and foreign markets.

3.6 LAW

Note: This section is dominated by the magistral studies of Dr. Niida Noboru. Item 3.6.7. on family law of course relates to family studies in sec. 8.4 below. Most of the remaining items relate to landownership in one way or another and are therefore closely relevant to the materials presented in sec. 7.6 on landholding and sec. 8.2 on village studies. Large studies of codes are at the end (3.6.19–21).

3.6.1 NIIDA Noboru 仁井田陞, **Chūgoku hōsei shi** 中國法制史 (History of the Chinese legal system), Iwanami Shoten, 1952, 363+indices and chronological table 25 pp. (Iwanami zensho, no. 165).

This authoritative socio-legal treatise describes China's legal institutions affecting criminal punishment, court procedure, mediation, social status (esp. slavery), feudalism, cities and gilds, persons, population records, the family and clan, land, exchange, property, and the like—in each case against the background of social structure and custom. The book opens, in fact, with an essay on the authoritarian nature of Oriental (*Tōyō* 東洋) Society and the influence of Neo-Confucian and other philosophers. Throughout, the author ranges widely over the field of theory, ancient and modern, concerning Chinese society and gives extensive bibliographical suggestions, all of which makes this a standard work. This heavily condensed volume is also useful as a preliminary guide to the whole range of Dr. Niida's researches. He is now director of the Tōyō Bunka Kenkyūjo, Tōkyō University. Reviewed by Fukushima Masao 福島 正夫 in **Shisō** 思想, 349 (July 1953), 900–905; and by Mano Senryū 間野潜龍 in **Shigaku zasshi**, 62.8 (Aug. 1953), 785–790.

3.6.2 NIIDA Noboru 仁井田陞, **Shina mibunhōshi** 支那身分法史 (History of Chinese law concerning social status), Zauhō Kankōkai 座右寶刊行會, 1942 (2nd printing, 1943, 997+index 17 pp.).

This massive volume takes up in great detail the traditional Chinese law concerning the clan and extended family, the household, marriage, parent-child relationships, slaves, and the like. Though dealing with traditional China as a whole, this large and comprehensive work is naturally of basic importance for the late-Ch'ing period as well. See review by Miyazaki Ichisada 宮崎市定 in **Tōyōshi kenkyū**, 7.2–3 (May–June, 1942), 177–178.

3.6.3 NIIDA Noboru 仁井田陞, "Shina kinsei gikyoku shōsetsu no sashie to keihō shiryō" 支那近世戲曲小説の挿畫と刑法資料 ("Illustrations from modern Chinese dramas and novels and historical materials on criminal law"), **Tōa ronsō,** 5 (Oct. 1941), 49–110, illus.

Dr. Niida here uses literary and pictorial sources, mainly of the Yuan and Ming periods, to obtain descriptions of (1) instruments used during judicial examination under torture and (2) for various punishments, together with (3) prisons, male and female, and (4) the execution of criminal sentences, up to and including the slow slicing process.

3.6.4 NIIDA Noboru 仁井田陞, "Min-Shin jidai no hitouri oyobi hitojichi monjo no kenkyū" 明清時代の人賣及人質文書の研究 (A study of documents on the sale of persons and mortgaging of persons in the Ming and Ch'ing periods), **Shigaku zasshi,** 46.4 (April 1935), 479–508; 5 (May 1935), 598–650; 6 (June 1935), 736–764.

This long research study first surveys the extensive literature of scholarly articles on this subject in Chinese and Japanese and then summarizes the general state of human traffic and mortgaging in China during the Ming and Ch'ing periods. The next sections survey the legal and governmental attitudes toward the problem, citing further examples, and then analyze textually many types and samples of contractual documents used for these purposes. Finally the author discusses the practice of palm-printing (*shou-mo* 手模) in Chinese legal documents.

3.6.5 NIIDA Noboru 仁井田陞, "Gen Min Shindai oyobi Reishi Annan no hoshō sei" 元明清代及び黎氏安南の保證制 ("Security System in the Dynasties of Yuan, Ming and Ts'ing and in Annam under Li"), **Shichō,** 5.3 (Oct. 1935), 32–62.

This comparative legal study (about one-third on the Ch'ing) describes the system of guarantees required with reference to commercial transactions, employment, and loans, and the obligations arising therefrom. This same topic is touched upon in Dr. Niida's later **Chūgoku hōseishi**, 1952, 3. 6. 1, pp. 346-351.

3. 6. 6 NIIDA Noboru 仁井田陞, "Shina no tochi daichō 'Gyo-rin zu-satsu' no shiteki kenkyū" 支那の土地臺帳「魚鱗圖冊」の史的研究 ("Historical study on Chinese Ledger of Land, *Yü-lin tu-ts'ê* [sic]"), **Tōhō gakuhō, Tōkyō,** 6 (Feb. 1936), 157-204, illus.

This factual legal analysis of land registers, chronologically by dynasties, continues the study of an important problem of legal history, previously discussed in **Shinkoku gyōseihō,** 1. 2. 1, and **Taiwan shihō,** 3. 6. 18. Only the last 20 pp. concern the Ch'ing period.

3. 6. 7 SHIGA Shūzō 滋賀秀三, **Chūgoku kazokuhō ron** 中國家族法論 (On Chinese family law), Kōbundō, 1950, 184 pp.

Developed from a thesis presented to the Law Faculty of Tōkyō University in 1943, this solid but somewhat formalistic study of "basic principles" (*kihon genri* 基本原理) makes careful and detailed use of the sources and scholarly literature of the subject. It is divided into chapters on inheritance (*shōkei* 承繼), reception of property (*juzai* 受財) and joint dwelling (*dōkyo* 同居), and under these heads covers a variety of sub-topics. The author goes into the religious side of family life and considers the religious concept of the ancestral clan (*tsung* 宗) as the basic principle of family law. He rather minimizes the historical aspect as well as class differentiations and tries to define a common legal concept which has been present in China everywhere and always. Though he acknowledges the constant advice of Dr. Niida Noboru, this book seems to have a different approach to this topic. The author is now at the Law Faculty, Tōkyō University.

3. 6. 8 KAINŌ Michitaka 戒能通孝, "Shina tochihō kankō josetsu, Hoku-Shi nōson ni okeru tochi shoyūken to sono gutaiteki seikaku" 支那土地法慣行序說──北支農村に於ける土地所有權と其の具體的性格 (An introductory note on the legal customs concerning the land system in China, the rights over landed property in North China villages and their concrete nature), pp. 85-176 in the author's **Hōritsu shakaigaku no shomondai** 法律社會學の諸問題 (Problems in legal sociology), Nippon Hyōronsha, 1943, 450 pp. (Also published in Tōa Kenkyūjo, ed., **Shina nōson kankō chōsa hōkokusho** 支那農村慣行調查報告書 [Reports of investigations of rural customs in China], series 1 第一輯, 1943, 329 pp.)

This is one of the pioneer analyses of the reports of the field investigation conducted by the SMR and the Tōa Kenkyūjo. The author participated in this research as a member of the Tōkyō team which was mainly in charge of the theoretical appraisal of the project and its results. With a heavily Germanic training in legal sociology (Gierke, Savigny and others) and also an experience of legal research in Japanese village life of the Tokugawa period behind him, the author analyses the seemingly "modern" aspects of the rights of property over land in North China as compared with those in Western and Japanese medieval society. He then appraises the characteristics of this situation in the context of the entire political and social structure of China, adding some hypothetical suggestions. This article is apparently intended to be a criticism of the then popular theory of "closed rural communities" (see further in sec. 8.2).

3.6.9 HAYASHI Megumi 林惠海, "Chūgoku nōka no kintō bunsan sōzoku no kenkyū" 中國農家の均等分産相續の研究 (A study of equal division of property inheritance among Chinese farm families), pp. 65–119 in Tōkyō Daigaku Shakai Gakkai 東京大學社會學會 (The association for sociology, Tōkyō University) ed., **Toda Teizō hakushi kanreki shukuga kinen rombun-shū, Gendai shakaigaku no shomondai** 戸田貞三博士還曆祝賀記念論文集, 現代社會學の諸問題 (Problems of contemporary sociology, a symposium commemorating the 60th birthday of Dr. Toda Teizō), Kōbundō, 1949, 552 pp.

Based on field work in the Yangtze delta area around Soochow, this sociological article is a long and careful study, with many charts, to show the actual process of land division.

3.6.10 WATANABE Kōzō 渡邊幸三, "Nankin fudōsan baikei no kenkyū" 南京不動産賣契の研究 ("A Study of the Deeds of Immovable Properties in Nanching"), **Tōhō gakuhō, Kyōto,** 22 (Feb. 1953), 95–122, maps.

This article, by a scholar who worked as a field-researcher attached to the SMR Shanghai branch, analyzes 19 legal documents used in the sale of land and houses in Nanking in the period 1792–1847. The author asserts the importance of studying legal documents from urban life, as distinguished from those used in rural districts.

3.6.11 SHIMIZU Kinjirō 清水金二郎, **Manshū chiken seido no kenkyū** 滿洲地券制度の研究 (A study of land-certificates in Manchuria), Jimbun Shorin 人文書林, 1946, 326 pp., illus.

This specialized study describes documents used in connection with land-holding (*ti-ch'i* 地契 or *ti-chao* 地照) in the four provinces of Manchuria (including Jehol), first in the Ch'ing period (pp. 13–68), then under the Republic and finally under Manchoukuo. The study is well researched and a great many examples of documents are given in printed form. The author is now at Kyūshū University.

3.6.12 SHIMIZU Kinjirō 清水金二郎, **Kei no kenkyū** 契の研究 (A study of contracts), Taigadō, 1945, 206+6 pp.

Subtitled "On customs and norms relating to land in Manchuria and China," this volume is drawn from field studies (including the investigations in North China by SMR and Tōa Kenkyūjo). It deals with the local procedures and documents involved in the sale and transfer of land, pawning of property, use of property as security, and the custom of *ya-tsu-ch'ien* 押租錢 or giving deposit money to the landlord at the time of making a tenancy contact. Local customary practices are compared with the pertinent legal statutes. The author's article, "Manshū ni okeru kei no kenkyū," **Tōa jimbun gakuhō,** 1.1 (Mar. 1941), 40–79, 3.6.14 (with Ishida Bunjirō as co-author), was a preliminary study of the same topics. His two other articles ("Hoku-Shi no ten kankō", 3.6.22, and "Manshū ni okeru ōsosen kankō," 3.6.13) are included in this book.

3.6.13 SHIMIZU Kinjirō 清水金二郎, "Manshū ni okeru ōsosen kankō" 滿洲に於ける押租錢慣行 (The custom of *ya-tsu-ch'ien* 押租錢 in Manchuria), **Tōa jimbun gakuhō,** 1.4 (Feb. 1942), 901–919.

A legal study, based on field survey reports by the Manchoukuo government, of the custom in Manchuria of giving deposit money to the landlord at the time of making a tenancy contract, studied against its historical background in China proper and Manchuria.

3.6.14 ISHIDA Bunjirō 石田文次郎 and Shimizu Kinjirō 清水金二郎, "Manshū ni okeru kei no kenkyū" 滿洲に於ける契の研究 (A study of contracts in Manchuria), **Tōa jimbun gakuhō,** 1.1 (Mar. 1941), 40–79.

A legal analysis of the procedures and documents involved in the making of various kinds of contracts concerning estates in Manchuria.

3.6.15 SUGINOHARA Shun'ichi 杉之原舜一, "Ten no hōteki seishitsu" 典の法的性質 (The legal nature of mortgages), **Hōritsu jihō 法律時報,** pub. by Nippon Hyōronsha, 19.1, whole no. 199 (Jan. 1947), 31–34.

A brief but highly suggestive discussion of mortgages as employed in North China villages, derived from the SMR field research which the author led the way in organizing. The author has serious doubts about the feasibility of using the ideas and terms of the modern Western legal system for the analysis of legal customs in China.

3.6.16 MAKINO Tatsumi 牧野巽, "Shinahō ni okeru gaiinshin no han-i no henka" 支那法に於ける外姻親の範圍の變化 (Changes in the extent of relatives on the wife's side in Chinese law), pp. 701–722 in **Katō hakushi kanreki kinen Tōyōshi shūsetsu,** 9.8.5.

A technical comparison of Ming and Ch'ing with earlier legal provisions as to the widening extent of affinity with the wife's relatives, taken as an aspect of the tendency to strengthen the joint-family relationship.

3.6.17 Mantetsu Sōmubu Shiryōka 滿鐵總務部資料課 (SMR, Research materials section of the general affairs department), comp., **Kantōchō no hōtei ni arawaretaru Shina no minji kanshū ihō 關東廳ノ法廷ニ現ハレタル支那ノ民事慣習彙報** (Reports on customary procedures in civil cases in China as seen in the courts of Kwantung), Dairen: SMR, 1934, 2 vols., total 2564 pp.+tables of contents 180 pp.

Based on the records of the local and higher courts of Kwantung during the 30 years from their inception up through 1931, and the records of a land research commission, this compilation by three legally trained researchers presents the findings of law and fact, and legal decisions and judgments, over the whole range of civil law cases, concerning persons, property, obligations, kinship relations, etc., and procedures in suits, commercial cases, and the like.

3. 6. 18 Rinji Taiwan Kyūkan Chōsakai 臨時臺灣舊慣調查會 (Temporary commission of the Taiwan Government-general for the study of old Chinese customs), **Dai-ichibu chōsa daisankai hōkokusho, Taiwan shihō,** 第一部調查第三回報告書，臺灣私法 (Third report of the investigation by the First Section, Private law in Taiwan), kan 卷 1, 2 vols., Kōbe, 1910, 741+471 pp.; kan 2, 2 vols., Tōkyō, 1911, 471+596 pp.; kan 3, 2 vols., Tōkyō, 1911, 517+477 pp. (this last supersedes the first edition of hen 編 3, 2 vols., Kōbe, 1909, 438+353 pp., which was published first of all and led to a revision of the whole work. Kan 1 and 2 were published only once, as above). **Taiwan shihō furoku sankōsho** 附錄參考書 (Private law in Taiwan, supplementary reference volumes), kan 1, 3 vols., Kōbe, 1910, 541 pp.; Tōkyō, 1911, 388 pp. and 437 pp.; kan 2, 2 vols., Tōkyō, 1910, 352 pp., and 1911, 360 pp.; hen 編 3, 2 vols., Kōbe, 1909, 294+172 pp.; kan 3, 2 vols., Tōkyō, 1910, 293+222 pp. (These last 4 volumes all contain materials relevant to kan 3 of the main work).

These monumental reports of the First Section 第一部 of this famous commission, in treating the law of immovable property, persons, movable property, commerce, and the like, take occasion to touch upon almost every conceivable aspect of Formosan society as it was in the early years of this century. Land-holding relations, the family system, and social and economic organizations of all kinds were investigated on the spot, and the results contain much of value for socio-historical research today. The research team for these reports was headed by Dr. Okamatsu Santarō 岡松參太郎, professor of civil law in Kyōto Imperial University, who was also the chief of the First Section of the commission. These volumes, like the companion publication, **Shinkoku gyōseihō,** 1. 2. 1, deserve extended appraisal in a special review article or monograph.

Dai-ichibu chōsa dai-ikkai hōkokusho 第一部調查第一回報告書 (First report of the investigation by the First Section), 2 vols., Kyōto, 1903, 470+323 pp.; supplementary vol., 480 pp. **Dai-ichibu chōsa dai-nikai hōkokusho** (Second report), kan 1, Kōbe, 1906, 780 pp.; supplementary vol., 389 pp.; kan 2, 2 vols., Kōbe, 1907, 791+341 pp.; supp. vol., 562 pp.

These are preliminary editions of the preceding work.

3. 6. 19 MURAKAMI Teikichi 村上貞吉, **Chūkaminkoku mimpō** 中華民國民法 (Civil law of the Chinese Republic), privately printed, 3 vols., preface 1930.

Under this title the author produced a commentary on the newly published code of civil law with the following subtitles: (1) **Sōsoku** 總則 (General rules), 111 pp. (2) **Saiken** 債權 (Law of obligations), pp. 112–475. (3) **Bukken** 物權 (Real rights), 309 pp. Like the further volumes listed below, these were printed for private circulation only, though they are solid legal studies by a competent specialist, who was a practicing lawyer of long residence in Shanghai (he gave his address as the Tōa Kōkyūkai 東亞(攷)攷究會 in Shanghai). The following studies by this author may also be noted:
(1) **Chūkaminkoku tegatahō** 手形法 (Law of bills of exchange), preface 1931, 393 pp.—a commentary on the code.

(2) **Chūkaminkoku kaishahō** 會社法 (Company law), preface 1931, 386 pp.—a commentary on the code.

(3) **Shina rekidai no keisei enkaku to genkō keihō** 支那歷代の刑制沿革と現行刑法 (The historical development of Chinese criminal law and the criminal law at present in force), preface 1932, 456 pp.—a historical survey (to pp. 254) and a translation of the penal code in force checked with the original text.

(4) **Shina ni okeru rikken kōsaku to kempō sōan shokō** 支那に於ける立憲工作と憲法草案初稿 (The work of making the constitution in China and the first draft constitution), preface 1934, 341 pp.—a historico-legal survey, with original texts of 23 drafts, official and private.

3.6.20 Kōain Kachū Renrakubu 興亞院華中連絡部 (Liason Office in Central China of the Asia Development Board), comp., **Chūkaminkoku hōrei sakuin, Kokuminseifu kōhō no bu** 中華民國法令索引——國民政府公報の部 (An index of laws and regulations of the Republic of China, section on the official gazettes of the Nationalist Government), Daidō Inshokan 大同印書館, 1942, 335 pp.

Compiled by Manabe Tōji 眞鍋藤治, this is a valuable and detailed, topically-arranged list of all laws, regulations, orders and decrees 各種法令 put out by the late Ch'ing and Republican governments in official gazettes from 1907 to 1937. These 2505 items are drawn from such compendia as **Cheng-chih kuan-pao** 政治官報 no. 1 to no. 1370, **Cheng-fu kung-pao** 政府公報 no. 1 to no. 1309, and **Kuo-min cheng-fu kung-pao** (daily) no. 1 to no. 1507. In addition to their topical arrangement (pp. 1–173), they are listed chronologically by day of issue (pp. 175–335).

3.6.21 Chūkaminkoku Hōsei Kenkyūkai 中華民國法制研究會 (Research society on the legal institutions of the Chinese Republic). ed., **Chūkaminkoku mimpō—sōsoku** 民法總則 (Civil law of the Republic of China, general rules), pub. by Chūō Daigaku 中央大學, 1931, 291 pp.

This small volume, commenting article by article on Section One (the general rules) of the Chinese Civil Code, was followed by an unusually comprehensive series of commentaries on the legal codes promulgated by the KMT Government. These were put out by the same society under the leadership formally of Dr. Matsumoto Jōji 松本烝治 and actually of Murakami Teikichi 村上貞吉 (d. 1940), a classmate of Matsumoto and a lawyer who stayed in Shanghai for more than thirty years. Those volumes in this uniform series that we have seen were published between 1933 and 1940, by the society at Chūō Daigaku, and all written by professors of the Faculty of Law, Tōkyō University.
They deal with the following topics:

(1) Wagatsuma Sakae 我妻榮, **Mimpō—saiken sōsoku** 債權總則 (Civil law, law of obligations, general rules), 1933, 520+5+9 pp.

(2) Wagatsuma Sakae, **Mimpō—saiken kakusoku** 各則 (Civil law, law of obligations, special rules), part 1, 1934; part 2 (with Kawashima Takeyoshi 川島武宜 as co-author), 1936; part 3 (with Kawashima), 1938; total 971 pp.+appendices (index, etc.).

(3) Wagatsuma Sakae and Kawashima Takeyoshi, **Mimpō—bukken** 物權 (Civil law, real rights) part 1, 1941, 329+3+4 pp.

(4) Tanaka Kōtarō 田中耕太郎 and Suzuki Takeo 鈴木竹雄, **Kaishahō 會社法** [公司法] (Law of companies and partnerships), 1933 (third printing 1940, 533+3+11 pp.).

(5) Tanaka Kōtarō and Suzuki Takeo, **Tegatahō 手形法** (Law of bills of exchange), 1934, 445+3+9 pages; with a section on historical (mainly medieval) background by Niida Noboru 仁井田陞 (26 pp.).

(6) Suzuki Takeo 鈴木竹雄 and Ishii Teruhisa 石井照久, **Kaishōhō 海商法** (Law of maritime commerce), part 1, 1936, 524 pp.; part 2, 1938, 373+6+14 pp.

(7) Ono Seiichirō 小野清一郎, **Keihō 刑法** (Penal law), vol. 1 (general rules 總則), 1933, 323+7 pp.; vol. 2 (special rules 分則, part 1), 1934, 385+4 pp.; vol. 3 (special rules, part 2), 1935, 436+3 pp.

(8) Ono Seiichirō and Dandō Shigemitsu 團藤重光, **Keiji soshōhō 刑事訴訟法** (Law of criminal procedure), part 1, 1938, 401 pp.; part 2, 1940, 480+21 pp.

(9) Miyazawa Toshiyoshi 宮澤俊義 and Tanaka Jirō 田中二郎, **Kempō sōan 憲法草案** (Draft constitution [of 1934]), 1935, 347+7 pp.

(10) Miyazawa Toshiyoshi and Tanaka Jirō, **Rikken shugi to Sammin shugi Gokenkempō no genri 立憲主義と三民主義・五權憲法の原理** (Constitutionalism and the principles of the *San-min chu-i* and of the Five-power Constitution), 1937, 170 pp.—this is not a commentary, but a comparative survey of the theoretical background.

There are additional volumes of commentaries (not seen by us) on such matters as the law of civil procedure, the law of trade-marks and the definitive draft of the constitution. Each volume translates the text of the code, point by point, giving also the original Chinese text and an extensive commentary, and compares the Chinese code with corresponding points in the Japanese code, together with notes and annotations, and references to German and other Western law. Indices and comparative tables of Chinese and Japanese terms are added to each volume. This series thus provides a fundamental perspective on the Chinese code. It is a virtual continuation of the earlier studies of Murakami Teikichi, 3.6.19, and the authors admit their heavy indebtedness to the writings and assistance of Mr. Murakami.

3.6.22 SHIMIZU Kinjirō 清水金二郎, "Hoku-Shi no ten kankō" 北支の典慣行 (Legal practices concerning mortgages [*tien* 典] on real estate in North China), **Tōa jimbun gakuhō,** 3.2 (Oct. 1943), 234–266.

Based on the field investigation reports of the famous research team of the South Manchurian Railway Co., this article provides an elaborate analysis of the practice, in North China, of receiving a definite sum of money in consideration of letting another party occupy and derive profit from specific real estate, which can be got back by repaying the same sum of money. Subsections discuss (1) contracts of *tien*, (2) rights and obligations of the party who has received such a mortgage, (3) rights and obligations of the party who has given the mortgage.

3.7 THE TREATY SYSTEM

Note: We insert at this point the treaty structure created by the foreign powers in their relations with China because we believe that this

structure must be regarded historically as a Chinese political institution—
no doubt a Sino-foreign creation, but an institution which formed part
of the Chinese polity nevertheless. On this subject the works of Ueda,
Irie, Saitō, and Hanabusa are the major items. Treaty texts are at the
end. Other aspects of foreign activity in China will be found in sec. 2
on political history, sec. 4 on Japanese and Russian expansion, sec. 6.6
on the Christian missionary movement, sec. 7.17 on the treaty ports and
sec. 7.18 on foreign investment and trade.

3.7.1 IRIE Keishirō 入江啓四郎, **Chūgoku ni okeru gaikokujin no chii**
中國に於ける外國人の地位 (The position of foreigners in China), Tōkyōdō
東京堂, 1937, 706 pp., indices 41 pp., bibliog. 7 pp.

A careful and penetrating historico-legal study of a major aspect of China's
foreign relations, based on the extensive Chinese, Western, Japanese and also
Russian literature on the treaty system. The author gives frequent and precise source references in his text, together with cross references, and although
the periodical literature seems to have been neglected, he quotes the basic
sources aptly and concisely, from an independent point of view. His chapters
deal with the historical development of the foreigner's status, pre-treaty and
post-treaty; the legal basis established for unequal Sino-foreign relations; the
foreigner's rights of residence and travel; settlements and concession areas;
spheres of interest and special areas; military, naval and police privileges;
consular jurisdiction; inland and coastal navigation; and missionary privileges.
The last chapters then take up the Chinese anti-foreign movement, foreigners'
defense of their position and the Kuomintang efforts to carry out a nationalistic economic policy. Although, as a whole, the construction of the book is
more fragmentary than systematic, its legal approach and pertinent references
to basic sources under every item make it a useful guide to further research.
The author, son of Murakami Teikichi, has been a journalist on international
affairs and is now a professor of the Aichi University.

3.7.2 UEDA Toshio 植田捷雄, **Shina gaikōshi ron** 支那外交史論 (On
the history of China's foreign relations), Ganshōdō, 1933 (4th printing,
1940, 244 pp.).

A widely-used survey of the Open Door and spheres of influence as twin
motifs in Chinese diplomatic relations, analyzing their development in specific
detail from the 1890s to 1931 on the basis of both Western and Japanese
secondary writings. The author stresses the need for a legal approach to the
study of China's diplomatic history; since every diplomatic issue in contempory
China has been in the nature of a dispute in the field of international law, it
follows that the traditional way of basing diplomatic history on a chronological
arrangement of facts can contribute little or nothing to the actual solution of
important diplomatic conflicts of the present day. This book can be considered
a general introduction to Dr. Ueda's later researches on leased territories, the
settlements and concessions, etc. He is now at the Tōyō Bunka Kenkyūjo,
Tōkyō University.

3.7.3 UEDA Toshio 植田捷雄, **Zai-Shi rekkoku ken'eki gaisetsu** 在支列
國權益概說 (A survey of the rights and interests of foreign powers in

China), Ganshōdō, 1939 (2nd printing, 1942, 455+82 pp.).

A useful guide to the subject, which has chapters on spheres of influence, leased territories, the settlements and concessions, Peking legation quarter, consular jurisdiction, foreign military rights, inland navigation rights, open ports, the maritime customs, indemnities, railroads, mines, posts, missionaries, the Kailan Mining Administration, and Hongkong. The appendix gives 21 useful tables, bibliography, index and a list of treaties.

3.7.4 UEDA Toshio 植田捷雄, **Shina ni okeru sokai no kenkyū** 支那に於ける租界の研究 (A study of the foreign concessions and settlements in China), Ganshōdō, 1941, 919 pp., 4 maps.

This definitive historical-and-legal study supersedes the author's earlier volumes, **Shina sokai ron** 支那租界論, 1934, and **Zōho Shina sokai ron** 増補, 1939. The fruit of many researches already published in books and articles, it summarizes the pre-treaty history of China's foreign trade centers and their development into treaty ports, and analyzes the growth of the Shanghai (pp. 58–301), Tientsin (pp. 302–363), Amoy and other settlements and concessions and the progressive revision of these arrangements after World War I. Part II then treats the legal bases, foreign administrative rights and institutional structures of these various foreign areas and gives a legal appraisal of the policies exercised toward domestic rebellion and international warfare. A bibliography (20 pp.) and index (31 pp.), as well as the notes provided in the text to both Western and Far Eastern literature, make this an important reference volume. Reviewed by Hirano Yoshitarō 平野義太郎 in **Tōagaku**, 4 (April 1942), 217–222.

3.7.5 UEDA Toshio 植田捷雄, "Shina ni okeru ryōjisaiban no kigen" 支那に於ける領事裁判の起源 ("Origins of the Consular Jurisdiction in China"), **Kokusaihō gaikō zasshi,** 40.10 (Dec. 1941), 877–907.

A summary, using mainly Western sources, of the long process of altercation, mainly at Canton, which preceded the British demand for extraterritoriality.

3.7.6 UEDA Toshio 植田捷雄, **Shina soshakuchi ron** 支那租借地論 (On the leased territories in China), Nikkō Shoin, 1943, 466 pp.

As distinct from Dr. Ueda's work on the foreign-administered and residential settlements and concessions in the treaty ports (see his **Shina ni okeru sokai no kenkyū,** 3.7.4), this volume surveys (1) the foreign diplomacy and (2) the international legal status of the leased territories secured by the imperialist powers in 1898 and afterward: Kiaochow, Kwantung, Weihaiwei, and Kwangchow-wan, including Japan's succession to the Russian and German positions. The second of these sections (pp. 137–466) studies quite extensively the many practical and theoretical legal questions involved, and evaluates conflicting legal interpretations.

3.7.7 SAITŌ Yoshie 齋藤良衞, **Shina keizai jōyaku ron, dai-ikkan, kaishijō no seishitsu** 支那經濟條約論，第一卷，開市場の性質 (A study of Chinese commercial treaties, vol. 1, the nature of the treaty ports), Gaimushō (Foreign Office), 1922, 1024 pp.

A documented legal study of the treaty ports and foreign-administered residential areas, which appears to cover all the major aspects of legal interest. Not followed by vol. 2 but by a larger work, 3.7.8.

3.7.8 SAITŌ Yoshie 齋藤良衞, **Gaikokujin no tai-Shi keizai katsudō no hōteki konkyo** 外國人の對支經濟活動の法的根據 (Legal bases for foreigners' economic activity regading China), Gaimushō (Foreign Office), 6 vols. in ca. 3346 pp., n. d. (preface to vol. 1 dated Nov. 1937).

This is the completely revised and enlarged edition of the author's former one-volume work entitled **Shina keizai jōyaku ron, dai-ikkan, kaishijō no seishitsu**, 3.7.7. These officially-sponsored volumes were constructed from a legal point of view, necessarity relying mainly on English-language documentation (which is extensively quoted) and apparently also on the Foreign Office archives. The 6 volumes present two major sections, on the rights and duties of foreigners in the treaty ports opened to trade, and on the rights and duties of foreigners in residential areas, respectively. This appears to be the most extensive work of its kind and the topics of the 13 chapters and numerous sub-sections may perhaps be left to the reader's imagination. A former diplomat, the author served as a Foreign Office adviser, particularly to Foreign Minister Matsuoka, and was active in the negotiations for the tri-partite alliance with Germany and Italy; purged but later cleared, he is now president of Aizu Tanki Daigaku 會津短期大學.

3.7.9 HANABUSA Nagamichi 英修道, **Chūkaminkoku ni okeru rekkoku no jōyaku ken'eki** 中華民國に於ける列國の條約權盆 (Treaty rights and interests of the great powers in the Chinese Republic), Maruzen, 1939, 848 pp., app. 136 pp, indices 65 pp.

A large, systematic legal-type treatise on the unequal treaties and their specific provisions, treaty by treaty and country by country, concerning trade, extrality, foreign settlements and concessions, coast trade and inland navigation, military rights, etc. A full, though formalistic, textual dissection of the treaty system.

3.7.10 NOGUCHI Kinjirō 野口謹次郎 and WATANABE Yoshio 渡邊義雄, **Shanhai kyōdō sokai to kōbukyoku** 上海共同租界と工部局 (The Shanghai International Settlement and the Municipal Council), Nikkō Shoin, 1939, 187 pp., charts.

This descriptive account of the various aspects of the foreign administration of Shanghai, by two members of the SMC staff, while carefully done, would appear to add little to materials in English except possibly a Japanese point of view.

3.7.11 UEHARA Shigeru 上原蕃, **Shanhai kyōdō sokai shi** 上海共同租界誌 (Gazetteer of the Shanghai International Settlement), Maruzen, 1942, 263 + app. 36 pp.

A descriptive account, written by a former Deputy Commissioner of Police, of the International Settlement's history, administrative organs, defense establishment, judical problems, external roads problem, etc., which though well done would appear to offer little data not available in English also.

3. 7. 12 HANABUSA Nagamichi 英修道, "Shanhai tochi eisoken ni kansuru jakkan no kōsatsu" 上海土地永租權に關する若干の考察 ("Some considerations about Perpetual Leases in Shanghai"), **Kokusaihō gaikō zasshi,** 41.10 (Oct. 1942), 971–992; 41.12 (Dec. 1942), 1172–1205.

A competent technical analysis of the special terminology, forms of documents, procedures, taxes, and other aspects of this once important subject.

3. 7. 13 IMAI Yoshiyuki 今井嘉幸, **Shina kokusaihō ron** 支那國際法論 (A treatise on international law concerning China), vol. 1 (no further volumes pub.), Maruzen, 1915, 484 pp.

A lucid pioneer work, subtitled "Foreign legal jurisdiction and foreign administrative areas", this volume surveyed the structure of foreign legal rights in China and Manchuria which grew out of foreign wars and treaties up to the time of the revolution of 1912. Superseded by works of W. W. Willoughby **(Foreign Rights and Interests in China,** 2nd ed., 2 vols., Baltimore, 1927) and Ueda Toshio (see his **Shina ni okeru sokai no kenkyū,** 3. 7. 4, 1941, and other writings). Dr. Imai (1878–1951) started his career as a judge, went to China in 1908 on the invitation of the Ch'ing government and taught law at *Pei-yang fa-cheng hsüeh-t'ang* 北洋法政學堂 in Tientsin. His research on the treaty system derived from this Tientsin experience. After the outbreak of the revolution, he became friendly with the revolutionaries and warned against Yuan Shih-k'ai's ambitions. At the time of the abortive revolution of 1916 he worked as an adviser to the revolutionary government in Yunnan. He was an intimate friend of Takayanagi Matsuichirō (see 7. 13. 14).

3. 7. 14 EGAWA Hidebumi 江川英文, for the Chūkaminkoku Hōsei Kenkyūkai 中華民國法制研究會, **Chūkaminkoku ni okeru gaikokujin no chii** 中華民國に於ける外國人の地位 (The status of aliens in the Republic of China), part 1, pub. by the Chūō Daigaku 中央大學, 1938, 197 pp. (No continuations published).

A brief legal survey by a professor of international private law of the Faculty of Law, Tōkyō University. Deals with (1) historical background, (2) trade and navigation, (3) residence and business, and (4) travel. Translations of some Ch'ing regulations and memorials are given.

3. 7. 15 ŌHIRA Zengo 大平善梧, **Shina no kōkōken mondai** 支那の航行權問題 (The problem of navigation rights in China), **Daitōa kokusaihō sōsho** 大東亞國際法叢書 (Greater East Asia international law series), vol. 4, Yūhikaku, 1943, 242+6 pp.,

As a wartime study, this volume discusses the history and legal features of foreign navigation rights in Chinese waters from the point of view of China's inclusion in the Greater East Asia Co-prosperity Sphere. While dated accordingly, it is nevertheless a useful and well documented piece of research. The author is now a professor of the Hitotsubashi University.

3. 7. 16 KOGA Motokichi 古賀元吉, **Shina oyobi Manshū ni okeru chigai-hōken oyobi ryōji saibanken** 支那及滿洲に於ける治外法權及領事裁判權 (Extraterritoriality and consular jurisdiction in China and Manchuria),

Nisshi Mondai Kenkyūkai 日支問題研究會, 1933, 255 pp.

This rather banal legal study by a lawyer analyzes the various freedoms conferred by extrality, the nature of the Chinese police power, the functioning of consular jurisdiction and police powers, and the application of these in treaty ports and foreign concession areas. The last section (from p. 185) is on Manchuria in particular.

3.7.17 SHINOBU Jumpei 信夫淳平, **Man-Mō tokushu ken'eki ron** 滿蒙特殊權益論 (On the special rights and interests in Manchuria and Mongolia), Nippon Hyōronsha, 1932, 541 pp.

A detailed exposition of Japan's legal and de facto position in Manchuria. Though published after the time of the Mukden incident, the draft was substantially completed before it with the object of throwing a factual light on the complexities of the so-called "special rights and interests" of Japan which had become intermingled with emotional considerations. This work may be compared with the contemporary Western studies of Dr. C. Walter Young and others on the Japanese position in Manchuria. No footnotes, bibliography or index. References are roughly given in the text.

3.7.18 Gaimushō Jōyakukyoku 外務省條約局, comp., **Ei Bei Futsu Ro no kakkoku oyobi Shinakoku kan no jōyaku** 英米佛露ノ各國及支那國間ノ條約 (Treaties between China and England, the United States, France and Russia, repectively), pub. by the treaty bureau of the Foreign Office, 1924, 2517 pp. + table of contents 136 pp.

This massive compilation prints the texts, mainly in Western languages (with Japanese translations which are not always reliable), of treaties, agreements, contracts, notes, etc., both by and concerning China, to which the four major powers were parties, from 1842 to 1922. A companion volume for other countries than Japan and these four (**Nichi, Ei, Bei, Futsu, Ro igaino kakkoku oyobi Shinakoku kan no jōyaku** 日, 英, 米, 佛, 露以外ノ各國及支那國間ノ條約) was published in 1926 (contents 61+1652+chronology 46 pp.) to complete the series. These volumes may be compared with the standard Western work by J. V. A. MacMurray, **Treaties and Agreements with and concerning China**, 2 vols., 1921.

3.7.19 Gaimushō Jōyakukyoku 外務省條約局, comp., **Nisshikan narabini Shina ni kansuru Nippon oyobi takoku kan no jōyaku** 日支間竝支那ニ關スル日本及他國間ノ條約 (Sino-Japanese treaties and treaties concerning China between Japan and other countries), pub. by the treaty bureau of the Foreign Office, 1923, contents 43+1192 pp., maps.

This compilation prints the texts in Japanese or occasionally Chinese, and frequently in English, of treaties, agreements, international regulations, and similar documents to which Japan was a party, from 1895 to 1922, covering a great variety of commercial and diplomatic subjects. These materials were brought together at the time of the Washington Conference of 1921–22 and cover a great deal more than the formal treaties.

3.8 CONSTITUTIONALISM

Note: This subject has been little studied. The chief work is that of Takahashi Yūji. Note also 3.6.19 and 3.6.21.

3.8.1 TAKAHASHI Yūji 高橋勇治, **Chūkaminkoku kempō** 中華民國憲法 (The constitutional law of the Chinese Republic), Yūhikaku, 1948, 318 pp. (Gaikoku kempō no kenkyū 外國憲法の研究, 5).

A systematic survey by a political scientist, running from the pre-revolutionary constitutional movement down through the succeeding documents which were put forward under the Ch'ing and the Republic and concluding with an 80-page reprint and ruthless analysis of the text of the constitution adopted under Kuomitang auspices at Nanking in 1946. Although rather schematic, this volume provides a handy survey of constitutional development against its social and economic background. Prof. Takahashi is now at the Shakai Kagaku Kenkyūjo 社會科學研究所, Tōkyō University.

3.8.2 TAKAHASHI Yūji 高橋勇治, "Chūgoku shinkempō to kihonteki jinken" 中國新憲法と基本的人權 ("Fundamental Human Rights in the New Chinese Constitution", **Shakai kagaku kenkyū,** 2 (1948), 1–28.

A detailed textual analysis of the provisions of the Chinese Nationalist Government constitution of 1946, referring to previous such documents and stressing the limitations on freedom of speech Note the same author's survey of the Chinese "renaissance", 6.1.12.

3.8.3 INADA Masatsugu 稻田正次, **Chūgoku no kempō** 中國の憲法 (The Chinese constitution), Seijikyōiku Kyōkai 政治教育協會, 1948, 206 pp.

Compiled as an aid to the historical understanding of the 1946 Nanking constitution, this study surveys the constitutional movement down to 1925, Sun Yat-sen's major constitutional devices, and the various steps taken by the Kuomintang government (to p.123). It then analyzes the provisions of the 1946 document in detail, adding the original text and a translation of it.

3.8.4 ISHIKAWA Tadao 石川忠雄, "Shimmatsu oyobi Minkoku shonen ni okeru rempōron to shōseiron" 清末及び民國初年における聯邦論と省制論 ("Arguments on Federalism and Sheng System in China: 1901–1913"), **Hōgaku kenkyū,** 24.9–10 (Sept.-Oct. 1951), 129–159.

This well informed study traces the discussion which arose over China's form of government, particularly the relation of the provinces to the central government, at the time of the early constitutional movement, before and immediately after 1911. Drawing widely on Chinese sources and studies in Chinese, Japanese and English, it analyzes the groups and views concerned down to the time when Yuan Shih-k'ai's despotic ambitions made the question academic. Prof. Ishikawa is now at Keiō University.

3.8.5 HARUMIYA Chikane 春宮千鐵, "Chūkaminkoku seikenshi no issetsu, En-yakuhō no seiritsu katei" 中華民國制憲史の一齣──袁約法

の成立過程 (A page in the history of constitution-making in the Republic of China, the process of drawing up the Yuan Shih-k'ai constitutional compact), **Shina kenkyū,** 50 (Mar. 1939), 125–158.

In this brief historical essay, the writer first traces the movement to make Yuan emperor, beginning with his presidential election law, and the break-up of the first Kuomintang, and then compares Yuan's constitutional compact with the Temple of Heaven draft 天壇憲法草案, concluding with the unfortunate Dr. Goodnow and the abortive restoration movement. Not a work of original research.

3.8.6 Matsumoto Hisashi 松本恒, "Chūkaminkoku seiji soshiki no ichikōsatsu" 中華民國政治組織の一考察 (A study of the political organization of the Republic of China), **Tōa jimbun gakuhō,** 1.4 (Feb. 1942), 1156–1175.

This analysis of the Wang Ching-wei 汪精衞 regime of the "true" (i.e., "puppet") Kuomintang at Nanking uses its organizational documents to describe the political structure set up there in 1939.

4. POWER POLITICS: JAPANESE AND RUSSIAN EXPANSION

Note: Our aim in this section is not to single out Japan and Russia as being more expansive or aggressive than other powers, but merely to call attention to the large and important body of Japanese research which has been devoted to the expansion of these two powers in Northeast Asia and the friction between them. The pioneer expansion of the British, as well as the Americans, toward China has been dealt with in sec. 2 above. On the whole, it came earlier and was primarily commercial rather than territorial. In surveying Japanese studies, it has seemed desirable to devote a major section to the growth of Japan's modern involvement on the continent, most of which has entered directly into the modern history of China. While treating the successive phases of the Japanese advance, and the stimulus it received from Russian encroachment, we do not attempt to deal with the domestic causes of Japanese expansion. We also eschew materials on the puppet state of Manchoukuo, classifying it as an aspect of Japanese history rather than of Chinese. Our sec. 4.8 is of course closely related to sec. 3.5 above; rather than group together the materials on Inner Asia, we have chosen to regard that region as generally peripheral and dependent on affairs elsewhere—at least in the way in which it has been studied, unrealistic as this may be in fact. Note that "Sino-Japanese cultural relations" are under sec. 6.8.

4.1 EARLY SINO-JAPANESE RELATIONS

Note: These rather miscellaneous bits and pieces offer some view of Sino-Japanese relations in the pre-treaty period when Sino-British relations have been so thoroughly studied. The most substantial item is that by Tabohashi Kiyoshi, 4.1.8.

4.1.1 ICHIMURA Sanjirō 市村瓚次郎, "Shina no bunken ni mietaru Nippon oyobi Nipponjin" 支那の文獻に見えたる日本及び日本人 (Japan and the Japanese as seen in Chinese records), **Rekishigaku kenkyū,** 109 (April 1943), 177–194.

> While concerned with pre-Ch'ing times, this essay and its quoted documents, by a scholar acquainted with the whole sweep of Chinese history, should provide useful background for understanding modern Sino-Japanese contact.

4.1.2 YANO Jin'ichi 矢野仁一, "Kan'ei Jōkyō jidai no Nagasaki no Shina bōeki 寛永貞享時代の長崎の支那貿易 (The Chinese trade of Nagasaki in the Kan'ei-Jōkyō periods), **Shigaku zasshi,** 38.10 (Oct. 1927), 925–952; 11 (Nov. 1927), 1035–1074.

> A factual study of the 17th century trade, using a German study by Oskar Nachod as well as Chinese and Japanese sources.

4.1.3 YANO Jin'ichi 矢野仁一, "Tokugawa Jidai ni okeru Nagasaki no Shina bōeki ni tsuite" 徳川時代に於ける長崎の支那貿易に就いて (On the Chinese trade of Nagasaki in the Tokugawa period), **Keizai ronsō,** 25.5 (Nov. 1927), 1000–1011; 25.6 (Dec. 1927), 1132–1152. "Nagasaki bōeki ni okeru dō oyobi gin no Shina yushutsu ni tsuite" 長崎貿易に於ける銅及び銀の支那輸出に就いて (On the export to China of copper and silver in the Nagasaki trade), ibid., 26.1 (Jan. 1928), 84–98; 26.2 (Feb. 1928), 316–334. "Jōkyō igo no Nagasaki no Shina bōeki ni tsuite" 貞享以後の長崎の支那貿易に就いて (On the China trade of Nagasaki after the Jōkyō period), ibid., 27.5 (Nov. 1928), 663–675; 27.6 (Dec. 1928), 802–828.

> These articles are based mainly on Dutch accounts, as well as on some Japanese records, and present data rather than conclusions. Further articles on the Chinese trade at Nagasaki, which do not appear to have been published by Dr. Yano in book form, will be found in **Tōa keizai kenkyū** as follows: (1) in 9.1 (Jan. 1925), 1–26; 9.2 (April 1925), 157–188; 9.3 (July 1925), 293–331. (2) in 12.1 (Jan. 1928), 1–25. (3) in 12.2 (April 1928), 54–74. (4) in 20.4 (Nov. 1936), 627–658; 21.1 (Jan. 1937), 37–60. (5) in 22.4 (Oct. 1938), 485–507.

4.1.4 KIMIYA Yasuhiko 木宮泰彦, **Nisshi kōtsū shi** 日支交通史 (History of communication between Japan and China), Kanezashi Hōryūdō 金刺芳流堂, vol. 2, 1927, 673+23 pp., illus.

> The last two chapters (vol. 2, pp. 476–584) of this lengthly survey deal with (1) the Nagasaki trade between Japan and the Ch'ing empire down into the middle of the 19th century, describing its procedures, commodities, etc. in detail; and (2) some 60 Chinese monks who visited Tokugawa Japan and the naturalized Chinese at Nagasaki who left numerous descendents. Pages 669–673 provide a brief chronology of early 19th century contact.

4.1.5 MARUYAMA Kunio 丸山國雄, "Kinsei makki ni okeru Nisshi bōeki" 近世末期に於ける日支貿易 (Sino-Japanese trade at the end of the Edo period), **Tōa ronsō,** 2 (Jan. 1940), 233–273.

> A study of the administration of the Japanese end of Sino-Japanese trade in the mid-nineteenth century, before and after the first Western treaties, with some attention also to the interesting question of Liu-ch'iu trade.

4.1.6 VAN GULICK (Van Kūrikku) 高羅佩, "Kenryū jidai ichi Shinajin no Nipponkan" 乾隆時代一支那人の日本觀 (A view of Japan by a Chinese in the Ch'ien-lung period), **Tōa ronsō,** 2 (Jan. 1940), 275–296, 4 plates.

> Contributed by the Dutch scholar R. H. Van Gulick under his Chinese name

(Kao Lo-p'ei), this article presents a Japanese translation from the **Hsiu-hai pien** 袖海編 of the Chinese painter Wang P'eng 汪鵬, as printed in the famous geographical collection **Hsiao-fang-hu-chai yü-ti ts'ung-ch'ao** 小方壺齋輿地叢鈔 of 1877. Wang went to and from Nagasaki annually for twenty years after 1764. His account of Nagasaki, though brief, should be of interest in comparison with Dutch accounts.

4. 1. 7 MORITA Yoshiko 森田美子, "Nisshin bōeki no ichisokumen" 日清貿易の一側面 ("A Phase of Sino-Japanese Trade"), **Shakai keizai shigaku,** 15. 3-4 (Oct. 1949), 72-86.

Morishima Chūrō 森島中良 (1754-1809) was a doctor trained in Dutch medicine and a dilletante essayist. He made a collection of 143 Chinese chops or firm labels of the Ch'ing period used in trade at Nagasaki. The present author describes the types, forms and uses of these chops or labels, giving examples and indicating the kinds of goods imported under them.

4. 1. 8 TABOHASHI Kiyoshi 田保橋潔, "Nisshi shinkankei no seiritsu, bakumatsu ishin ki ni okeru" 日支新關係の成立――幕末維新期に於ける (The establishment of a new relationship between Japan and China, in the late Tokugawa and Restoration periods), **Shigaku zasshi,** 44. 2 (Feb. 1933), 163-199; 3 (Mar. 1933), 314-338.

Based on Japanese and Chinese sources including the newly published **Ch'ou-pan i-wu shih-mo** 籌辦夷務始末, this article describes in detail the painstaking Sino-Japanese negotiations to establish a diplomatic relationship on a modern and strictly equal basis, from the early soundings by Tokugawa officials in 1862 down to the signing of a treaty in 1871, including the abortive effort on the part of Japan to revise the unratified treaty and the final exchange of ratifications in 1873.

4. 1. 9 OKITA Hajime 沖田一, "Bakufu dai-ichiji Shanhai haken kansen Senzai-maru no shiryō" 幕府第一次上海派遣官船千歳丸の史料, ("The Historical Documents relating to the Senzai-Maru, the First Government Ship despatched to Shanghai by the Shogunate"), **Tōyōshi kenkyū,** 10. 1 (Dec. 1947), 48-58; 10. 3 (July 1948), 198-212.

A bibliographical article concerning records left by members of this first modern mission to China. Note the published diaries of the same period (4. 1. 10).

4. 1. 10 NŌTOMI Sukejirō 納富介次郎, and HIBINO Teruhiro 日比野輝寛, **Bunkyū ninen Shanhai nikki** 文久二年上海日記 (Diaries of 1862 in Shanghai), Osaka: Zenkoku Shobō 全國書房, 1946, 165 pp., pub, by the Tōhō Gakujutsu Kyōkai, preface 23 pp. by Toyama Gunji 外山軍治, illus.

First-hand records of historical interest. Note also the article by Okita Hajime, 4. 1. 9.

4.2 JAPAN'S CONTINENTAL EXPANSION: GENERAL AND BIOGRAPHICAL

Note: General studies of the course of empire are followed here by studies by or about individual groups, adventurers, and statesmen. This latter subject is continued in sec. 4.6 below.

4.2.1 ICHIKO Chūzō 市古宙三, **Kindai Nippon no tairiku hatten** 近代日本の大陸發展 (Continental development of modern Japan), Keisetsu Shoin 螢雪書院, 1941, 448 pp.

This wartime volume by one of the most scholarly of modern Japanese sinologists provides a useful survey of the development of Japan's continental policy from 1868 to 1922, centered first around Korea and then around Manchuria (to p. 345). An appendix prints 90 pages of key treaties and documents and an 11-page list of articles from **Gaikō jihō** 外交時報 and various other journals. The text is concise and plentifully supplied with sub-heads to facilitate reference.

4.2.2 UEDA Toshio 植田捷雄, **Nikka kōshōshi** 日華交渉史 (A history of Sino-Japanese relations), Nomura Shoten 野村書店, 1948, 279 pp.

This little volume, subtitled "Japanese continental expansion and the process of its collapse", offers a handy but commonplace summary of early contact (37 pp.), modern relations up to 1931 (to p. 147), and the Manchurian and China incidents up to the Japanese surrender. The last chapter concerns the fate of Japanese interests in China; though published before the Military Tribunal records were fully available, this is a well-informed account. No footnotes, bibliography or index.

4.2.3 KASHIMA Morinosuke 鹿島守之助, **Teikoku gaikō no kihon seisaku** 帝國外交の基本政策 (Basic policies in the foreign relations of the Japanese Empire), Ganshōdō, 1938 (3rd printing, 1942, 424+sources 8 and index 16 pp.).

A standard research study, mainly based on Western sources, of Japanese relations with the powers from the triple intervention of 1895, through the development of the Anglo-Japanese alliance and Russo-Japanese treaties, etc., down to the Washington Conference and Russo-Japanese treaty of 1925, with chapters on the Nine-power Treaty (pp. 363–395) and on Japan's continental policy in general (416–484). A new post-war printing by the same publisher has a different title (**Nippon gaikō seisaku no shiteki kōsatsu** 日本外交政策の史的考察, 1951). Dr. Kashima was once a foreign service official, and then became a leading specialist in diplomatic history, writing two other major books on the origins of World War I and on the diplomacy of Bismarck. He is now an influential businessman and a Diet member.

4.2.4 WATANABE Ikujirō 渡邊幾次郎, **Nippon kinsei gaikō shi** 日本近世外交史 (History of Japanese modern foreign relations), Chikura Shobō 千倉書房, 1938, 494 pp.

While the major content of this standard and rather old-style text is beyond our scope, its treatment of the Sino-Japanese War of 1894–5 (pp. 123–282) and of the 21 Demands of 1915 (pp. 475–486) may be noted as representing a respectably conservative Japanese view.

4. 2. 5 SHINOBU Jumpei 信夫淳平, **Meiji hiwa nidai gaikō no shinso** 明治秘話二大外交の眞相 (Secrets of the Meiji period, the facts about two great diplomatic negotiations), Banrikaku Shobō 萬里閣書房, 1928, 519 + index 7 pp.

·Using the German, British and other war-origin documents, Dr. Shinobu here explores the diplomatic history of the Anglo-Japanese Alliance and of the Russo-Japanese War. Sources are cited briefly in the text.

4. 2. 6 TABOHASHI Kiyoshi 田保橋潔, "Meiji gonen no 'Maria Rusu' jiken" 明治五年の「マリア・ルス」事件 (The *Maria Luz* incident of 1872), **Shigaku zasshi,** 40. 2 (Feb. 1929), 230–246; 3 (Mar. 1929), 364–375; 4 (April 1929), 483–508.

A thoroughly documented reexamination of a famous case in international law, that of the unfortunate Peruvian bark *Maria Luz* carrying Chinese coolies.

4. 2. 7 KUZUU Yoshihisa 葛生能久, **Nisshi kōshō gaishi** 日支交渉外史 (An unofficial history of Sino-Japanese relations), pub. by the Kokuryūkai 黑龍會, 1938–39, 2 vols., 648 + 776 pp.

Published by the so-called Black Dragon Society (more accurately, "Amur [River] Society"), these stout volumes show a good deal of patriotic and chauvinist fervor as well as sympathy for the early revolutionary movement in China. They recount the phases of Japanese relations with China since the Restoration of 1868. Formosa, Liu-ch'iu, Korea, the war of 1894–95, Russia in Manchuria, the Boxers, the war of 1904–5, and American interest in Manchuria, are all dealt with in volume one. Volume two begins with the 1911 revolution and runs through Yuan Shih-k'ai, the "so-called" Twenty-one Demands, warlord relations, the Paris and Versailles conferences, the nationalist anti-Japanese movement, and background and development of the Manchurian and China incidents, etc, etc. Covering so broad a front, the narrative is unencumbered with source references and may be regarded as chiefly of value for its full expression of the super-patriotic attitudes which have been connected with Japan's continental expansion.

4. 2. 8 SATŌ Saburō 佐藤三郎, "Kōakai ni kansuru ichikōsatsu" 興亞會に關する一考察 ("A study on the Kōakai [The Asia Reconstruction Society]"), **Yamagata daigaku kiyō (Jimbun-kagaku)** 山形大學紀要（人文科學）("Bulletin of the Yamagata University [Cultural Science]"), pub. by the university, Yamagata, no. 4 (Aug. 1951), 399–411.

An account of a Sino-Japanese association founded in Tōkyō in 1880 to rebuild Asia against Western aggression, with the participation of Japanese leaders and members of the Chinese legation. Renamed "Ajia Kyōkai" in 1883, it eventually became inactive, though existing until 1900; however, its Chinese language school, Kōa Gakkō, trained many persons for continental service, and its brief history expressed a pan-Asian ideal.

4.2.9 TAKEUCHI Zensaku 竹内善作, "Meji makki ni okeru Chū-Nichi kakumei undō no kōryū" 明治末期における中日革命運動の交流 (Interaction of the Chinese and Japanese revolutionary movements in the late Meiji period), **Chūgoku kenkyū,** 5 (Sept. 1948), 74–95.

This lecture describes an incident of 1907 in which the later KMT stalwart Chang Chi 張繼, with Chang T'ai-yen 章太炎 and other Chinese in Tōkyō, joined in forming a Sino-Japanese society named the "Asiatic Humanitarian Brotherhood" 亞洲和新會 under which the author believes there occurred an early introduction of socialism. This reminiscent lecture was given by one of the Japanese participants in this movement, and so has the merit of being a first-hand account.

4.2.10 Tai-Shi Kōrōsha Denki Hensankai 對支功勞者傳記編纂會 (Commission for the compilation of biographical memoirs of those who rendered service regarding China), **Tai-Shi kaikoroku** 對支回顧錄 (A record looking back on China), by the commission, 1936, 2 vols., 782+1520 pp.; **Zoku tai-Shi kaikoroku** 續 (Supplement), Dai Nippon Kyōka Tosho 大日本敎化圖書 co., vol. 1, 1942, 588 pp.; vol. 2, 1941 (sic), 1316 pp.

Compiled by an agency set up by the Tōa Dōbunkai 東亞同文會 in 1934 and headed by Nakajima Masao 中島眞雄, the first volume of this work surveys the course of Sino-Japanese relations in modern times, including all the incidents, wars and treaties down to 1931. It also has sections on the growth of transport, banking, loans, cultural relations (as fostered by Japanese societies), the press and education between the two countries. These latter especially provide much useful information. Vol. 2 gives biographies of some 830 Japanese individuals then deceased, who had been active regarding China in the period since 1868. Letters and documents are quoted extensively and publications noted. The supplement, vol. 1, continues the summary of Japan's continental exploits, which had outstripped the editors, down to 1937, while vol. 2 adds 214 more biographies, most of them made quite lengthy by the quotation of personal letters and other documents. These two volumes of biographies, although written from an ultra-patriotic angle, give a panorama of many varieties of persons—statesmen, diplomatic-consular and other civilian officials, officers, military interpreters, secret-service men, businessmen, scholars and teachers, journalists, lawyers, and the like, and even though many minor figures are included, much useful data is here provided.

4.2.11 Kokuryūkai 黑龍會 (Amur [River, or "Black Dragon"] Society), **Tōa senkaku shishi kiden** 東亞先覺志士記傳 (Memoirs of pioneer East Asian patriots), pub. by the society, vol. 1, 1933, 886 pp.; vol. 2, 1935, 898 pp.; vol. 3, 1936, 804 pp., many illus.

This monument to Japanese chauvinist expansion contains a history of the incidents, issues and efforts connected with the Japanese forward movement into Korea, Manchuria and China together with biographies of about 1000 individuals who were active in the process. The historical section (to p. 143 of vol. 3) includes accounts of the Nisshin Bōeki Kenkyūjo (1. 396–413) and Tōa Dōbunkai (1. 605–611), as well as almost every other aspect of Japanese activity in war and revolution in China. As a kind of companion work to **Tai-Shi kaikoroku,** 4.2.10, this strongly partisan compilation provides much debatable but often unique material for historians.

4.2.12 MIYAZAKI Tōten (Torazō) 宮崎滔天 (寅藏), **Sanjūsan-nen no yume** 三十三年の夢 (The thirty-three-years dream), Bungei Shunjūsha 文藝春秋社, 1943, 342 pp., illus. (Original edition dated 1902).

This famous autobiography of one of Sun Yat-sen's closest Japanese collaborators has been studied in the recent volume by Dr. Marius Jansen (Harvard University Press). Its colorful narrative comes down to the abortive Hui-chou 惠州 putsch of 1900. This new edition includes a reminiscent essay (pp. 292-342) by the author's son, Miyazaki Ryūsuke 宮崎龍介 in which are included a memoir, once published in the press, of the author's wife and several letters by Miyazaki Torazō himself.

4.2.13 GOTŌ Shimpei 後藤新平, **Nippon shokumin seisaku ippan, Nippon bōchō ron** 日本植民政策一班，日本膨脹論 (A general view of Japanese colonial policy. On the expansion of Japan), edited by Nakamura Akira 中村哲, Nippon Hyōronsha, 1944, 252 pp. (Meji bunka sōsho).

Of these reprints of two essays by Count Gotō, the first (originally a lecture given in 1914) sets forth his ideas of colonial development in the Japanese empire based particularly on his aims and experience in Formosa and Manchuria. Gotō's ideas are of particular interest because of the important part he played in sponsoring research in both regions (e.g. **Shinkoku gyōseihō,** 1.2.1, and other work when he was in Formosa, and the research of the South Manchurian Railway Co.). The editor, a specialist in constitutional law and political science, adds an appraisal of these writings, a biography and a bibliography (pp. 3-40).

4.2.14 Gaimushō 外務省, comp. and 藏版 copyrighted by, **Komura gaikō shi** 小村外交史 (A history of the foreign policy of Marquis Komura), Shimbun Gekkansha 新聞月鑑社, 1953, 2 vols., 459+533 pp., illus. (Published separately from the **Nippon gaikō bunsho** 日本外交文書別冊).

This authoritative study of the career and foreign policy of one of the chief architects of Japan's continental policy of the early 20th century was first drafted by Dr. Shinobu Jumpei 信夫淳平, working in the Foreign Office archives, in the 1920s. It was revised before publication by Usui Katsumi 臼井勝美 with the aid of certain members of the diplomatic corps, and now provides the fullest available account of the part played by Komura Jutarō 小村壽太郎 from the 1890s to 1911 as minister at several capitals and foreign minister during two periods. Quite aside from Komura's influence in shaping and renewing the Anglo-Japanese alliance, he played a key role in Japan's China policy, as minister at Peking in connection with the Boxer crisis, and later. A chronology of his career (pp. 493-501) and another of events (pp. 505-533) are included in volume 2. Reviewed by Hattori Shisō 服部之總 in **Shigaku zasshi,** 62.9 (Sept. 1953), 876-879.

4.2.15 YAMAURA Kan'ichi 山浦貫一, comp., **Mori Tsutomu (Kaku)** 森恪 (Life of Mori Tsutomu [Kaku]), Mori Kaku Denki Hensankai 森恪傳記編纂會, 1940, 55+1124+chronology 18 and index 13 pp., ca. 25 pp. photos.

An elaborate record of the career and ideas of a leading expansionist (1883–1932) who was active both in promoting Japan's continental economic development and then after 1912 in politics and councils of state, becoming chief secretary of the Cabinet and one of the fathers of Manchoukuo. The narrative is based more on oral statements, given by his friends for this compilation, than on written sources. Ca. 60 letters of Mori and some other writings of his, mainly on China problems, are also printed.

4.2.16 Shanhai Zasshisha 上海雜誌社, comp., **Hakusen Nishimoto-kun den** 白川西本君傳 (Biography of Mr. Nishimoto Hakusen), privately printed, 1934, 150 pp., illus.

A brief biography together with many reminiscent notes by relatives and friends of Nishimoto Shōzō 西本省三 (pen-name Hakusen 白川) [1878–1928], an old China hand who was at first a teacher of Chinese language at the Tōa Dōbun Shoin and then worked for many years as a journalist in Shanghai. He was an intimate friend of Cheng Hsiao-hsü 鄭孝胥 and advocated the restoration of the Ch'ing dynasty.

4.3 ACQUIRING LIU-CH'IU AND OPENING KOREA

Note: These studies concern Japan's expansion in the 1870s, omitting the Formosa question.

4.3.1 UEDA Toshio 植田捷雄, "Ryūkyū no kizoku o meguru Nisshin kōshō" 琉球の歸屬を繞る日清交渉 ("Sino-Japanese negotiations disputing the sovereignty over the Lewchew Islands"), **Tōyō bunka kenkyūjo kiyō,** 2 (Sept. 1951), 151–201.

Mainly based on the recently published volumes of **Nippon gaikō bunsho** 日本外交文書 as well as on the Chinese documents (**Ch'ing-chi wai-chiao shih-liao** 淸季外交史料 and Li Hung-chang's papers 李文忠公全集), this article follows in detail the Sino-Japanese dispute over the Liu-ch'iu Islands, from Japan's scheme of annexation in the early Meiji period, through the period of the mediation of General Grant and the unratified treaty of 1880, down to the Sino-Japanese War of 1894–95. See also the article by Mikuniya Hiroshi 三國谷宏 on "Grant's Mediation concerning the Subjection of Loochoo Islands" in **Tōhō gakuhō, Kyōto,** 10, part 3 (Oct. 1939), 29–64.

4.3.2 TABOHASHI Kiyoshi 田保橋潔, "Ryūkyū hammin bangai jiken ni kansuru kōsatsu" 琉球藩民蕃害事件に關する考察 (An investigation of the outrage suffered by the natives of Ryūkyū at the hands of the aborigines [in Formosa]), pp. 663–688 in **Ichimura hakushi koki kinen Tōyōshi ronsō** 市村博士古稀記念東洋史論叢 (Collected essays in Oriental history to commemorate the 70th anniversary of Dr. Ichimura), Fuzambō, 1933, 1214 pp.

A brief study of the Liu-ch'iu dispute between Japan and China in 1872–73 making some use of the newly published Chinese (IWSM) documents.

4.3.3 MIURA Hiroyuki 三浦周行, "Meiji jidai ni okeru Ryūkyū shozoku mondai" 明治時代に於ける琉球所屬問題 ("Das Problem der politischen Zugehörigkeit der Liukiuinseln im Meiji-Zeitalter"), **Shigaku zasshi,** 42.7 (July 1931), 711–724; 12 (Dec. 1931), 1351–1385 (this latter part was posthumously edited by the author's student, Nakamura Naokatsu 中村直勝).

Mainly based on the Japanese sources then available (including a manuscript indicating the opinion of the famous French legal consultant, Boissonade), this study by an eminent scholar of Japanese history describes the development of the Sino-Japanese dispute over the Liu-ch'iu Islands. This has presumably been superseded by Dr. Ueda's recent study, 4.3.1.

4.3.4 OKUDAIRA Takehiko 奧平武彥, **Chōsen kaikoku kōshō shimatsu** 朝鮮開國交涉始末 (A complete account of the negotiations in the opening of Korea), Tōkō Shoin, 1935, 195+15 pp. (pub. by Keijō Imperial University Law Society).

This research monograph deals with the period from before the Japanese treaty of 1876 down through the Shufeldt treaty of 1882 and the signing of the British and other treaties which followed. It is based on United States, British, Japanese and Korean archives, and Chinese and other publications, making it the most authoritative account yet available on this topic, even though it is less comprehensive in scope than the work of Tabohashi Kiyoshi, **Kindai Nissen kankei no kenkyū,** 4.4.2.

4.3.5 TABOHASHI Kiyoshi 田保橋潔, "Kindai Chōsen ni okeru kaikō no kenkyū" 近代朝鮮に於ける開港の研究 (A study of the opening of the ports in modern Korea), pp. 599–639, in **Oda sensei shōju kinen Chōsen ronshū** 小田先生頌壽記念朝鮮論集 (Collected writings on Korea to commemorate the 60th birthday of Professor Oda), Ōsakayagō Shoten 大阪屋號書店, 1934, 1074 pp., illus.

A detailed and documented account of the role of the ports as originally envisioned in the Japanese treaty of 1876, the choice of sites, and Japanese-Korean negotiations concerning Genzan (Wonsan 元山) and Jinsen (Inchon 仁川).

4.3.6 AKIHO Ichirō 秋保一郎, "Fuzan kaikō o meguru jakkan no rongi" 釜山開港を繞る若干の論議 ("On the Opening of Port Pusan in 1876 and the Disputations over the Opening of Korea to Foreign Intercourse"), **Tōyō bunka kenkyū,** 11 (May 1949), 65–78.

A brief and somewhat interpretative study using the **Nippon gaikō bunsho** 日本外交文書 documents, and stressing the discrepancies between the assumptions of Japanese and Korean officials.

4.4 JAPAN AND CHINA IN KOREA: THE FIRST SINO-JAPANESE WAR

Note: The major work here is that of Tabohashi Kiyoshi, 4.4.2.

4.4.1 TABOHASHI Kiyoshi 田保橋潔, **Kindai Nichi-Shi-Sen kankei no kenkyū** 近代日支鮮關係の研究 ("A Critical Study on the Diplomatic Relations of Japan with China and Korea from the Treaty of Tientsin [1885] to the Outbreak of the Chino-Japanese War [1894]"), **Keijō teikoku daigaku hōbungakubu kenkyūchōsa sasshi dai-sanshū** 京城帝國大學法文學部研究調査册子第三輯 ("Bulletin of the Faculty of Law and Letters, Miscellaneous Series, Vol. 3, Keijō Imperial University"), Seoul, 1930, 252 pp., photos.

This monograph on the diplomatic history of the Sino-Japanese conflict over Korea was the first version of the masterly work in 2 volumes later published in 1940 as **Kindai Nissen kankei no kenkyū**, 4.4.2.

4.4.2 TABOHASHI Kiyoshi 田保橋潔, **Kindai Nissen kankei no kenkyū** 近代日鮮關係の研究 (A study of modern Japanese-Korean relations), Keijō: Chōsen Government-general, 1940, 2 vols., 1133+969 pp., many illus.

These enormous volumes grew from Professor Tabohashi's earlier study, **Kindai Nichi-Shi-Sen kankei no kenkyū**, 4.4.1, which, though acclaimed by critics, seemed to touch on official secrets; distribution of its copies was prohibited subsequent to publication. The publication of this larger work was made possible only as a secret printing sponsored by the Korean Government General for a quite limited circulation; certain professors of the Faculty of Letters of Tōkyō Imperial University refused to accept it as a doctoral dissertation because it infringed upon the national dignity. As finally published the work would seem to justify such concern only in the sense that it is by far the most complete and authoritative account of the origins of the Sino-Japanese War of 1894. Major sections concern: 1) Korea's modern history, especially the rise to power of the Tai Won Kun 大院君; 2) inauguration of Korean-Japanese relations, growth of Japanese policy and negotiation of the treaty of Kōka (Kiang-hua 江華) of 1876, including Ch'ing policy concerning it (pp. 133–556); 3) opening of ports, incidents and further negotiations, the Chinese intervention, and Sino-Japanese treaty of Tientsin of 1885; 4) Sino-Japanese rivalry, Korean movements of reform and rebellion, incidents and international negotiations up to the war of 1894 (vol. 2 to p. 638); 5) a separate study of Japanese-Korean relations concerning Tsushima. No brief note can do justice to Professor Tabohashi's thorough, systematic, and detailed use of sources including Japanese and Korean archives, Chinese and Western published documents, Japanese memoirs, and the like. Bibliography 35 pp. These volumes were continued in his **Nisshin-sen'eki gaikōshi no kenkyū**, 4.4.3. Early versions of several sections appeared in the Keijō journal, **Seikyū gakusō**; see especially 21 (Aug. 1935), 1–105, on the Korean revolt of 1882.

4.4.3 TABOHASHI Kiyoshi 田保橋潔, **Nisshin sen'eki gaikōshi no kenkyū**

日清戰役外交史の研究 ("A Diplomatic History of the Sino-Japanese War 1894–1895"), Tōkō Shoin, 1951, 556 pp., list of the author's works, 5 pp., English summary, 5 pp., (Tōyō bunko ronsō 東洋文庫論叢, 32).

This impressive and scholarly monograph describes the origin and course of the war and particularly the international negotiations before, during, and after it, up to the ratification of the Treaty of Shimonoseki. The exemplary use of Western, Chinese and Japanese materials and also of Korean government documents makes this well balanced volume presumably the last word on its subject. Reviewed by Ubukata Naokichi 幼方直吉 in **Rekishigaku kenkyū**, 154 (Nov. 1951), 51–52.

4.4.4 SHINOBU Seizaburō 信夫清三郎, **Mutsu gaikō, Nisshin sensō no gaikōshiteki kenkyū** 陸奥外交——日清戰爭の外交史的研究 (The foreign policy of Mutsu, an historical study of the diplomacy of the Sino-Japanese war), Sōbunkaku 叢文閣, 1935, 617 pp.

A narrative account, from the Tong-hak 東學 rebellion through the various stages of fighting, negotiation, intervention, reform efforts and incidents leading up to the outbreak of the war of 1894–5. The author uses a considerable body of Japanese, Chinese and Western materials, including contemporary newspapers and magazines. The famous memoirs of the Japanese foreign minister, Mutsu Munemitsu 宗光, **Kenkenroku 蹇々錄**, are also fully used. This big volume is based on the Marxist interpretation of Japan's modern history provided by Yamada Moritarō 山田盛太郎, Hattori Shisō 服部之總, and Hirano Yoshitarō 平野義太郎. Reviewed by Fujisawa Ichirō 藤澤一郎 in **Rekishigaku kenkyū**, 3.2 (Dec. 1934), 186–188. Prof. Shinobu, son of Dr. Shinobu Jumpei, is now at Nagoya University.

4.4.5 TATSUMI Raijirō 巽來治郎, **Nisshin sen'eki gaikō shi** 日清戰役外交史 (Diplomatic history of the Sino-Japanese war), Tōkyō Semmon Gakkō Shuppanbu 東京專門學校出版部, 1902, 1227 pp., documentary appendix 59 pp.

Although now of ancient vintage, this account benefitted by the author's experience as a historiographer under the General Staff of the Army and may still have value for reference to contemporary events and opinions. The contemporary importance of this book lay in its discussion of the application of international law to each phase of the conflict. The book was officially banned at one time.

4.4.6 UMETANI Noboru 梅溪昇, NISHIMURA Mutsuo 西村睦男, KITAMURA Hirotada 北村敬直, KANG, Z. E. (Kyō Zai-gen) 姜在彦, "Nisshin Sensō" 日清戰爭 (The Sino-Japanese war), **Shirin**, 35.4 (Oct. 1953), 335–374.

A symposium of four papers given at the 1952 annual meeting of the Shigaku Kenkyūkai 史學研究會 (The Historical Society), followed by a briefly recorded discussion: Umetani, "Sino-Japanese War as seen from the Japanese Viewpoint"; Nishimura, "The Industrial Revolution and the Sino-Japanese War"; Kitamura, "Sino-Japanese War as seen from the Chinese Viewpoint"; Kang, "Sino-Japanese war as seen from the Korean Viewpoint". Sources are cited in the course of the respective presentations. Prof. Kitamura's suggestive report analyses the background rivalry between Li Hung-chang's peace party (who generally monopolized the industrial enterprises patronized by the state) and the war party led by reformers such as K'ang Yu-wei and Chang Chien (who had a program for establishing private-enterprise industries).

4. 4. 7 MINAMI Tokuko 南さく子, "Nisshin Sensō to Chōsen bōeki" 日清戦争と朝鮮貿易 ("Sino-Japanese War and Korean Trade"), **Rekishigaku kenkyū,** 149 (Jan. 1951), 43–46.

This brief note analyzes the Japanese exports to Korea in the pre-war period, and criticizes the conventional Marxist interpretation (see, e.g., Shinobu Seizaburō, **Mutsu gaikō,** 4. 4. 4) which maximizes the desire of the Japanese industrial interests, especially the cotton industry, to expand their market, as a decisive factor in bringing on the war.

4.5 MODERN HISTORY OF MANCHURIA AND KOREA

Note: We confine this section mainly to bibliographical guidance, the subject matter being peripheral to China and too extensive to pursue in detail.

4. 5. 1 Rekishigaku Kenkyūkai 歷史學研究會, comp., **Manshūshi kenkyū** 滿洲史研究 (Researches in Manchurian history), Shikai Shobō 四海書房, 1936, 312 pp.+bibliography 97 pp. (first published as a special number, **Rekishigaku kenkyū,** 5.2 [Dec. 1935], 312+97 pp.)

This volume's list of publications and sources in all languages includes the contents of Chinese gazetteers for the Three Eastern Provinces, and reference sections in major Chinese compendia, but most of the bibliography (52 pp.) lists Japanese works. The 13 research articles include the following on the modern period:
(1) Mishima Hajime 三島一, "Manshūshi kenkyū josetsu" 滿洲史研究序說 (Introduction to the study of Manchurian history), 1–6.—A brief reference article.
(2) Kawakubo Teirō 川久保悌郎, "Shimmatsu ni okeru Kichirinshō seihokubu no kaihatsu 清末に於ける吉林省西北部の開發 (Exploitation of the northwestern part of Kirin province in the late Ch'ing period), 147–184.—A solidly based article on an aspect of Chinese colonization in Manchuria, typically exemplifying the erosion created on bannermen's lands by Chinese peasants.
(3) Ōkami Suehiro 大上末廣, "Kindai ni okeru Manshū nōgyō shakai no henkaku katei" 近代に於ける滿洲農業社會の變革過程 (The process of transformation in Manchurian agrarian society in the modern period), 185–212.—An interpretative analysis of the modernizing process in Manchuria since the opening of Newchwang (Ying-k'ou 營口) as a treaty port in 1862.
(4) Eguchi Bokurō 江口朴郎, "Tsāri to Manshū mondai" ツァーリと滿洲問題 (The czar and the Manchurian problem), 213–236.—A study by a specialist in the diplomatic history of modern Europe, which traces the relationship between the czarist despotism and Russian Far-Eastern policy, with special attention to Witte and the pre-history of the Russo-Japanese war.

(5) Nohara Shirō 野原四郎, "Shindai ni okeru Man-Shi no keizaiteki yūgō" 清代に於ける滿支の經濟的融合 (The economic union of Manchuria and China proper in the Ch'ing period), 237–246.—A brief interpretative survey, mainly on the 19th century after the opening of Newchwang.

(6) Momose Hiromu 百瀬弘, "Wagakuni ni okeru Manshū kinseishi kenkyū no dōkō" 我國に於ける滿洲近世史研究の動向 (Trends of research in Japan on the modern history of Manchuria), 279–292.—An excellent historiographical reference survey.

(7) Suzuki Shun 鈴木俊, "Manshū jiken to Shinajin no Manshū kenkyū" 滿洲事件と支那人の滿洲研究 (The Manchurian Incident and research on Manchuria by the Chinese), 299–312.—A reference survey, stressing the sudden heightening of Chinese scholarly concern as stimulated by the Incident.

4.5.2 Nagao Sakurō 永雄策郎, "Mantetsu o chūshin to suru gaikō" 滿鐵を中心ことする外交 (Negotiations centering on the South Manchurian Railway Co.), **Keizai shiryō** 經濟資料 (Economic source materials), pub. by the East Asia Economic Investigation Office 東亞經濟調查局 of the South Manchurian Railway, 14.5 (May 1928), 1–465, map.

This is the second part of a study of railways in colonial regions of the world under the grandiose title of **Shokuminchi tetsudō no sekai keizaiteki oyobi sekai seisakuteki kenkyū naishi shokuminchi tetsudō no gaiteki kenkyū** 植民地鐵道の世界經濟的及世界政策的研究乃至植民地鐵道の外的研究 (A study of colonial railways, from the view point of world economy and world politics, or an external study of colonial railways), see ibid. 14.4. It deals with the efforts of Harriman in railway enterprise in Manchuria, negotiations over the Fa-ku-men 法庫門 and various other lines, and eventual collapse of the Knox neutralization scheme of 1910, relying mainly on Western sources. The author was on the SMR research staff.

4.5.3 Chōsen Sōtokufu Chōsenshi Henshūkai 朝鮮總督府朝鮮史編修會 (Chōsen Government-general, commission for the compilation of Chōsen history), **Kindai Chōsenshi kenkyū** 近代朝鮮史研究 (Researches in modern Korean history), Keijō: Chōsen Government-general, 1944, 625 pp.

This officially sponsored volume was published as the first of a series but the subsequent volumes have never appeared. The book contains three sections: by Tabohashi Kiyoshi 田保橋潔 on political reform in modern Korea; by Teratani Shūzō 寺谷修三, a Korean scholar, on the conflict of new and old in the Seisō 成宗 period; and by Tagawa Kōzō 田川孝三 on the modern rural society of North Korea and the problem of vagrancy. Tabohashi's section traces in detail the successive systems of administrative organization (pp. 1–190) and adds an appendix of key documents (pp. 191–302). This volume is a by-product of the officially-sponsored Korean historiography discontinued after the collapse of Japanese colonialism, and is one of the relatively few pieces of first-class scholarship dealing with Korea. Mr. Tagawa is now in the Tōyō Bunko.

4.5.4 Hatada Takashi 旗田巍, **Chōsen shi** 朝鮮史 (History of Korea), Iwanami, 1951, 299+11 pp. (Iwanami zensho, 154).

This convenient little volume surveys Korean history bibliographically, in addition to its narrative summary, period by period. About one hundred pages deal with the modern century. The critical, general bibliography, chronological table and index are all usefull, and the book is full of lucid interpretations based on a sophisticated, socio-economic historical analysis. Reviewed by Suematsu Yasukazu 末松保和, Sudō Yoshiyuki 周藤吉之, and Yamabe Kentarō 山邊健太郎 in **Rekishigaku kenkyū**, 156 (March 1952), 41–49. The special issue of **Rekishigaku kenkyū** on **Chōsenshi no shomondai** 朝鮮史の諸問題 ("Problems on the History of Korea"), (July 1953), 118 pp., contains articles by Hatada and others (some critical of him) and bibliography.

4.6　JAPANESE CENTERS IN CHINA

Note: The first four items relate to the important Tōa Dōbun Shoin in Shanghai, which has been responsible for more of the studies recorded in this volume than any other single research agency or center. Subsequent items record the growth of Japanese communities in China.

4.6.1　INOUE Masaji 井上雅二, **Kyojin Arao Sei** 巨人荒尾精 (Arao Sei, a great man), 1910 (2nd printing, Tōa Dōbunkai, 1936, 220 pp.)

A biography of one of the key personalities in the history of Japanese expansion towards China. Arao Sei (1858–1896) went to China in 1886 as an intelligence officer of the Japanese Army General Staff, and worked for 3 years in Hankow at the center of an intelligence network, in charge of a number of espionage agents and researchers. In 1890, he founded in Shanghai a school called Nisshin Bōeki Kenkyūjo 日清貿易研究所 (The Research Institute for Sino-Japanese Trade), which was the forerunner of the Tōa Dōbun Shoin. **Shinkoku tsūshō sōran**, 7.2.2, was a product of this early intelligence-research work and set an example of digging into the Chinese scene which was influential in later Japanese studies of China. This biography is factually written and many documents are quoted. **Shina**, 25.10 (Oct. 1934), frontispiece and pp. 234–292, provides supplementary information on Arao Sei, and an appendix of 77 pp. reprints two of his writings of 1894 and 1895.

4.6.2　Tōa Dōbunkai 東亞同文會, comp., "Konoe Kazan kō kinenshi" 近衞霞山公記念誌 (Commemorative writings on Prince Konoe Kazan), **Shina**, 25.2–3 (Feb. 1934), 1–213, illus.

This special issue prints many reminiscent articles commemorating the 30th anniversary of the death of the late Prince Konoe Atsumaro 篤麿, father of the wartime premier, who was a leading statesman and one of the key personalities of Sino-Japanese relations in the Meiji period. He was an influential patron of Japanese "patriots" who were keenly concerned in China problems. In 1900 he sponsored in Nanking (with important help on the part of Liu K'un-i 劉坤一, Governor-general of the Liang-kiang provinces) a school called Nankin Dōbun Shoin 南京同文書院, which was removed to Shanghai the next year and rechristened Tōa Dōbun Shoin. The school became the alma mater of many scholars, teachers, journalists, merchants, diplomatic and consular officials, interpreters and intelligence agents, many of whose works are noted in this bibliography.

4.6.3 Tōa Dōbun Shoin Koyū Dōsōkai 東亞同文書院滬友同窓會, comp., **Sanshū Nezu sensei den** 山洲根津先生傳 (Biography of Nezu Hajime 根津一 [pen-name "Sanshū"]), Nezu Sensei Denki Hensambu 根津先生傳記編纂部 (in Tōa Dōbunkai), printed for limited circulation, 1930, 490 pp., many photos.

A biography of the patriarchal figure of the Tōa Dōbun Shoin, including his own writings and autobiographical notes. Nezu Hajime (1860–1926) started his career as an army officer, and helped Arao Sei to found and run the Nisshin Bōeki Kenkyūjo. He then became the president of the Shanghai Tōa Dōbun Shoin in 1901, and remained in this position until 1923. This work was compiled by Otake Fumio and Yamada Kenkichi 山田謙吉 (pen-name 岳陽). See also Munekata Kingo 宗像金吾, comp., **Tōa no senkakusha Sanshū Nezu sensei narabini fujin** 東亞の先覺者山洲根津先生竝夫人 (The East Asian pioneer, Professor Nezu [Sanshū], and his wife), pub. by the compiler, for limited circulation, Kyōto, 1943, 364 pp.

4.6.4 (Shanhai) Tōa Dōbun Shoin (上海)東亞同文書院, **Sōritsu sanjisshū-nen kinen Tōa Dōbun Shoin shi** 創立三十周年記念東亞同文書院誌 (A record commemorating the thirtieth anniversary of the founding of the T'ung-wen College), Shanghai: pub. by the same, 1930, 116 pp., map, many photos.

A history of this important training and research center, arranged by periods from the founding of the preceding Nisshin Bōeki Kenkyūjo 日清貿易研究所 in 1890, the establishment of the Tōa Dōbunkai in Tōkyō in 1898, the Nankin Dōbun Shoin 南京同文書院 in 1900, and the Tōa Dōbun Shoin in 1901, down to 1930 (to p. 64). There follows a description of the institution and its work as of that date with data on sponsors, staff members, field trips, publications (**Shina kenkyū**, 9.9.21, from 1920), etc., numbers of graduates and their later lines of work (914 in business, 110 officials, 55 journalists, out of 1847 Japanese graduates). This volume was prepared by Otake Fumio.

4.6.5 OKUDA Otojirō 奧田乙治郎, **Meiji shonen ni okeru Honkon Nippon-jin** 明治初年に於ける香港日本人 (Japanese at Hongkong in the early Meiji period), Nettai Sangyō Chōsakai 熱帶產業調査會 of the Taiwan Government-general, printed for limited circulation, Taihoku, 1937, 331+29 pp.

Written by a consular officer stationed at Hongkong, from the consulate archives, this account of Japanese activities there mainly in the period 1873–1889 was first printed in instalments in the Japanese press (香港日報) at Hongkong. It records the founding of the consulate and then faithfully summarizes major developments there and in the local community on a year by year basis down to 1889, and also mentions briefly the later period under several topics. A stenographic record of two conferences held by Japanese pioneers in Hongkong is printed as an appendix. The author is a graduate of the Tōa Dōbun Shoin.

4.6.6 YONEZAWA Hideo 米澤秀夫, "Shanhai hōjin hatten shi" 上海邦人發展史 ("L'histoire du développement des japonais à Shanghai"), **Tōa keizai kenkyū,** 22.3 (July 1938), 394–408; 23.1 (Jan.–Feb. 1939), 112–126.

Beginning with the vessel sent by the Shogunate in 1862, this piece summarizes the growth of the Japanese community and its institutions, including the early Japanese steamer service (begun in 1875), founding of the Nisshin Bōeki Kenkyūjo, see 4.6.1–3, and early business firms, listed by date. The author was on the staff of the Japanese chamber of commerce in Shanghai.

4.6.7 Kikuchi Akishirō 菊地秋四郎 and Nakajima Ichirō 中島一郎, **Hōten nijūnenshi** 奉天二十年史 (A history of twenty years of Fengtien [Mukden]), Fengtien: Hōten Nijunenshi Kankōkai 刊行會, 1926, 770+124 +172 pp., illus., maps.

This semi-official commemorative volume summarizes factually the history of Japanese military and political activities, migration, administration (judicial, police, etc.), railway, commercial and fiscal developments, foreign relations, industries, social services and the like in the city of Mukden and adds a 124-page who's who of Japanese pioneers and another 172-page who's who of the local Japanese leadership.

4.6.8 Tenshin Kyoryūmindan 天津居留民團, comp., **Tenshin kyoryūmindan nijisshūnen kinenshi** 天津居留民團二十周年記念誌 (Commemorative record of twenty years of the Tientsin Residents Association), Tientsin: pub. by the association for private distribution, 1930, 30+698 pp., 40 pp. photos., maps.

This massive tome records the growth of the foreign treaty port of Tientsin, its population, administration, economy, communications, education and religion, and social life in great detail (to p. 334), and then recounts the similar aspects of the development of the Japanese community and concession area, with equal detail.

4.7 RUSSIAN EXPANSION IN MANCHURIA

Note: The most valuable item in this section is that by Yano Jin'ichi, 4.7.1. Note also 2.7.8 and 2.7.13.

4.7.1 Yano Jin'ichi 矢野仁一, **Nisshin ekigo Shina gaikō shi** 日淸役後 支那外交史 ("A History of the Post-bellum Diplomacy of China after the Sino-Japanese War"), Kyōto: Tōhō Bunka Gakuin, Kyōto Kenkyūjo (The Academy of Oriental Culture, Kyōto Institute), Memoirs, vol. 9, 1937, 709 pp., index and bibliography 21 pp., English summary 17 pp.

An important study of Sino-Russian diplomacy in the fateful decade between the Sino-Japanese and Russo-Japanese wars. Dr. Yano utilized the Western documents on the origin of World War I, extensive Russian literature, and volumionus Chinese sources of all sorts to produce an outstanding work in diplomatic history. The manifold events and interests involved in this story are subordinated to the main theme of the negotiations by Li Hung-chang and by Russia, through which China eventually permitted Russian expansion into

Manchuria and thus, in the author's view, set the stage for the war of 1904–5. This meticulous and painstaking volume, the basic work on its subject, should be of value to Western diplomatic historians. See review by Mikuniya Hiroshi 三國谷宏, **Tōyōshi kenkyū**, 3.2 (Nov.–Dec. 1937), 137–141.

4.7.2 YANO Jin'ichi 矢野仁一, **Manshū kindai shi** 滿洲近代史 (Modern history of Manchuria), Kōbundō, 1941, 520 pp.

This detailed history of Russian encroachment on Manchuria first considers the institutional situation of Manchuria in the early Ch'ing period, and then covers the subject from the first Russian descent of the Amur through their successes in 1850–60 and 1900, and on down to the Russo-Japanese War and its aftermath. The prelude to that war receives the amplest treatment (some 200 pp.). Use of Chinese documents casts light on the policy of Li Hung-chang. No footnotes, bibliography or index, but references are roughly given in the text.

4.7.3 YANO Jin'ichi 矢野仁一, "Pekin no Rokoku kōshikan ni tsuite" 北京の露國公使館に就いて (On the Russian legation in Peking), **Geimon**, 6.9 (Sept. 1915), 884–897: 10 (Oct.), 1065–1091.

A detailed account of the location of the early Russian mission (*O-lo-ssu kuan* 俄羅斯館) among other tribute hostels in Peking, and the early Russian tribute missions.

4.7.4 HANAWA Sakura 塙作樂, "Roshia teikoku no Kyokutō shinshutsu" 露西亞帝國の極東進出 ("Eindringen nach dem fernen Osten des russischen Kaiserreichs"), **Rekishigaku kenkyū**, 10.9, whole number 81 (Sept. 1940), 894–923; 10.10 (Oct. 1940), 1025–1065; continuation not found.

A survey of Russian penetration of Manchuria down to the treaty of Peking of 1860, which makes good use of the Chinese (IWSM) and other documents, although it relies on well-known Western-language accounts for the Russian side of the story. Attention is centered mainly on Muraviev.

4.7.5 AKIHO Ichirō 秋保一郎, "Ro-Shin mitsuyaku to Chokureiwan mondai" 露清密約と直隸灣問題 ("The Sino-Russian Secret Treaty of 1896 and the Pechihli-Bay Problem"), **Kokusaihō gaikō zasshi**, 36.10 (Dec. 1937), 948–979; 37.2 (Feb. 1938), 149–170; 37.3 (Mar. 1938), 201–218; 37.6 (July 1938), 518–543.

This study uses the published Chinese documents and published Western documents and literature to work out the course of the negotiations which resulted in the Li-Lobanov treaty of 1896 that invited the Russian bear into the Manchurian parlor. The author deals at length with the premises on which the deal was made and the conclusion of the secret alliance, which soon collapsed although Russian treaty rights continued. The Pechihli-Bay problem concerned the seizures of Port Arthur and Weihaiwei.

4.7.6 MIYAZAKI Masayoshi 宮崎正義, Minami Manshū Tetsudō Kabushiki Kaisha, Shachōshitsu Chōsaka 南滿洲鐵道株式會社社長室調查課 (South Manchurian Railway Company, Research Department of the

Director General's Office), **Kindai Ro-Shi kankei no kenkyū, en Kokuryū chihō no bu** 近代露支關係の研究——沿黑龍地方之部 (A study of modern Russo-Chinese relations, section on the region along the Amur), Dairen, pub. by the Research Department, 1922, 400 pp. (Research reports, vol. 17).

Published as the first of a trilogy (vols. 2 and 3, which have not been published, were to deal with Russo-Chinese relations in Manchuria and Mongolia), this early research study made extensive use of Russian and Chinese sources (the bibliography, pp. 395–400, lists 45 of the former, 26 of the latter). It covers the early Russian incursion into the Amur valley and contact with the Chinese (to p. 116), the Treaty of Nerchinsk and subsequent friction down to the treaties of 1858–60 (to p. 324), and problems of boundary, navigation, migration and the like thereafter. Names printed in Russian are given *kana* readings. Completed a decade or more before the availability of the Chinese documents, this study has presumably been largely superseded.

4.7.7 YOSHIDA Kin'ichi 吉田金一, and MOMOSE Hiromu 百瀨弘, **Rokoku no Kyokutō shinryaku ni kansuru chōsasho** 露國の極東侵略に關する調査書 (Research report on Russian aggression in the Far East), Tōa Keizai Chōsa Kyoku 東亞經濟調査局 (East Asia Economic Investigation Office [of the SMR]), 1933, 3 vols. mimeo. for limited distribution, 156+147+229 pp. (Vols. 1 and 3 by Yoshida, vol. 2 by Momose).

Of these draft volumes, the first deals with the Nerchinsk and Kiakhta treaties, the second with the Russian absorption of the Amur region and Maritime Province down to 1860, and the third with the Ili question and Russian relations on China's northwest frontier in the latter part of the 19th century. While relying mainly on Western-language sources, and using Russian sources only at second-hand, these mimeographed volumes also use the Chinese (e.g., IWSM) documents.

4.7.8 YUMBA Seikichi 弓場盛吉, comp., for the Harbin office, South Manchurian Railway Co., **Tō-Shi tetsudō o chūshin to suru Ro-Shi seiryoku no shōchō** 東支鐵道を中心とする露支勢力の消長 (The extension of Russo-Chinese power, especially over the Chinese Eastern Railway), pub. by the SMR, 1928, 2 vols., 1457 pp.

This research report surveys the early development of Russo-Chinese relations over the Chinese Eastern Railway in Manchuria down to 1917 (to p. 276) and thereafter, including the ramified negotiations concerning it in the 1920s. The story is long and detailed.

4.7.9 SHIMODA Reisuke 下田禮佐, "Ro-Shin kankei no kenkyū" 露清關係の研究 (A study of Russian-Ch'ing relations), pp. 403–452 in **Ogawa hakushi kanreki kinen shigaku chirigaku ronsō** 小川博士還曆記念史學地理學論叢 (Collected writings in history and geography to commemorate the 60th birthday of Dr. Ogawa), Kōbundō, 1930, 1064 pp.

A survey of 200 years of Russo-Chinese relations, particularly concerning trade, down to the middle of the 19th century. Using well known sources in

French and English and a few Ch'ing documents, the author concentrates on describing the commercial contact, especially its machinery as recorded in **Hui-tien shih-li 會典事例.**

4. 7. 10 IWAMA Tōru 岩間徹, **Rokoku Kyokutō seisaku to Witte** 露國極東政策とウイッテ (Russian Far Eastern policy and Witte), Hakubunkan, 1941, 284 pp.+index 10 pp., map.

A scholarly monograph on Russo-Japanese relations over Manchuria and Korea in the period from the beginning of the Trans-Siberian railway to the Portsmouth treaty of 1905. The author is critical of Witte's memoirs and relies considerably on the works of B. A. Romanov, **Rossija v Manczurii** and Kurt Krupinski, **Russland und Japan** (Berlin 1940), together with Japanese and Western memoirs and documents.

4. 7. 11 SAITŌ Yoshie 齋藤良衞, **Sovieto Rokoku no Kyokutō shinshutsu** ソヴイエト露國の極東進出 (Soviet Russia's advance into the Far East), Nippon Hyōronsha, 1931, 376 pp.

While this volume seems somewhat more journalistic than academic, as though written for the public at the time, it nevertheless presents a well organized summary of the Soviet take-over in Siberia (to p. 77) and the successive phases in the development of Soviet policy toward and in China, from the various missions of Yurin and Paikes, Joffe, Karakhan and others to the work of the Comintern and Borodin, and the various negotiations over the Chinese Eastern Railway. Few sources are cited.

4. 7. 12 SAITŌ Seitarō 齋藤淸太郎, "Rokoku no Tōa seisaku to rekkoku" 露國の東亞政策と列國 (Russia's Far Eastern policy and the powers), in **Tōzai kōshō shiron,** 1939, (9. 8. 11), pp. 1003–1050.

This essay provides a broad survey of the many facets of Russian policy in East Asia down to the First World War.

4. 7. 13 FUSE Katsuji 布施勝治, **Sowēto Tōhō saku** ソウエート東方策 (Soviet Eastern policy), Ōsakayagō Shoten, 1926 (2nd printing 1927, 479 pp.).

The author, a journalist for more than a decade, traveled in Russia and elsewhere in Asia and interviewed at one time or another Lenin, Stalin, Trotsky, Chicherin, Wu P'ei-fu, Feng Yü-hsiang, Chang Tso-lin and similar figures, so that this volume contains many first-hand impressions of the process of "bolshevization" 赤化. Its aim is to analyze the Soviet policy behind this process in Asia, which it does by presenting an historical retrospect, a description of the Soviet policy for national minorities (including self-government and cultural revival) and an account of Soviet policy regarding Turkey, Persia, Afghanistan, India, Mongolia, and the Peking and Canton regimes. The latter sections deal with the KMT-CCP alliance, anti-imperialism, and Soviet propaganda in China (pp. 235–411).

4.8 INNER ASIAN RELATIONS

Note: See sec. 3.5 above, of which this is a brief continuation. We have found rather few studies of modern international relations in Inner Asia (as opposed to Manchuria), although several of the general works in sec. 1 above include surveys. On Mongolia, note the bibliography by Iwamura and Fujieda, 9.2.10. We advisedly refrain from entering the rich field of Mongol studies in Japan.

4.8.1 IRIE Keishirō 入江啓四郎, **Shina henkyō to Ei-Ro no kakuchiku** 支那邊疆と英露の角逐 (The Chinese frontier and Anglo-Russian rivalry), Na-u-ka-sha ナウカ社, 1935, 604+3 pp., map.

This important study of the modern power relations concerning Mongolia, Sinkiang and Tibet utilizes a tremendous body of English, Russian, Chinese and other literature to give a comprehensive treatment of the subject. An introductory general section surveys China's nineteenth century relations with Inner Asia and the advance of Russian and English interests there (also English-French rivalry over Yunnan). Succeeding sections on the three major areas of Mongolia (pp. 145–340), Sinkiang, and Tibet (pp. 445–601) take up their geography, societal structure, and recent political development, leading in each case to a detailed study of their modern international relations. The treatment is clear, concise and carefully annotated, although necessarily superficial on many aspects of so large a subject. As this field of research has been almost unexplored in Japan, this book published in 1935 is still a useful guide.

4.8.2 YANO Jin'ichi 矢野仁一, **Kindai Mōkoshi kenkyū** 近代蒙古史研究 (Researches in modern Mongolian history), Kōbundō, 1925 (6th printing, 1940, 468 pp.).

These 35 chapters treat in factual detail a variety of topics, covering the period down to the 1920's, including Ch'ing institutional relations with the Mongol tribes, problems of Chinese settlement and cultivation in Mongolia, the Lama church, and Russian relations with Mongolia. Sources are noted in the text in great variety, as well as spellings of many names.

4.8.3 SAKAMOTO Koretada 坂本是忠, "Mongoru minzokushugi no ippanteki kōsatsu" モンゴル民族主義の一般的考察 ("A general inquiry into the Mongolian nationalism"), **Tōyō bunka,** 10 (Aug. 1952), 37–60.

This well-balanced and analytic survey article discusses the growth of Mongol nationalism in the period from the 19th century down to the mid-20th century, including the influence of Japan. The author acknowledges a considerable debt to Owen Lattimore.

4.8.4 NISHIDA Tamotsu 西田保, **Sa Sō-dō to Shin-kyō mondai** 左宗棠と新疆問題 (Tso Tsung-t'ang and the Sinkiang question), Hakubunkan 1942, 302 pp., map.

A convenient account of the geographico-political situation of Chinese Turkestan, Anglo-Russian great-power interest in the region, the revolt of

Yakub Beg, Russian occupation of Ili, and Tso Tsung-t'ang's eventual reconquest, using Chinese-Japanese-Western sources, including Gideon Ch'en and W. L. Bales. Tso Tsung-t'ang figures mainly in the last chapter (86 pp.).

4.8.5 NOHARA Shirō 野原四郎, "Chūgoku ni okeru Kaikyō seisaku" 中國における回敎政策 (The government policy toward the Moslem religion in China), in **Kindai Chūgoku kenkyū,** 1948, (9.8.6), pp. 299–323.

A brief look at the Moslem policy of Sun's Three Principles and of the Kuomintang, compared with the announced Chinese Communist policy, and with some reference to the Sinkiang situation. The author is now at the Chūgoku Kenkyūjo.

4.9 JAPANESE EXPANSION AND THE CHINESE REPUBLIC

Note: These materials (arranged chronologically by subject) provide some insight into successive phases of Japanese expansion. Japanese relations with modern China are further explored in sec. 5 below.

4.9.1 YOSHINO Sakuzō 吉野作造, **Nikka kokkō ron** 日華國交論 (Essays on Sino-Japanese relations), Shinkigensha 新紀元社, 1947, 298 pp. (Yoshino Sakuzō hakushi minshushugi ronshū, 6.)

The title chapter of this volume of reprinted writings by a leading China specialist is an especially well informed critical narrative of Japan's relation to the Chinese revolutionary movement, from the early days of aid to Sun Yat-sen down through the 1920s (to p. 139; this essay was originally published in 1930 as a separate book entitled **Tai-Shi mondai** 對支問題). Later chapters on aspects of Japanese policy and Chinese foreign relations are also mainly from the 1920s. See Dr. Yoshino's more substantial **Shina kakumei shi,** 5.2.1, and **Chūgoku kakumei shiron,** 5.2.2.

4.9.2 UEDA Toshio 植田捷雄, "Dai-ichiji taisen ni okeru Nippon no sansen gaikō" 第一次大戰に於ける日本の參戰外交 (The negotiations concerning Japan's entrance into the First World War), in **Kindai Chūgoku kenkyū,** 1948, (9.8.6), pp. 327–361.

This study uses the published British, American, Chinese and Japanese materials (especially the biography of Katō Takaaki 加藤高明). The author confirms the usual interpretation that the determining factor in motivating Japan's leaders to enter the war was their expansionist concern over China. Repercussions in China of Japan's attitude are also mentioned (pp. 349–359).

4.9.3 MATSUMOTO Tadao 松本忠雄, **Nisshi shinkōshō ni yoru teikoku no riken** 日支新交涉に依る帝國の利權 (The rights and interests of the Empire in reference to the recent Sino-Japanese negotiations), Genshinsha 元眞社, 1915, preface by Katō Takaaki 加藤高明, 9 pp. (enlarged 3rd ed., 1921, 258+65 pp).

This volume sets forth the basis for the Twenty-one Demands, item by item, giving a pro-government view, and also includes 50 pages of documents and the author's retrospect on the question in 1921 (last 65 pp.), which includes quotation of contemporary domestic and foreign opinion. The author, a graduate of the Tōa Dōbun Shoin, is said to have been a follower of Katō, and was later a member of the Diet.

4.9.4 YOSHINO Sakuzō 吉野作造, **Nisshi kōshō ron** 日支交渉論 (On the Sino-Japanese negotiations), Keiseisha Shoten 警醒社書店, 1915, 288 pp., maps.

A detailed exposition and justification (in view of the foreign competition faced by Japan in China), of the Twenty-one Demands, written for the edification of the Japanese public and based on current news sources and interviews.

4.9.5 SHINOBU Jumpei 信夫淳平, **Taishō gaikō jūgonenshi** 大正外交十五年史 (A history of fifteen years of foreign relations in the Taishō era), Kokusai Remmei Kyōkai 國際聯盟協會, 1927 (3rd printing, 1929, 253+ chronology 11 pp.)

While this little volume for general readers by a competent scholar is concerned with Japanese policy, its several chapters on relations with China over the Twenty-one Demands, the Shantung question, the Washington conference, extraterritoriality and customs reform, provide an authoritative statement of the contemporary Japanese view. Dr. Shinobu, a former diplomat, has been a professor of international law and diplomatic history in Waseda University and acted for many years as a legal consultant to the Navy.

4.9.6 SHŌTA Kazue 勝田主計, narrator (kōjutsu 口述), **Kiku no newake, Nisshi keizaijō no shisetsu ni tsuite** 菊の根分け──日支經濟上の施設に就て (The division of the chrysanthemum's roots, on the institution of a Sino-Japanese economy), printed for private distribution by the Jijokai 爾汝會, 1918 (2nd printing, 1933, 199 pp.).

The author was the Minister of Finance in the Terauchi 寺内 cabinet, and was the promoter of the so-called "Nishihara 西原 Loans". This small volume outlines in considerable detail the various forms which Japanese economic activity had taken in and toward China—banks and development companies; loans of all sorts, economic, political, and military, including the famous loans abovementioned (pp. 88–171); and problems connected with enterprises in China in which Japan had an interest, such as economic advisors, the Boxer indemnity, the Maritime Customs, and the Han-yeh-p'ing Co., these latter being touched on only briefly.

4.9.7 Shanhai Kyoryūmindan 上海居留民團 (Shanghai Japanese residents association), comp., **Shanhai jihen shi** 上海事變誌 (A record of the Shanghai incident), Shanghai: privately printed by the association for limited circulation, 1933, 20+816+62+91 pp., maps, illus.

This account of the 1932 Shanghai "undeclared war", with prefaces by Vice-admiral Nomura Kichisaburō 野村吉三郎, ex-minister to China Shigemitsu Mamoru 重光葵 and others, aims to record the deeds particularly of the Japanese resident community, which subscribed to pay for its compilation. It describes in detail (1) the anti-Japanese movement, incidents, and Japanese counter-measures preceding the hostilities; (2) the course of the conflict itself, from the military, diplomatic and civilian angles (this last in much detail), and public support sent from Japan; and (3) the aftermath. The appendix gives a chronology and martyrology.

4. 10   AMERICAN POLICY AND THE INTERNATIONAL ORDER

Note: These fragmentary studies give Japanese views of the problems presented to Japan in connection with China by the United States and the League of Nations in particular. Note especially 4. 10. 8, a recent study of the origins of the Pacific War. Studies of Japanese-American relations are beyond our scope.

4. 10. 1   TACHI Sakutarō 立作太郎, **Beikoku gaikōjō no shoshugi** 米國外交上の諸主義 (Some principles of American foreign policy), Nippon Hyōronsha, 1942, 378 pp.+index 13 pp.

Professor Tachi (1874–1943), for many years a leader in the field of international law and diplomatic history at Tōkyō University, here analyzes the reasons for and expressions of American isolationism, the Monroe doctrine, neutrality, freedom of the seas, Pan-Americanism and similar principles, including with reference to China the ideas of the Open Door, territorial integrity, and the Stimson doctrine (pp. 193–291). This book is a product of careful legal analysis tinged with a not unnatural, conscious or unconscious tendency to justify Japanese foreign policy.

4. 10. 2   OGAWA Heiji 小川平二, "Kyūkakoku jōyaku no seiritsu, Beikoku Kyokutō seisakushi no issetsu" 九ケ國條約の成立──米國極東政策史の一齣 (The conclusion of the Nine-power Treaty, a page in the history of American Far Eastern policy), **Tōa kenkyū shohō,** 14 (Feb. 1942), 35–88.

A survey article on the origin of the Open Door and its subsequent diplomatic use, relying chiefly on Western sources.

4. 10. 3   SHIMMI Yoshiji 新見吉治, "Beikoku no Tōyō seisaku" 米國の東洋政策 (American Far Eastern policy), in **Tōzai kōshō shiron,** 1939, (9. 8. 11), pp. 1243–1292.

A survey of American policy in the periods up to 1894, 1905, 1917, and 1931. The few sources noted do not appear exhaustive.

4. 10. 4   TAMURA Kōsaku 田村幸策, **Daitōa gaikōshi kenkyū** 大東亞外交史研究 (A study in the diplomatic history of Greater East Asia), Dai Nippon 大日本 pub. co., 1942, 431 pp.

This wartime research study of the Washington Conference treaties is mainly based on the American and other published official documents, and provides a documented, step by step analysis of how the conference came into being, how the Nine-Power Treaty was put together, the Anglo-Japanese Alliance succeeded by the Four-Power Treaty, and the Five-Power Naval Treaty concluded. A supplementary chapter surveys the course of American Far Eastern policy since 1898. The volume is dedicated to the proposition that the seeds of the war of 1941-45 were sown in 1921-22, and not by Japan, but the author does not appear to have sought out more than the official record as the basis for his wartime conclusions.

4.10.5 SAEGUSA Shigetomo 三枝茂智, **Kyokutō gaikō ronsaku** 極東外交論策 (A policy discussion of Far Eastern international relations), Shibun Shoin 斯文書院, 1933, 682 pp.

Written at the height of Japan's controversy with the League of Nations, this volume by an ex-Foreign Office official and Doctor of Law at Tōkyō University seeks to provide an historical understanding of the whole background of the dispute. It is therefore primarily devoted to the development of Japanese policy and its relation to the Open Door, the League, and the like, in the context of international politics, and may be noted here as an example of the very extensive literature (which we do not seek to explore) on the Japanese side of the Far Eastern scene. Note may also be taken of the author's earlier volume, **Shina no gaikō zaisei** 支那の外交・財政 (China's foreign relations and fiscal administration), pub. by the Tōa Dōbunkai in 1921, 472 pp., which is largely concerned with the positions and interests of the powers in China.

4.10.6 HANABUSA Nagamichi 英修道, **Manshūkoku to monko kaihō mondai** 滿洲國と門戸開放問題 (Manchoukuo and the Open Door question), Nippon Kokusai Kyōkai 日本國際協會, 1934, 325 pp., index 12 pp.

A study of the origin and development of the Open Door and equal-opportunity doctrines, Japanese-American rivalry, Japan's maintenance of these doctrines in relation to Manchoukuo (sic) and American dissatisfaction, up to the Amau (Amō) 天羽 statement of 1934. Although a partisan study of a contested subject, this volume is of interest as a reasoned and documented presentation of the Japanese view. The author succeeded Tanaka Suiichirō as professor at Keiō University in the field of Far Eastern international relations.

4.10.7 HANABUSA Nagamichi 英修道, **Monko kaihō kikai kintō shugi** 門戸開放機會均等主義 (The Open Door and equal-opportunity principle), Nippon Kokusai Kyōkai, 1939, 363+27 pp.

This is a reprint of the author's **Manshūkoku to monko kaihō mondai**, 4.10.6, with the addition of two chapters on the China Incident and a new conclusion.

4.10.8 Nippon Gaikō Gakkai 日本外交學會, ed., **Taiheiyō Sensō gen'in ron** 太平洋戰爭原因論 (On the origins of the Pacific War), Shimbun Gekkansha 新聞月鑑社, 1953, 766+chronology 22 pp.

Edited by Ueda Toshio 植田捷雄, this large study by 15 authors, most of whom are leading specialists on China problems, takes up the relationship to the war of the emperor and his chief ministers (by Inada Masatsugu 稻田正次), the bureaucracy (by Tsuji Kiyoaki 辻清明), and the political parties (by Yabe

Sadaji 矢部貞治); the phases in the pre-war period including the suppression of social movements (by Mukōyama Hiroo 向山寛夫), the Manchurian incident (by Takahashi Yūji 高橋勇治), Japan's withdrawal from the League of Nations (by Hanabusa Nagamichi 英修道), the Tripartite Pact (by Ueda Toshio), and Japanese-American relations (by Kumakiri Nobuo 熊切信男); and other factors or aspects such as the legal aspects of the outbreak of the war (by Ōhira Zengo 大平善梧), Japan's right of self-defence (by Ichimata Masao 一又正雄), the Japanese economy (by Muramatsu Yūji 村松祐次), Chinese boycotts against Japan (by Uchida Naosaku 内田直作), the question of oil (by Itagaki Yoichi 板垣與一), and of Germany's war responsibility (by Irie Keishirō 入江啓四郎). The main points of the papers are usefully summarized by the editor (pp. 3-21). The use of the many post-war memoirs, records of the Military Tribunal for the Far East and other sources, by a group of writers competent in diplomatic, political, and economic history, makes this an important contribution. The result should be noteworthy, at least, as providing Japanese scholarly points of view based on the Japanese sources now accessible, which may eventually provide a basis for criticism of the Tribunal's records.

4.10.9 IIZUKA Kōji 飯塚浩二, "Ajia no kaihō to nashonarizumu" アジアの解放とナショナリズム ("The Emancipation of Asia and its Nationalism"), **Tōyō bunka**, 10 (Aug. 1952), 1–11.

A scholarly and somewhat ironical brief essay by an expert in human geography, commenting on the papers and discussions of the Institute of Pacific Relations conference at Lucknow in October 1950. Prof. Iizuka much appreciated the opening speech of Mr. Nehru, and stresses the psychological aspect of the problem of Asian nationalism which he believes has hardly been understood nor sympathized with by Westerners.

4.10.10 IRIE Keishirō 入江啓四郎, "Kokusai Rengō to Chūgoku no chii" 國際聯合と中國の地位 ("China and the United Nations"), **Tōyō bunka kenkyū**, 5 (Aug. 1947), 18–27.

A legal analysis of various aspects of the relationship between the UN and China.

5. REPUBLICAN CHINA

Note: Japan's emergence as one of the imperialist powers in China, at the turn of the century, was accompanied by an acceleration of study and compilation concerning the current Chinese scene, which we have noted above in sec. 1. 2, 1. 3 and 4. 6. By the time of the Republican Revolution this activity began to show results in the writing of journalists and scholars whose greater cultural proximity to the Chinese realities enabled them to describe the politics and personalities of the new republic more intimately than Western observers were able to do. Late-Ch'ing China having just gone to school in Japan, and being still unnationalistic, Japanese observers of the early republican era seem to have had unexcelled access to Chinese leaders. We believe the works noted in this section provide a major key to the understanding of the Chinese Republic's crucial first decade (1912–1921), when the Western liberal-constitutional order was half-heartedly emulated, only to be rejected in the end.

5.1 GENERAL ACCOUNTS OF THE REVOLUTIONARY PROCESS

Note: These materials fall in between the broader studies of modern China in general, listed under sec. 1. 1 above, and the more specialized materials on the Chinese Communist movement in particular, listed under sec. 5. 6 below. The first item below is a recent appraisal of the new regime. Items in the remaining sections are related to their various historical periods. We have not tried to deal with the extensive flow of Japanese materials on China-since-1949.

5. 1. 1 TAKEUCHI Yoshimi 竹內好, **Gendai Chūgoku ron** 現代中國論 (A discussion of contemporary China), Kawade Shobō, 1951 (6th printing, 1953, 201 pp., Shimin 市民 bunko, 58).

This popular vest-pocket volume discusses the Japanese view of China, the Chinese resistance, major movements (literary, youth, and women), politics (the spirit of the new China), literature (relations between Japanese and Chinese literature) and thought and history. The last chapter, "What is the modern age?", is the same as the author's chapter in **Tōyōteki shakai rinri no seikaku,** 6. 1. 7. Among recent Japanese appraisals of the current Chinese scene, this small volume is remarkable for its combination of unrestrained sympathies and independent criticism, and for its vigorous exposition of the phychological and intellectual climate of opinion, Chinese and Japanese. The author seems especially good at exposing the mentality which lies beneath the various Japanese approaches to China problems. He has specialized on Lu Hsün. Writing as a specialist in literature rather than history, he gives a rather impressionistic account, which in view of its popularity would seem to demand the attention of Western students of contemporary Sino-Japanese relations.

5.1.2 TACHIBANA Shiraki 橘樸, **Chūkaminkoku sanjūnen shi** 中華民國三十年史 (A thirty-year history of the Chinese Republic), Iwanami Shoten, 1943, 207 pp. (Iwanami shinsho, 96).

Actually written, so it is said, by Matsumoto Shin'ichi 松本愼一, this little booklet is widely regarded as one of the best-balanced brief surveys of the Chinese revolution. It summarizes succinctly the Western invasion, the Revolution of 1911, its frustration 革命の停滞 in the period of Yuan Shih-k'ai and the warlords, the Nationalist Revolution, and the disorder attendant upon the KMT-CCP split and Japanese aggression after 1931.

5.1.3 TACHIBANA Shiraki 橘樸, **Chūgoku kakumei shiron** 中國革命史論 (An historical study of the Chinese revolution), Nippon Hyōronsha, 1950, 407 pp., preface 10 pp. by Hirano Yoshitarō 平野義太郎, commentary postscript of 19 pp. by Ōtsuka Reizō 大塚令三.

The author died in Mukden in October 1945 at the age of 65, having published two important socio-historical analyses of China a decade earlier **(Shina shakai kenkyū,** 8.1.11, and **Shina shisō kenkyū,** 6.1.3, both pub. 1936). The present posthumous volume is mainly an account of the Kuomintang-Communist cooperation and rivalry during the 1920s, followed by a study of Chinese Communist policy and doctrine which events since the author's death have perhaps made somewhat academic. Chapter 3 (108 pp.) gives an illuminating analysis of warlords, including Chiang Kai-shek as a warlord of a new type. Although the author followed the development of the Chinese revolution with keen interest and sympathy, he rather minimized the role of the proletarian movement and the possible consequences of Chinese communism and was bitterly critical of the mob violence during peasant uprisings. He was also optimistic about the potentialities of the left-wing Kuomintang. See the critical review by Hanamura Yoshiki 花村芳樹, in **Tōyō bunka,** 3 (July 1950), 111–115. Tachibana, as an SMR research staff member, lacked academic connections but nevertheless had a number of disciples indirectly. While influenced by Marxist theory, he picked it up rather late, after 40, and accepted less of it than contemporaries like Suzue Gen'ichi.

5.1.4 NOZAWA Yutaka 野澤豐, "Chūgoku no minzoku kaihō undō 中國の民族解放運動 (The national liberation movement of China), pp. 116–128 in Rekishigaku Kenkyūkai, ed., **Rekishi ni okeru minzoku no mondai** 歷史における民族の問題 (Problems of nations in history), Iwanami Shoten, 1951, 185 pp.

This report on the May Fourth Movement, its origin, development and intensification, is followed by a 35-page transcript of a discussion of modern nationalism in Japan, Ireland, India, and Spain as well as China, on which other papers were presented in the 1951 annual meeting of this society and are here printed. Like the volume for 1950, **Kokka kenryoku no shodankai** 國家

權力の諸段階 (Phases of state power), Iwanami Shoten, 1950, 135 pp., all history is surveyed here, in somewhat general terms and on a basically Marxist framework of ancient, feudal, and modern or imperialist periods. The same periodization is applied in the 1952 annual volume, **Minzoku no bunka ni tsuite** 民族の文化について (On national cultures), Iwanami Shoten, 1953, 205 pp., which includes a 6-page report on Chinese culture during the anti-Japanese united front period by Saitō Akio 齋藤秋男. Discussed in this framework, the phenomena of the Chinese revolution appear as Chinese examples of a set of universal truths.

5.1.5 Anonymous (Suzue Gen'ichi 鈴江言一), for the SMR, Research section of the general office 庶務部調査課, **Chūgoku musankaikyū undō shi** 中國無産階級運動史 (History of the Chinese proletarian movement), Dairen: pub. by the SMR, 1929, 677 pp., classified confidential.

This analytic survey of the revolutionary movement takes up the youth movement (including the May Fourth, 1919, and May Thirtieth, 1925, movements and subsequent developments, pp. 1–41), the CCP (its founding, its youth corps, activities and policies), the holders of power and the proletarian movement (feudal warlords, the KMT, the Northern Expedition, etc., 222–342), the labor union movement (centered about the four congresses of the labor organization), and the peasant movement (in Kwangtung, Hunan and Wuhan, 609–677). This is a substantial compilation of current reports and appraisals, and was one of the most important early writings on the CCP by a Japanese scholar. A new printing has been published by a commercial publisher: Suzue Gen'ichi, **Chūgoku kaihō tōsō shi** 中國解放鬪爭史 (A history of the struggle for emancipation in China), Ishizaki Shoten 石崎書店, 1953, 568 pp. (including 4-page index of CCP documents cited)+index of names 3 pp., with 5-page preface by Itō Takeo 伊藤武雄.

5.1.6 SUZUE Gen'ichi 鈴江言一, **Shina kakumei no kaikyū tairitsu** 支那革命の階級對立 (The confrontation of classes in the Chinese revolution), Taihōkaku 大鳳閣, 1931, 504+9 pp.

Reflecting on his experience in the midst of revolution in China, the author presents an analysis of the various class relationships between and among warlords and bureaucrats; landlords, local bullies, rotten gentry, and peasants; the bourgeoisie and imperialism; the intelligentsia; and labor. The author's factual and often colorful treatment is heavily tinged with Marxist-Leninist theory. This volume is a condensed popular edition of the author's earlier work, **"Chūgoku musankaikyū undō shi"**, 5.1.5.

5.1.7 OIKAWA Tsunetada 及川恒忠, **Shina seiji soshiki no kenkyū** 支那政治組織の研究 (Studies in Chinese political organization), Keiseisha 啓成社, 1933, 1051+place-name dictionary 地名字典 81 pp.

This volume is heavily based on the studies which were originally collected from a number of specialists like Hatano Ken'ichi for a projected "China yearbook" (**Shina nenkan** 支那年鑑) which, however, was never completed. The book covers geography, the political history of the Republic (pp. 47–121), of the republican constitutions, of the parliaments (239–280), and of political parties (281–318); the government structure, judicial system (577–622), army, navy (771-790), and financial history at some length (pp. 791–1051), especially government loans. Key documents are included. While sources are not given, a good deal of information seems to have been brought together, but the result is a pedestrian compilation rather than a work of original understanding.

5.1.8 ISHIDA Mikinosuke 石田幹之助, editorial supervisor, **Tōa no gensei** 東亞の現勢 (The contemporary state of East Asia), the seventh and last vol. in the series **Tōyō bunkashi taikei** 東洋文化史大系 (A general cultural history of the Orient), Seibundō Shinkōsha 誠文堂新光社, 1939, 379 pp., photos and illus., maps.

Like other volumes in this popular and flossy series (see 2.1.13), this one surveys all aspects of developments in Republican China, especially along cultural lines, and summarizes also the positions of the various powers in the Far East, down through the founding of Manchoukuo and the outbreak of war in 1937. The text, which is supplied with *kana* readings throughout, is enlivened by a truly unique collection of photographs on every sort of subject.

5.1.9 SHIMIZU Yasuzō 清水安三, **Shina tōdai shinjimbutsu** 支那當代新人物 (New personages of present-day China), Ōsakayagō Shoten, 1924, 298 pp., preface by Yoshino Sakuzō 吉野作造, illus.

Rather full essays on some 25 leading figures, including warlords and scholars, by the editor of the **Pekin Shūhō** 北京週報, who had spent 7 years in China as a missionary. Biographies include Li Yuan-hung 黎元洪, Ts'ao K'un 曹錕, Chang Tso-lin, Wu P'ei-fu, Feng Yü-hsiang, K'ang Yu-wei, Liang Ch'i-ch'ao, Hu Shih, Ch'en Tu-hsiu, Li Ta-chao, Sun Wen, Ts'ai Yuan-p'ei.

5.1.10 SHIMIZU Yasuzō 清水安三, **Shina shinjin to reimei undō** 支那新人と黎明運動 (China's new men and the movement for a new day), Ōsakayagō Shoten, 1924, 387 pp., preface by Yoshino Sakuzō.

These essays by a Japanese missionary, a graduate of Dōshisha Daigaku, who had been in Peking during the golden age of the new thought movement, reflect various features of it. Topics include anti-Confucianism, the literary reform, the student, womens' and mass movements, anti-Japanism, the new literature and new education, and the anti-Christian movement. This is a companion volume to the same author's biographical studies, 5.1.9. These two volumes were once widely read in Japan.

5.2 THE REVOLUTION OF 1911 AND YUAN SHIH-K'AI

Note: These are mainly contemporary accounts and materials. In general, they give more intimate detail than is available in Western writings. See also Naitō's **Shina ron** (1.4.1).

5.2.1 YOSHINO Sakuzō 吉野作造 and KATŌ Shigeshi 加藤繁, **Shina kakumei shi** 支那革命史 (History of the Chinese revolution), Kyōto: Naigai Shuppan Kabushiki Kaisha, 1922, 457 pp.

From Sun Yat-sen's founding of the Hsing-Chung-hui 興中會 in 1895 to his resignation as provisional president in 1912, this compact narrative in 43 chapters recounts the sequence of events, incidents, conferences, putsches and organizational efforts of the revolutionists down to 1911 (pp. 1–176) and then presents the "annals" of the 1911 revolution, province by province and step by step. Translations of documents and occasional references are given in the text. Yoshino Sakuzō went to China in 1906 to teach, on the recommendation of Dr. Ume Kenjirō 梅謙次郎, professor of civil law at Tōkyō University; he then took service under Yuan Shih-k'ai at Tientsin and later taught there; after travel in the West, he eventually became a professor of political history at Tōkyō, publishing several volumes on European history and also compiling the famous **Meiji bunka zenshū** 明治文化全集 in 14 volumes. On the work of Katō Shigeshi, see his collected studies, **Shina keizaishi kōshō**, 9. 8. 8.

5. 2. 2 YOSHINO Sakuzō 吉野作造, **Chūgoku kakumei shiron** 中國革命史論 (An historical discussion of the Chinese revolution), Shinkigensha, 新紀元社, 1947, 340 pp. (vol. 7 in Dr. Yoshino Sakuzō's **Minshu shugi ronshū** 民主主義論集).

This volume reprints two realistically written popular accounts of contemporary Chinese politics: a "Short history of the Chinese revolution" 中國革命小史 first published in 1917, and an appraisal of developments in "China after the third revolution" 第三革命後の中國 (i.e. the revolt begun in Yunnan in 1915), published in 1922. The author gives a well-informed account, stressing the role played by Japan in Chinese revolutionary developments of the period. Written by an expert in political science, these accounts provide an interesting analysis of the strategies of both the revolutionary and the counter-revolutionary forces in the Chinese political scene.

5. 2. 3 IWAMURA Michio 岩村三千夫, **Minkoku kakumei** 民國革命 (The Republican Revolution), **Shakai kōseishi taikei**, 9. 8. 7, series 8, 1950, 76 pp.

An interpretative study of the Revolution of 1911 under (1) its background (the impact of capitalism, the racial nature of Manchu rule, rise of new classes, etc.), (2) development and structure of the revolutionary forces (analyzed in the successive phases of the movement), and (3) the counter-revolution and its social basis (Yuan and the Peiyang army, basis of warlordism, etc.). Orthodox communist studies of Wu P'o 武波 (i.e. Fan Wen-lan 范文瀾), Hua Kang 華崗 and others are listed for reference as well as the factual studies of Yoshino Sakuzō 吉野作造, Katō Shigeshi 加藤繁, Hirakawa Seifū 平川清風 and Hatano Ken'ichi 波多野乾一. Reviewed by Nozawa Yutaka 野澤豐 in **Shichō**, 44 (May 1951), 65–66. The author is now at the Chūgoku Kenkyūjo.

5. 2. 4 NOZAWA Yutaka 野澤豐, "Shingai kakumei no kaikyū kōsei, Shisen bōdō to shōshin kaikyū" 辛亥革命の階級構成──四川暴動と商紳階級 (The class structure of the 1911 Revolution, the Szechwan outbreak and the merchant-gentry class), **Rekishigaku kenkyū**, 150 (Mar. 1951), 84–91.

A brief documented account and interpretation of the Szechwan movement against the central government's railroad building program, the peasant uprisings, the eventual outbreak of revolution, and its class angle. See critique by

Kitamura Hirotada 北村敬直 in **Rekishigaku kenkyū,** 152 (July 1951), 49–51. Mr. Nozawa is now at the Tōkyō Kyōiku Daigaku.

5. 2. 5 TANAKA Masayoshi 田中正美, "Shimmatsu Shisen bōdō no hottan ni kansuru ichishiken" 清末四川暴動の發端に關する一試見 ("The Beginning of the Riot in Ssu-ch'uan at the End of Ch'ing Dynasty"), **Shichō,** 44 (Mar. 1951), 31–41.

This brief note uses 26 references to **Ta-Ch'ing Hsüan-t'ung cheng-chi 大清宣統政紀** and similar late Ch'ing sources, as well as contemporary Japanese observations. The author emphasizes, in the background of the riots, the imposition of special contributions (amounting to 3% of the crop) upon the peasantry in the effort to get funds for the railroad nationalization program. Mr. Tanaka is now at the Tōkyō Kyōiku Daigaku.

5. 2. 6 TAKEUCHI Katsumi 竹内克己 and KASHIWADA Tenzan (Chūichi) 柏田天山 (忠一), **Shina seitō kessha shi 支那政黨結社史** (A history of political parties and societies in China), Hankow: pub. by the author (Takeuchi), 1918, 2 vols., 342+430 pp., charts.

This pioneer study of the political factions of the early Republican period first surveys the events of the revolution and its aftermath through the "third revolution" of 1915–1916 and then goes into the details of the innumerable cliques, clubs, parties and groupings as they formed and reformed from year to year. The book is extraordinarily full of names and summaries of political pronouncements; while source references are not given, it is evident that the authors worked mainly from periodical and press files (see vol. 1, p. 8, for list).

5. 2. 7 ISHIYAMA Fukuji 石山福治, **Shinchō oyobi sono kakumeitō 清朝及其革命黨** (The Ch'ing dynasty and its revolutionary party), Tōkyōdō 東京堂, 1911, 20+190 pp.

Published Nov. 7, 1911, with prefaces dated October, this quickly written volume gave the Japanese public a background view of Sun Yat-sen and his movement in the context of Ch'ing decline. A score of incidents, assassinations, and other revolutionary acts of the preceding decade are recounted with useful contemporary detail.

5. 2. 8 EGUCHI Bokurō 江口朴郎, "Shina no taisen sanka o meguru kokusai kankei" 支那の大戰參加を繞る國際關係 ("Die internationale Beziehung betreffs des chinesischen Teilnahme an dem Weltkrieg"), **Rekishigaku kenkyū,** 9. 10, whole number 71 (Nov. 1939), 1002–1025.

This study uses mainly the documents in the **Foreign Relations of the United States** to work out the stages by which China's entry into World War I was eventually effected. The author particularly notes the shift caused by the war in the international balance of power affecting China and the influence of this upon the Chinese political scene. Prof. Eguchi is now at the College of General Education (教養學部), Tōkyō University.

5. 2. 9 YAMAJI Aizan 山路愛山 (Yakichi 彌吉), **Shina ron 支那論** (Essays on China), Minyūsha 民友社, 1916, 327 pp.

Written by an unusual private historian of Japanese history and prefaced by a laudatory Chinese dedication to Li Yuan-hung 黎元洪 extolling Sino-Japanese cooperation, the 46 suggestive essays in the first part of this volume expatiate on things Chinese in a quick survey of the chief episodes of modern history, from the Taipings down to Yuan Shih-k'ai's failure to become emperor. The author expresses a rather cynical and grudging but nevertheless unavoidable admiration for the role played by Yuan. The latter part contains essays on representative government in China and on Koxinga (Cheng Ch'eng-kung, 鄭成功, pp. 229–288) and a chronology. The entire text is supplied with *kana* readings of all characters and is phrased in very simple, though rather sentimental, terms. The author produced a dozen or more volumes, mainly on Japanese history but including one on the history of Chinese thought, another on the comparison of Chinese and Japanese culture, and another on Confucius.

5. 2. 10 INABA Iwakichi (Kunzan) 稲葉岩吉 (君山), **Shina teisei ron 支那帝政論** (On the imperial regime in China), Hōkōkai 奉公會, 1916, 202+ 40 pp.

With a preface on "China's future government" by Naitō Torajirō and an appendix which translates the famous memorandum of Dr. Goodnow suggesting China's need of imperial rule, this volume has all the earmarks of an historian's tract for the times. Under 20 headings the author describes various aspects and phases of Yuan Shih-k'ai's effort to assume the imperial title and also notes the movement against it.

5. 2. 11 NAITŌ Juntarō 内藤順太郎, **Seiden En Sei-gai 正傳袁世凱** (An authentic biography of Yuan Shih-k'ai), Hakubunkan, 1913, 256 pp., illus.

With the guidance of many contemporaries, beginning with Liang Shih-i 梁士詒, the author recounts Yuan's brilliant career from his original abduction of the Tai Won Kun in Korea through his service in North China, retirement, and reemergence to become president. A 7-page chronology of Yuan's life and a 4-page chronology of the revolution begin and close the volume. This informed contempory account is favorable to its subject.

5. 2. 12 SEKIYA Jūrō 關矢充郎, **Kaiketsu En Sei-gai 快傑袁世凱** (The amazing Yuan Shih-k'ai), Jitsugyō no Nipponsha, 1913, 410 pp., preface by Count Ōkuma Shigenobu 大隈重信.

A somewhat laudatory biography of President Yüan which takes him in rather simple terms through the phases of his career as Li Hung-chang's proconsul in Korea (pp. 7–143), his training of the new Ch'ing army and service as Shantung governor and Chihli governor general, etc. (to p. 285), and his final emergence in 1911–12. This is a highly circumstantial, old-style narrative, undocumented, and supplied with *kana* readings throughout the text.

5. 2. 13. SAITŌ Hisashi 齋藤恒, "Shintei no taii to En Sei-gai" 清帝の退位と袁世凱 (The Manchu emperor's abdication and Yuan Shih-k'ai), **Shigaku zasshi,** 27. 12 (Dec. 1916), 1278–1304.

A lecture, based on personal observation, by an Army lieutenant colonel (major, at the time of the lecture), who was on the spot when the Manchu dynasty collapsed.

5.2.14 MAKINO Yoshitomo 牧野義智, **Shina gaikōshi** 支那外交史 (History of Chinese foreign relations), Kinkōdō 金港堂, 1914, 573+36 pp., illus.

> The first section of this book is an old-style survey which nevertheless had the merit in its day of covering China's relations with Japan as well as with the West. The second and third sections of the volume (pp. 284–576) deal in detail with contemporary China in the first years of the Republic and give some critical comments: first, the establishment of the Republic and of Yuan Shih-k'ai in power, and their relations with the consortium and concerning foreign railways; second, the position and policies of each of the major powers in East Asia (Britain, U.S.A., France, Germany, Russia and Japan). The appendix is a useful who's who of 44 contemporary Chinese. No footnotes, bibliography or index.

5.2.15 KITA Kazuteru ("Ikki") 北一輝, **Shina kakumei gaishi** 支那革命外史 (An unofficial history of the Chinese revolution), Daitōkaku 大鎧閣 1921, 416 pp.; **Zōho Shina kakumei gaishi** 增補支那革命外史 (An unofficial history of the Chinese revolution, revised and enlarged), Uchimi Bunkōdō 內海文宏堂, 1937, (reprinted 1938, 490 pp., illus., a mere reprint with the addition of three documents, pp. 437–490).

> An interestingly iconoclastic and chauvinist work which argues that neither the Japanese, the Americans nor Sun Yat-sen had a correct understanding of the problem of governing China. Conceiving that Western liberal dogma has misled both Japan and China, the author urges Sino-Japanese cooperation and speculates widely and rather wildly on power politics (anti-British and pro-American) while commenting on events from 1911 to 1919. The book is based on the author's own impressions from his ten years' association with Chinese revolutionists such as Sung Chiao-jen 宋教仁, Ch'en Ch'i-mei 陳其美 and T'an Jen-feng 譚人鳳. Unlike Naitō Torajirō, he considers the pre-revolutionary stage of China as "medieval", corresponding to the French ancien régime and the Tokugawa period in Japan, and his heroes are Napoleon Bonaparte, Emperor Meiji, Saigō Takamori and Lenin. The author later became a doctrinaire thinker of the rightist movement and the idol of simple-minded younger military officers. He participated as an oracular leader in the abortive coup d'état of Feb. 26, 1936, and was executed thereafter.

5.3 THE WARLORD PERIOD

Note: The last three items are specifically on the warlord triumvirate of Chang, Feng, and Wu. Of the earlier items, several are based on newspaper files, which, in the absence of the files, gives them considerable second-hand value.

5.3.1 HIRAKAWA Seifū 平川淸風, **Shina kyōwa shi** 支那共和史 (A history of the Chinese republic), Shanghai: Shunshinsha 春申社, 1920, 868 pp.

> The author of this large volume was a journalist with the **Ōsaka Mainichi,** stationed for some time in Shanghai, where he compiled this chronicle evidently

from newspaper files. The result is a detailed running account of political manoeuvres and week to week developments in the hectic decade after 1911 in which many documents and statements of leaders are quoted in Japanese translation. The author proceeds systematically through the origins of the 1911 Revolution (pp. 1–101), events from October 1911 through the unsuccessful Second Revolution of 1913 (to p. 265), Yuan Shih-k'ai's regime until his death (to p. 456), the conflict between the government and parliament (to p. 559) and between north and south (to p. 743), and so to the collapse of the Tuan Ch'i-jui 段祺瑞 crowd. Chronology, pp. 819–868. Replete with names and dates, this compilation should be of reference value.

5.3.2 MATSUI Hitoshi 松井等, **Shina gendai shi** 支那現代史 (Modern history of China), Meizendō 明善堂, 1924, 351 pp.

This careful account of contemporary history was based of necessity largely on press and periodical files, although it lists an extensive bibliography (pp. 9–14), and was probably the best such work of its day. The narrative in 24 chapters deals with the coming of the 1911 Revolution, the vicissitudes of politics at Peking, activities of the powers, Yuan Shih-k'ai, etc., down through the Washington Conference. There is a 30-page index and the confusing monotony of politics is relieved by a few pages on contemporary Chinese thought (ch. 24).

5.3.3 TAMURA Kōsaku 田村幸策, **Saikin Shina gaikō shi** 最近支那外交史 (History of China's recent foreign relations), Gaikō Jihōsha 外交時報社, vol. 1, 1938, 1110 pp.; vol. 2, 1939, 1116 pp. (The originally projected third volume has not been published.)

These massive volumes provide probably the most detailed account yet available of the Chinese Republic's foreign relations from the time of its birth up to the Washington Conference, in the decade 1911–1922. Domestic policies were of course closely geared to foreign relations, e.g. financially, and this is therefore a major account of a much neglected decade. The chapters deal with (1) foreign relations aspects of the 1911 Revolution (vol. 1, pp. 1–440), including rearrangements concerning the foreign position in China, foreign loans, recognition, Outer Mongolia and Tibet, etc. (2) China and World War I (pp. 441–1110), including her entrance and the peace settlement. (3) Post-war foreign relations (vol. 2, pp. 1–472), with Russia, Japan and other powers, financial, military and otherwise. (4) China and the Washington Conference (pp. 473–1116). This makes four normal-sized volumes. Unfortunately, sources are mainly in Western languages, including British bluebooks, **U.S. Foreign Relations,** conference documents, some memoirs and monographs by Willoughby, Pollard, et al. Some Japanese works are also used. But the author relies heavily upon the American documents and certain secondary works as the chief basis of his narrative and the result is a factual account rather than a critical one broadly based on the bibliography of the subject.

5.3.4 KASHIWAI Kisao 柏井象雄, "Minkoku shoki no zaisei seisaku to sono seika" 民國初期の財政政策とその成果 (The fiscal policy of the early Republican period and its results), **Tōa jimbun gakuhō,** 3.3 (Jan. 1944), 576–611.

Using mainly Chinese and Japanese sources, including the large compilation of Chia Shih-i 賈士毅, this useful study discusses (1) the ideals of the fiscal authorities after 1911, (2) their attempt at reforms and its results, (3) the fiscal difficulties of the time and increase of public loans and 4) two counter-measures which were attempted—the idea of international control and the "fiscal adjustment commission" (Ts'ai-cheng cheng-li-hui 財政整理會).

5.3.5 HYŪGA Saburō 日向三郎, "Shina gumbatsu no honshitsu to rek-koku tono kankei" 支那軍閥の本質と列國との關係 (The fundamental nature of the Chinese warlords and their relations with the foreign powers), **Tōa ronsō,** 1 (July 1939), 123–146.

A brief comment on the place of the warlords in the Chinese political structure, their rise after the 1911 revolution, relations with imperialist powers, and recrudescence in the Kuomintang, written in rather general and doctrinaire terms.

5.3.6 SONODA Kazuki 園田一龜, **Shina shinjinkokki** 支那新人國記 (A record of the new men and country of China), Ōsakayagō Shoten, 1927, 674+index 14 pp., illus.

This who's who was produced by a journalist who wrote 218 columns for the Mukden paper **Hōten shimbun** 奉天新聞 between July 1926 and March 1927. Arranged by provinces, his columns discuss both leading personalities, two or three at a time, and also the general political configuration of which they form a part. Given the personal politics of the warlord period, this approach seems promising, and while the compiler offers fewer statistics than a standard who's who, he also offers a good deal more contemporary lore and insight.

5.3.7 HEMMI Jūrō 逸見十郎, **Chūkaminkoku kakumei nijisshūnen kinen shi** 中華民國革命二十周年記念史 (A history commemorating twenty years of the Chinese national revolution), Keijō: Hemmi Kōenkai Hombu 逸見後援會本部, vol. 1, 1931, 800 pp., illus.

This survey of the Chinese scene, patriotically compiled to guide the author's countrymen, summarizes the history of the revolution and warlord politics, with some 300 brief biographic entries (to p. 356), surveys the Nanking Government (to p. 467), and then works over the Chiness railroads (to p. 776). The author was an old China hand and a behind-the-scenes politician who had some connection with Wu P'ei-fu. This book is based on his personal observations as well as on miscellaneous secondary writings.

5.3.8 SONODA Kazuki 園田一龜, **Tōsanshō no seiji to gaikō** 東三省の政治と外交 (The politics and foreign relations of the Three Eastern Provinces), Hōten Shimbunsha 奉天新聞社, 1925, 344+37 pp.

The author, a journalist with a Mukden paper **(Seikyō jihō** 盛京時報), traces the course of events in Manchuria from the 1890s and then contributes an interesting though spasmodic account of events during the rise of Chang Tso-lin 張作霖 from the time of the 1911 revolution, and the coincident development of the Japanese position. The book has the first-hand raw-material value of being based on miscellaneous news items from a relatively unstudied era and region. Tables list events from 1903 to 1925 and the succession of Chinese officials in the three provinces.

5.3.9 FUSE Katsuji 布施勝治, **Shina kokumin kakumei to Fū Gyoku-shō** 支那國民革命と馮玉祥 (The Chinese national revolution and Feng Yü-hsiang), Ōsakayagō Shoten, 1929, 500 pp., illus.

Based on a diary kept by an experienced correspondent for the **Ōsaka Mainichi** and the **Tōkyō Nichi-nichi** who spent five years (1924–1929) in Peking, this first-hand account takes the "Christian General" as its focal point and colorfully and anecdotally recounts the warlord manoeuvres, marches, counter-marches, coups d'état, and betrayals of the period as they related to Feng as a central figure. There are chapters on his trip to Russia, his betrayal of Wu P'ei-fu 吳佩孚, relations with Chang Tso-lin 張作霖, Chiang Kai-shek 蔣介石, and Yen Hsi-shan 閻錫山, among others, and a general appraisal of Feng's personality and policies during perhaps his best years, based on interviews, with a score of interesting photographs. The result is a volume of considerable research value.

5.3.10 OKANO Masujirō 岡野增次郎, ed., **Go Hai-fu** 吳佩孚 (Wu P'ei-fu), printed for private circulation, Banseikaku 萬聖閣, 1939, 563+1339 pp., photos.

This two-thousand-page memorial to the scholarly warlord contains, among many things, biographical data, writings and speeches, extensive accounts by his biographer of Wu's policies and activities, his assistants, his relations with Chang Tso-lin, summaries of innumerable interviews, his relations with Japanese, and the like. Whether or not the Marshal's Japanese aide has thus been able to outdo Boswell, he presents a good deal of material. According to the compiler-author, he had forty (sic) years experience of personal intercourse with Marshal Wu, from the time of the Russo-Japanese War when Mr. Okano served as an intelligence agent under the Japanese Army.

5.4 SUN YAT-SEN

Note: See also the accounts of Miyazaki Tōten (4.2.11) and Yoshino Sakuzō (5.2.1). It may be suggested that the Japanese view of Dr. Sun is in many cases quite different from the generally accepted Chinese view.

5.4.1 SUZUE Gen'ichi 鈴江言一, **Son Bun den** 孫文傳 (Life of Sun Yat-sen), Iwanami Shoten, 1950, 555+index 14 pp., illus.

This volume is a revised form of the biography of the same title put out by the author in 1931 under the Chinese pen-name of Wang Shu-chih 王樞之. Originally planned as part of a study of the entire modern revolution in China, the volume devotes its first 100 pages to developments before Sun entered the scene, and another section (pp. 337–415) to his ideas. The last chapter (pp. 433–501) surveys the development of the revolution after his death down to 1930, and is followed by a chronology (24 pp.), bibliography (4 pp.), an appraisal of the author and his times by Itō Takeo 伊藤武雄 (20 pp.), and an index of personal names. This book is an attempt at political biography with a Marxist interpretation. The narrative, based on easily available sources, is soberly factual but is illuminated by a vivid, though rather schematic, analysis. The author received rigorous training in Marxist theory as well as in Chinese classics from Nakae Ushikichi 中江丑吉. He lived in China from 1920 to 1944 and collaborated with Chinese left-wing nationalists and communists in the 1920s, but his later activities were purposely limited to scholarship. He was attached twice to the South Manchurian Railway Co. research staff but was persecuted by Japanese military authorities in the 1940s and he died in 1945 of an illness caused by these hardships. See reviews by Niida Noboru 仁井田陞, in **Tōyō bunka,** 4 (Nov. 1950), pp. 125–129; and by Takahashi Yūji 高橋勇治 in **Chūgoku kenkyū,** 14 (June 1951), 94–96.

5. 4. 2 KAYANO Nagatomo 萱野長知, **Chūkaminkoku kakumei hikyū 中華民國革命秘笈** (Secrets of the Chineses national revolution), Kōkoku Seinen Kyōiku Kyōkai 皇國青年敎育協會, 1941, 431 pp.+app. ca. 150 pp., photos.

This inside story of the career of Sun Yat-sen and his party describes their vicissitudes down to the time of his death and records a wealth of lore and information concerning the persons and the incidents in the revolutionary movement, including the author's own experiences. Some 70 photographs and specimens of calligraphy of early party workers, and 640 party members' oaths, are reproduced as an appendix. This volume indicates how largely Japan, and certain Japanese in particular, figured in early Kuomintang history. The author was born in 1873, in Tosa 土佐 (Kōchi prefecture), a center of the *Jiyū minken undō* 自由民權運動 (Liberal people's rights movement), as the son of a progressive samurai. He went to China in 1891 and worked as a press correspondent. In 1905, he became acquainted with Sun Yat-sen and his followers at Hongkong and after that consistently devoted himself to the revolutionary movement as an adviser or private secretary of Sun over a period of some thirty years.

5. 4. 3 TAKAHASHI Yūji 高橋勇治, **Son Bun 孫文** (Sun Yat-sen), Nippon Hyōronsha, 1944, 228 pp. (Tōyō shisō sōsho, 17).

This account, written in a fluent style for the general public, is condensed from the authors's unpublished scholarly report presented to the Tōhō Bunka Gakuin. Although this book is intended to be and really succeeds in being, a critical reappraisal of Sun's life, the author seems to have been unable to free himself from the conventional approach which stresses the problem whether Sun was idealistic or materialistic. The volume includes a survey of Chinese constitutionalism. See review by Deguchi Yūzō 出口勇藏 in **Jimbun kagaku,** 1.1–2 (July 1946), 106–116. In this connection see also the article by Nagai Kazumi 永井算己, "Kō-Chū-kai no seiritsu o meguru ichikōsatsu" 興中會の成立をめぐる一考察 ("A Study about the establishment of Sun Yat Sen's Kochukai"), **Shinshū daigaku kiyō 信州大學紀要** (Memoirs of Shinshū University), 3 (May 1953), 25–36. This is a critical reappraisal of Sun's famous petition to Li Hung-chang in 1894 and the establishment of the Hsing-Chung-hui. The author seeks to differentiate Sun's tactical statements and his private intentions, and concludes that Sun was already anti-Manchu and for a democratic republic at that time.

5. 4. 4 YAMAMOTO Hideo 山本秀夫, "Son Bun ni okeru kakumei seishin keisei no katei" 孫文における革命精神形成の過程 (Stages in the formation

of the revolutionary spirit in Sun Yat-sen), **Chūgoku kenkyū,** 1 (June 1947), 1–17.

    Subtitled as centering on Sun's book on Psychological Reconstruction (**Hsin-li chien-she 心理建設**), this brief article seeks to draw conclusions from Sun's invocation of the principle "Knowledge is difficult, action easy" and similar expressions of that period. The author sees this aspect of Sun's thought as a comprehensive key to understanding his basic attitude.

5.4.5 YAMAMOTO Hideo 山本秀夫, "Sombunshugi tochi kakumei riron no hatten kōzō" 孫文主義土地革命理論の發展構造 (The development and structure of the doctrine of agrarian revolution in Sun Yatsenism), pp. 139–165 in Niida Noboru, ed., **Kindai Chūgoku no shakai to keizai,** 5.7.1.

    As a research on the *San-min chu-i*, this article traces Sun's doctrines of the "equalization of land rights" 平均地權 and "land to the tiller" 耕者要有 其田, and their controversial interrelationship. The writer points out the inconsistencies between them, and the revolutionary character of the latter principle as it emerged in the latest phase of development of the *San-min chu-i*.

5.4.6 IWAMURA Michio 岩村三千夫, "Son Bun Samminshugi no hatten katei" 孫文三民主義の發展過程 (Stages in the development of Sun Yat-sen's Three Principles of the People), **Chūgoku kenkyū,** 2 (Nov. 1947), 44–74.

    In tracing the growth and reformulation of Sun's doctrine, the author defines three stages: 1905–1911, 1912–1921, and 1922-24, and follows the changing interpretations of each of the Three Principles through these stages. While not heavily documented, this article is obviously well informed. It argues for the importance of a developmental study of Sun's thought, on the ground that, after the death of Sun, the last stage of his ideological development was purposely neglected by his right-wing successors. See also the author's book entitled **Samminshugi to gendai Chūgoku 三民主義と現代中國** (The Three Principles of the People and contemporary China), Iwanami Shoten, 1949, 146 pp. (Iwanami shinsho, 6).

5.4.7 DEGUCHI Yūzō 出口勇藏, **Son Bun no keizai shisō 孫文の經濟 思想** (Sun Yat-sen's economic ideas), Kyōto: Kōtō Shoin 高桐書院, 1946, 159 pp.

    This rather slight volume is based in part on articles written in 1942–43 and published in **Tōa jimbun gakuhō,** 2.1 (March 1942), 71–98, on the *min-sheng* 民生 principle; and 3.3 (Jan. 1944), 543–575, on the *min-tsu* 民族 principle. The author made this study at the Kyōto Institute of Humanistic Sciences and expresses a debt to Ojima Sukema 小島祐馬 and Kōsaka Masaaki 高坂正顯. Stressing the Western origin of Sun's thought, the author poses a thesis that the Chinese revolution is a revolution "from the side", as compared with so-called revolutions "from above" or from "below". The results of his research, however, seem rather conventional and pedestrian.

5.4.8 ISHII Masuo 石井壽夫, **Son Bun shisō no kenkyū 孫文思想の研究** (A study of Sun Yat-sen's thought), Meguro Shoten 目黑書店, 1943, 318 pp. (Tairiku 大陸 sōsho).

A rather conventional interpretation, tinged with ultra-patriotic sentiments for a wartime audience, of Sun's ideas, grouped under his Three Principles, with a special appendix giving statements of his concerning Japan. While Dr. Sun is constantly quoted, with page references to the Foreign Office translation of his works, it seems a bit difficult to unscramble his ideas from those of the author, probably because the author seeks to identify himself with the inner struggles and aspirations of Dr. Sun as a patriot.

5.4.9 ONO Noriaki 小野則秋, **Son Bun** 孫文 (Sun Wen), Taigadō, 1948, 257 pp., illus.

The author, librarian of Dōshisha 同志社 University, recounts Sun's career from first to last in 27 chapters and adds a 5-page chronology. Extensive quotations are translated in the text, though without very clear attribution, and each section is replete with the Chinese names connected with the subject, to which *kana* readings are attached. The volume is, however, based on secondary works and not intended for more than the general reader.

5.4.10 BABA Kuwatarō 馬場鍬太郎, "Shina sangyō kaihatsusaku to shite no Son Bun gakusetsu" 支那産業開發策としての孫文學說 ("Les doctrines de Sun Yat-sen considérées au point de vue de leur efficacité pour faire progresser la production en Chine"), **Tōa keizai kenkyū,** 15.1–2 (April 1931), 88–110.

By one of the chief compilers of the Tōa Dōbun Shoin in Shanghai, this is a summary of Sun's vision of the industrialization of China through developing waterways and establishing big commercial ports, expressed in his **The International Development of China** (first published as articles in **The Far Eastern Review,** Mar. 1919 to Nov. 1920).

5.5 THE NATIONALIST REVOLUTION AND GOVERNMENT

Note: These are mainly contemporary accounts, some of them quite well-informed. Arrangement is chronological, after the first item.

5.5.1 HATANO Ken'ichi 波多野乾一, **Chūgoku Kokumintō tsūshi** 中國 國民黨通史 (A general history of the China Kuomintang), Daitō Shuppan-sha 大東出版社, 1943, 629 pp.

The author graduated from the Tōa Dōbun Shoin in Shanghai in 1912, studied in Peking and spent almost 20 years in journalism before entering government work. This volume draws on his extensive background experience and knowledge of the documentation to put down a full record of the origins, composition and activities of the successive organizations that, when strung together, form the history of the Kuomintang—from 1894 to Wang Ching-wei's ill-starred defection in wartime Nanking. Especially useful are the quotation of numerous party documents and pronouncements and the expression of a critical view necessarily different from that of Chou Lu 鄒魯 and other KMT historians. See also his studies of Chinese Communism, 5.6.1 and 5.6.2. Mr. Hatano is now in Tōkyō and connected with the Sangyō Keizai 産業經濟 Shimbunsha.

5.5.2 SHIMIZU Tōzō 清水董三, "Chūgoku Kokumintō" 中國國民黨 (The Chinese Nationalist Party), **Shina kenkyū,** 12 (Dec. 1926), 199–300.

This long contemporary account deals briefly with the history of Dr. Sun and the party, and then describes its reorganization with the inclusion of communists, its regulations, organization, party government principle, propaganda, army, mass movement, etc., giving a good many names and details in the process and a long bibliography.

5.5.3 TAKAKU Hajime 高久肇, for the Shanghai office, SMR, **Kokkyō ryōtō no teikei yori bunretsu made, kokumin kakumei no gensei, sono ichi** 國共兩黨の提携より分裂まで──國民革命の現勢其一 (The present condition of the Nationalist Revolution, part one, the KMT and CCP from coalition to disruption), pub. by the SMR Shanghai office 上海事務所 and classified confidential, 250 pp., charts, preface dated Aug. 1927.

The writer collected materials and held interviews on the spot in Hunan and at Wuhan. He describes the CCP movement and KMT-CCP coalition (to p. 96), and the factors and events leading up to the 1927 split (pp. 97–250), giving a wealth of circumstantial detail and documentation. This appears to be a useful report from an independent observer. The author was a leading specialist on the KMT, and wrote several other reports (not seen by us) for the SMR, one of which was entitled **Hokubatsu kansei made no Kokumintō** 北伐完成迄の國民黨 (The KMT up to the consummation of the Northern Expedition), without the name of the writer but actually written, so it is said, by Mr. Takaku; all of them are considered valuable for study of KMT history.

5.5.4 IDE Kiwata 井出季和太, **Shina no kokuminkakumei to Kokumin-seifu** 支那の國民革命と國民政府 (China's national revolution and the Nationalist Government), pub. by the research section of the Taiwan Government-general, 2 vols., 238 and 276 pp., prefaces dated May and August, 1928.

This contemporary compilation gives the history of the revolution and of Dr. Sun in a nutshell and then does the same for the KMT, including a section on the Canton period after 1924 (pp. 195–238). Volume 2 then goes into a good deal of detail on the organization, activities, and policies of the KMT during the Nationalist Revolution. While the assiduous compiler seldom notes sources, he evidently worked from full news files, particularly noting anything in lists.

5.5.5 MIYAWAKI Kennosuke 宮脇賢之介, "Han-Shō undō o chūshin to shite Shina jikyoku no tembō" 反蔣運動を中心として支那時局の展望 ("Un coup d'oeil sur la situation de la Chine autour du mouvement anti-Chiang"), **Tōa keizai kenkyū,** 14.4 (Oct. 1930), 600–629.

This rather detailed recital of current events, mainly concerning Wang Ching-wei's flirtation with Yen Hsi-shan, is cited as an example of the often factual reporting which appears in this journal, as also in **Mantetsu chōsa geppō, Gaikō jihō** 外交時報 and similar publications, and which appear likely to interest the historian.

5.5.6 Tōa Keizai Chōsakyoku 東亞經濟調查局, comp., **1930-1931 Shina seiji keizai nenshi** 一九三〇年一九三一年支那政治經濟年史 (A year's history of Chinese politics and economics for 1930-1931), Senshinsha 先進社, 1932, 821 pp.

This volume, developed from the current commentaries (the columns entitled "Tōa jihō" 東亞時報) in the monthly **Tōa,** covers events in China, Mongolia and Manchuria from January 1930 to September 1931,—i.e. the historical background of the Manchurian Incident. The period is cogently divided into phases and aspects—domestic politics, the rights recovery movement, domestic capitalist developments, imperialism of the powers, the silver problem, Sino-Japanese relations, etc.—and a large number of documents are translated at length.

5.5.7 CHŌ Shū-tetsu (Chang Hsiu-che) 張秀哲, **Kokumin seifu no gaikō oyobi gaikō gyōsei** 國民政府の外交及外交行政 (The foreign relations of the National Government and its administration of foreign affairs), Nisshi Mondai Kenkyūkai 日支問題研究會, 1935, 438 pp.

Developed from a dissertation done for the Faculty of Law of Tōkyō Imperial University, this study by a student from Taiwan summarizes KMT history, the structure of the Canton, Wuhan and Nanking governments and their handling of foreign relations (to p. 264). It then surveys the activity of the KMT and the Nationalist Government regarding foreign affairs, the late-Ch'ing and Republican organs for handling the same, and the establishment of Chinese missions abroad (to p. 384), and concludes with a 35-page who's who of leaders and diplomats.

5.5.8 NAKANISHI Isao 中西功, "Saikin Shina seiji no sandankai to Kokumintō gumbatsu no dōkō" 最近支那政治の三段階と國民黨軍閥の動向 (Three stages of recent Chinese politics and the tendencies of the Kuomintang warlords), **Mantetsu chōsa geppō,** 16.5 (May 1936), 59-80.

A critical analysis of (1) the KMT failure in its program of "unification" (1928-31), (2) its unsuccessful campaigns against the Chinese Communists (1931-34), combined with the seizure of Manchuria, and (3) the dictatorial trend at Nanking as the Communists escaped westward and the Japanese moved to take over North China (1934-35). This contemporary comment appears to give the CCP a higher survival value than the KMT.

5.5.9 KAZAMA Takashi 風間阜, **Kinsei Chūgoku shi** 近世中國史 (History of modern China), Sōbunkaku 叢文閣, 1937, 369 pp.

The author's residence as a journalist in Shantung from 1917 to 1923 and from that date to 1937 in Peking gave him a first-hand experience of warlord and Kuomintang China. His narrative covers from 1911 to the abortive Fukien revolt of 1934 in systematic chronological sections and its lack of an academic footnote-apparatus seems fully compensated by its well-informed treatment of Chinese politics on the basis of first-hand Legation Quarter gossip and day to day observation. The bibliography (pp. 367-9) is not very comprehensive but the author has been able to rely on the earlier narratives of Hirakawa Seifū (1920), 5.3.1, Matsui Hitoshi (1924), 5.3.2, and others. This seems to be one of the more reliable summaries of the superficial events of the period.

5.5.10 HIBIKI Mitsuo 日疋滿雄, "Shina hokubatsu kokumin kakumei ronkyū" 支那北伐國民革命論究 (An inquiry into the national people's revolution during the Northern Expedition in China), **Tōa ronsō,** 5 (Oct. 1941), 363-396.

The author treats the Northern Expedition in two parts: from Canton to Wuhan and Shanghai in 1926–27 and from Nanking to Peking in 1927–28. He describes the worldwide popular response to the first victories of the earlier period, the relations with the powers and between the Wuhan and Nanking governments, and after 1928 the eventual relapse into continued domestic strife. Rather commonplace sources are cited.

5.5.11 YAMAUCHI Kiyomi 山內喜代美, **Chūgoku Kokumintō shi** 中國 國民黨史 (History of the China Kuomintang), Ganshōdō Shoten, 1941, 356+bibliog. 3 pp.

Written in wartime when the "true" or "orthodox" 正統 Kuomintang had been installed at Nanking under Wang Ching-wei 汪精衞 (Wang Chao-ming 兆銘), this is a straightforward, undocumented, factual and pro-Wang account of the Kō-Chū-kai (Hsing-Chung-hui 興中會), Dō-mei-kai (T'ung-meng-hui 同盟), and successive forms of the KMT and its history, winding up with Wang's new central government. Appendix reprints writings of Sun, Wang and Chou Fu-hai 周佛海.

5.6 HISTORY OF CHINESE COMMUNISM

Note: This section includes works specifically on the Chinese Communist movement, rather than the Chinese revolution as a whole. We begin with the publications of Hatano Ken'ichi and the Foreign Office, followed by those of Ōtsuka Reizō. Of the remaining items many, as might be expected, are party-line accounts. We also add some examples of the contemporary Japanese reports on Chinese Communism in the late 1920s and early 1930s, published before Edgar Snow called the attention of the Western world to the latest phase of the Chinese revolution (**Red Star over China,** 1937); these reports, which could be listed much more extensively from Japanese periodicals of the period, should have considerable research value for those who are now interested in finding out what happened to China. Broader studies of the Chinese revolution, including some of the best treatments of the Communist movement, will be found in sec. 5.1 above.

The latest annotated bibliography in this field is by Ichirō Shirato, edited by C. Martin Wilbur, **Japanese Sources on the History of the Chinese**

Communist Movement, An annotated bibliography of materials in the East Asiatic Library of Columbia University and the Division of Orientalia, Library of Congress (East Asian Institute of Columbia University, 433 West 117th Street, New York, November, 1953, 69 pp. including index).

5.6.1 HATANO Ken'ichi 波多野乾一, for the Gaimushō Jōhōbu 外務省情報部 (Information Bureau, Ministry of Foreign Affairs), **Shina kyōsantō shi 支那共産黨史** (History of the communist party in China), printed by the Ministry for restricted circulation and classified "confidential" (**hi** 祕), dated on cover July 1932, preface dated Oct. 12, 1932, 694+index 8 pp.

This officially-sponsored study, based on the author's private materials as well as on Foreign Ministry sources, is not only one of the first but also one of the most valuable researches into the early history of the Chinese Communist Party from the period of its founding down to the middle of 1932. A 5-page chronology and 11-page bibliography are followed by a 36-page descriptive list of 182 communist source items compiled by Ōtsuka Reizō 大塚令三. Many of these latter are secret party publications, evidently available to the compiler. Hatano's text is heavily interlarded with quoted documents, and the six annual volumes, 5.6.2, which he superintended in subsequent years similarly brought together a great deal of material.

5.6.2 HATANO Ken'ichi 波多野乾一, for the Gaimusho Jōhōbu 外務省情報部 (Information Bureau, Ministry of Foreign Affairs), **Chūgoku Kyōsantō sen-kyūhyaku-sanjū-ni-nen shi 中國共産黨一九三二年史** (History of the Chinese Communist Party for 1932), printed by the Ministry for restricted circulation, preface Feb. 1933, 348+index 5 pp.; the same for 1933, preface Feb. 1934, 937+index 3 pp.; 1934, preface Feb. 1935, 978+index 5 pp.; 1935, preface Mar. 1936, 766+index 3 pp.; 1936, printed Feb. 1937, 766+index 4 pp.; and 1937, preface June 1938, 1090 pp.

These six volumes form a continuation and supplement to the **Shina Kyōsantō shi** similarly produced by Hatano and the Ministry, 5.6.1. Compiled by an acknowledged expert in the subject, they provide probably the most thorough single account for most of the periods covered. News sources and communist publications are cited; classified government sources were also used but are not divulged. A chronology of events is added to each volume, together with a selection of documents, articles, and reports translated into Japanese from various sources (including foreign correspondents such as Edgar Snow and Agnes Smedley). These documents in fact form the bulk of the publication: i.e., pp. 227–348 in the volume for 1932; pp. 311–937 for 1933; pp. 471-978 for 1934; pp. 169–766 for 1935. Both the documents and Hatano's prefatory surveys are arranged with reference to major topics, incidents and problems of current interest.

5.6.3 HATANO Ken'ichi 波多野乾一, **Gendai Shina no seiji to jimbutsu 現代支那の政治と人物** (Politics and personalities of contemporary China), Kaizōsha 改造社, 1937, 562 pp.

The author, having been a journalist before he became a leading specialist on Chinese communism, produced some 150 articles on China for popular journals in the period 1932–37, 50 of which are selected for this book and grouped under major topics such as the aftermath of the Shanghai undeclared war of 1932, the Fukien revolt, Chiang and Chinese fascism, the North China situation, the CCP army, or the anti-Japanese united front and Sian incident. These articles show a considerable grasp of Chinese political situations and personalities, who are described and analyzed in great detail.

5. 6. 4 HATANO Ken'ichi 波多野乾一, **Mō Taku-tō to Chūgoku no kōsei** 毛澤東と中國の紅星 (Mao Tse-tung and China's red stars [a Who's who of the Chinese Communist Party]), Teikoku Shoin 帝國書院, 1946, 179+ index 6 pp. ("Freshman's Library").

After a brief sketch of Chinese Communist Party history and of Mao's life, this volume presents thumb-nail biographies of some 110 leaders (pp. 65–179), not excluding Ch'en Tu-hsiu 陳獨秀 (pp. 147–154).

5. 6. 5 HATANO Ken'ichi 波多野乾一, **Mō Taku-tō** 毛澤東 (Mao Tse-tung), Fukuchi Shoten 福地書店, 1949, 179+6 pp.

This volume is identical with Hatano's **Mō Taku-tō to Chūgoku no kōsei,** 5. 6. 4, of 1946, except for a new preface.

5. 6. 6 Gaimushō Chōsakyoku 外務省調査局 (Bureau of Research and Documentation of the Foreign Office), comp., **Mō Taku-tō shuyō genronshū** 毛澤東主要言論集 (A collection of important speeches and essays of Mao Tse-tung), printed by the Foreign Office for limited circulation, 1948, 550 pp.

Based on materials from the collections of Hatano Ken'ichi 波多野乾一, Senda Kyūichi 千田九一 and Kayahara Nobuo 萱原信雄, this volume prints translations of 27 items dated between 1934 and 1948. While some of these are well known in English translations also, some appear to be rather rare.

5. 6. 7 Gaimushō Chōsakyoku 外務省調査局 (Bureau of Research and Documentation of the Foreign Office), **Chūkyō gairon** 中共概論 (A General account of the Chinese Communists), printed by the Foreign Office for limited circulation, 1949, 462 pp., charts.

Written by Kayahara Nobuo 萱原信雄 and a number of other members of the research staff, this is a companion volume to those on Mao's utterances (**Mō Taku-tō shuyō genronshū,** 5. 6. 6) and China's postwar politics (**Sengo ni okeru Chūgoku seiji,** 5. 7. 2). Building on the immense backlog of Foreign Office materials on this subject (see under Hatano Ken'ichi, 5. 6. 1-3), the authors describe, as of the era of the Communist takeover, the history and structure of the CCP, its political and economic policies, cultural program, and armed forces, adding a 77-page appendix of documents and 23-page chronological table.

131

5.6.8 YOSHIKAWA Shigezō 吉川重藏, **Chūkyō sōran** 中共總覽 (A general survey of the Chinese Communists), Jiji Tsūshinsha 時事通信社 ("Jiji Press"), 1950, 381 pp., tables.

The chief author is a Foreign Office official with experience in China and abroad, who in 1950 became head of the first section of the Bureau of Research and Documentation. This volume was first drafted for office use using F.O. materials (see our separate note on it under the title **Chūkyō gairon, 5.6.7**). It was then developed for publication with F.O. backing. It takes up CCP history (pp. 3–29), party organization (30–67), political program and system (68–146), economics (147–260), international relations (261–278), mass movements (279–287) and culture (to 313). Chronology 34 pp. and index 20 pp. While necessarily a rapid survey of a changing situation, this sober volume appears to have reliability as a reference work.

5.6.9 ŌTSUKA Reizō 大塚令三, "Chūgoku Kyōsantō bunken kō" 中國共產黨文獻考 ("Reference about the Communist Party of China"), **Mantetsu Shina gesshi** 滿鐵支那月誌 ("The Shanghai S.M.R. Co's Monthly"), 7.4 (April 1930), 64–77; 5 (May), 45–60; 6 (June), 70–87.

This long article lists and describes in detail 182 pieces of CCP publications, almost all of which were of course secret or illegal, which appeared in China before June 1930—one of the most complete lists available. The first 56 items all pre-date the Canton commune of December 1927. See the author's CCP history of 1940, 5.6.11.

5.6.10 ŌTSUKA Reizō 大塚令三, "Chūgoku Kyōsantō nenshi kō" 中國共產黨年誌稿 (A draft chronology of the CCP), **Mantetsu chōsa geppō**, 12.12 (Dec. 1937), 62–115.

Evidently not published by the author in book form, this extensive date-list from 1920 to the end of 1926 is heavily annotated with source references to contemporary publications, many now unobtainable.

5.6.11 ŌTSUKA Reizō 大塚令三, **Shina Kyōsantō shi** 支那共產黨史 (History of the Chinese Communist Party), Seikatsusha, 1940, 2 vols., 220+266 pp.

The author was a member of the South Manchurian Railway Co. research staff. He states that this summary of Chinese Communist Party history to 1940 is based on more than ten years research, mainly in contemporary news sources, and sections are dated as composed at various times during the 1930s. One part was presented as a Japanese data paper at the Sixth (Yosemite) Conference of the Institute of Pacific Relations in 1936. The author deals with personalities and defections from the party as well as with shifts of strategy. He conceives Japan's task to be the saving of China from communism, and his second volume stresses the Chinese soviets' affinity for Soviet Russia and the growth of the Chinese Red Army. While documented only occasionally, these various accounts factually summarize much of the contemporary story of CCP developments.

5.6.12 OJIMA Sukema 小島祐馬，**Chūgoku Kyōsantō** 中國共産黨 (The Chinese Communist Party), Kōbundō, 1950, 78 pp. (A-te-ne library アテネ文庫, no. 112).

This tiny pocket-size volume summarizes the growth of the communist movement during successive phases of the Chinese revolution. The author stresses the decisive importance of the leadership of the Comintern and his concluding comments (pp. 67–72) express grave doubts as to the validity and feasibility of Marxist theory as now applied to China.

5.6.13 MIYAMOTO Michiharu 宮本通治，"Shina ni okeru saikin no sekishoku rōdō kumiai undō ni tsuite" 支那に於る最近の赤色勞働組合運動に就て ("On about the Red Unionist movement in recent China"), **Mantetsu Shina gesshi** 滿鐵支那月誌 ("The Shanghai S. M. R. Co's Monthly"), 7.2 (Feb. 1930), 89–119.

An early study, concerned mainly with the Fifth All-China Labor Conference held in November 1929 at Shanghai, using contemporary reports.

5.6.14 MURATA Shirō 村田孜郎，**Shina no sayoku sensen** 支那の左翼戰線 (The left front in China), Banrikaku Shobō 萬里閣書房, 1930, 328 pp., illus.

By a journalist with some China experience, this popularized and rather gossipy compilation, built up from his news articles, retails the story of the abortive "March revolution" 三月革命 of 1927 in Shanghai and then flashes back to survey the origin and organization of the Chinese Communist Party, its success at Canton, activities of the Comintern, and various aspects of the split between Wuhan and Chiang. The author also includes vignettes of Chiang and others, and various aspects of Shanghai life during revolution and terror. This is a journalistic work in both the good and the bad senses of the term.

5.6.15 HIMORI Torao 日森虎雄，"Shina Sekigun oyobi Sověto kuiki no hatten jōkyō" 支那赤軍及ソヴェート區域の發展情況 (Circumstances in the development of the Chinese Red Army and Soviet areas), **Mantetsu chōsa geppō,** 12.8 (Aug. 1932), 41–133；12.9 (Sept. 1932), 56–164, map.

Under 40 headings this contemporary report translates or summarizes a large body of current documentation on the anti-CCP campaigns and the CCP itself, in one of its least known periods.

5.6.16 Gunseibu Gunji Chōsabu 軍政部軍事調査部 (Military affairs research department of the military government board), comp., **Manshū Kyōsanhi no kenkyū** 滿洲共産匪の研究 (Researches on the Communist bandits in Manchuria), pub. by the Gunseibu Komonbu 顧問部 (Adviser's office of the military government board), 1937 (dated K'ang-te ssu-nien 康德四年, although 三年 on cover and title page), 924+146 pp., photos, maps., charts.

Labelled volume 輯 one, secret, and a combined intelligence report 綜合情報, this massive tome was compiled from reports gathered from all parts of the Manchoukuo military and police network as well as of the Japanese consular service in order to describe the Soviet Russian-backed program of banditry, disorder and resistance to the Japanese dispensation. It gives a history of the communist-bandit movement, stemming from the CCP in China, and describes its activities in the eastern, southern and northern parts of the country, region by region, giving historical summaries and details of incidents and organizations. With an appendix of documents (146 pp.) on the party, army and mass organizations, this appears to be a major source on the subject.

5.6.17 HIRANO Yoshitarō 平野義太郎, **Chūgoku ni okeru Shimminshu-shugi kakumei** 中國における新民主主義革命 (The New Democratic revolution in China), Chūō Kōronsha, 1949, 233 pp.

The leading figure in the Chūgoku Kenkyūjo begins this volume by relating the Chinese Communist cause to that of world peace and then discusses its relation to Japan and the rest of Asia. Part two is on the political theory and institutions of the New Democracy in practice, part three on land reform and economic policies, and part four on recent developments under the CCP leadership in the civil war.

5.6.18 NAKANISHI Isao 中西功, **Chūgoku Kyōsantō shi** 中國共産黨史 (A history of the Chinese Communist Party), Hakutosha 白都社 (2nd ed., 1949, 271 pp.)

Written in prison during wartime without many sources available, by a communist member of the Diet (at the time of publication), and dedicated to 5 anti-militarist Japanese comrades, the sections of this somewhat garrulous book deal with (1) the problem of the Chinese soviet revolution (describing KMT warlordism, imperialism and the agrarian crisis, etc.), (2) the Chinese soviet strategy (including the decisions of the Comintern and of the CCP Sixth Congress of June 1928), and (3), as an appendix, the advance from New Democracy to socialism as of 1949. See also by the same author, **Chūgoku kakumei to Chūgoku Kyōsantō**, Jimminsha 人民社, 1946, 95 pp.; and **Chūgoku Kyōsantō to minzoku tōitsu sensen** 民族統一戰線, Taigadō 大雅堂, 1946, 154 pp., the latter with Nishisato Tatsuo 西里龍夫 as co-author. Reviewed by Iwamura Michio 岩村三千夫 in **Chūgokn kenkyū**, 1 (June 1947), 94–101.

5.6.19 NAKANISHI Isao 中西功 and NISHISATO Tatsuo 西里龍夫, **Bukan ni okeru kakumei to hankakumei** 武漢に於ける革命と反革命 (Revolution and counter-revolution in Wuhan), Minshu Hyōronsha 民主評論社, 1948, 217 pp.

Feeling that the events of 1927 in China should have inspirational value for the coming revolution (as of 1948) in Japan, the authors revised this Wuhan part of Nakanishi's **Chūgoku Kyōsantō shi**, 5.6.18, written in prison in 1944. They begin with the "rising tide" 高潮 of the Great Revolution as defined by the Seventh Plenum of the Comintern in late 1926 and, for events from then until the end of the abortive "Canton commune" of 1927, proceed to offer a straight party-line interpretation, stressing the wise adaptation of the party leadership to the changing situation.

5.6.20 OKAMOTO Saburō オカモトサブロウ (岡本三郎), "Kō-Nichi minzoku tōitsu sensen no keisei katei" 抗日民族統一戰線の形成過程 (The process of formation of the anti-Japanese national united front), **Rekishi-gaku kenkyū,** 138 (Mar. 1949), 2–18.

An apparently orthodox party-line account of this aspect of the Chinese revolutionary process from 1928 through the Manchurian Incident and the successive stages in the growth of the anti-Japanese movement thereafter, up to late 1937.

5.6.21 IWAMURA Michio 岩村三千夫, "Mō Taku-tō no shisō to sono hatten" 毛澤東の思想とその發展 (The thought of Mao Tse-tung and its development), Aoki Shoten 青木書店, 1951, 211 pp. (Aoki bunko 青木文庫, 6.)

This vest-pocket size volume is a thorough revision of the author's **Mō Taku-tō no shisō** published in 1950 (Sekai Hyōronsha 世界評論社), and is based on wider documentation, although the author says he still had not seen **Mao Tse-tung hsüan-chi 選集**. His well organized analysis is divided into (1) sources of Maoism (Marxism and theories of Lenin and Stalin regarding China), (2) theory of the agrarian revolution (from Hunan and the Chinese Soviet period), (3) theory of the anti-Japanese war (the united front, etc.), (4) theory of the New Democracy and (5) its economic order, (6) the Party and its purification (*cheng-feng*), and (7) Mao's position on realism and art and letters and (8) on patriotism and internationalism. A biographical summary (pp. 184–207), and bibliographies of Mao's works and of reference sources are added, the last recommendation being J. Stalin's essays on the Chinese revolution.

5.6.22 YAMAMOTO Shōjirō 山本正治郎, "Chūgoku kakumei ni okeru riron to jissen, Mō Taku-tō shisō no hatten" 中國革命における理論と實踐——毛澤東思想の發展 (Theory and practice in the Chinese revolution, the development of Mao Tse-tung's thought), **Keizaigaku zasshi 經濟學雜誌**, pub. by the Economic Research Society of the Osaka Commercial Science University 大阪商科大學經濟研究所, 27.6 (Dec. 1952), 38–77.

Beginning with a statement of 1902 by V. I. Lenin, the author asserts the need for revolutionary theory and practice to go hand in hand, etc. His examination of the exemplification of this principle by Mao Tse-tung in the periods to 1927, 1937, 1945, and 1949, seems (on the basis of statements quoted from Mao and the official account by Hu Ch'iao-mu 胡喬木) to come to the same conclusion, etc.

5.6.23 HOSOI Shōji 細井昌治, "1925–27 nen no Chūgoku kakumei to rekkoku no taido" 1925–1927 年の中國革命と列國の態度 ("The Attitude of the Powers at the Time of the Chinese Revolution of 1925–27"), **Rekishigaku kenkyū,** 162 (March 1953), 12–27.

This posthumous article uses commonplace Marxist assumptions to construct a quick chronological summary of China's foreign relations during the troubled period of the May 30th Incident, the Northern Expedition, the Nanking Incident, etc, down to the Canton Commune. There is some citation of Western press sources but little evidence of intensive research.

5. 7 POST-WAR CHINA

Note: Although the contemporary scene is largely outside the scope of this volume, these items represent current scholarship (or in some cases, propaganda) concerning it. On post-war foreign trade, see 7.18.12.

5. 7. 1 NIIDA Noboru 仁井田陞, ed., for the Mombushō Chūgoku Gendai Shakai Kenkyū Iinkai 文部省中國現代社會研究委員會 (Ministry of Education, Research Committee on Modern Chinese Society), **Kindai Chūgoku no shakai to keizai** 近代中國の社會と經濟 (Modern China's society and economy), Tōkō Shoin 刀江書院, 1951, 350 pp.

We have noted certain items of the 12 chapters of this symposium separately by author and title, when their subject matter so required. The remaining chapters not noted elsewhere, which refer primarily to the contemporary situation, are the following:

(1) Hirano Yoshitarō 平野義太郎, "Shimminshu shugi ni okeru chihō gyōsei oyobi shihō no henkaku" 新民主主義における地方行政および司法の變革 (The revolution in local government and the judicial system under the New Democracy), pp. 1–49.

(2) Nakamura Jihei 中村治兵衞, "Kahoku nōson no sompi, gendai Chūgoku no chihō zaisei no ichikenkyū 華北農村の村費——現代中國の地方財政の一研究 (Official expenditures of villages in North China, a study of local finance in contemporary China), pp. 83–111.—Mainly based on the reports of field investigations by the SMR and the Tōa Kenkyūjo.

(3) Ubukata Naokichi 幼方直吉, "Pan, dōkyōkai, dōgyōkōkai to sono tenka" 幫・同鄉會・同業公會とその轉化 (*Pang*, regional associations, professional associations, and their transformation), pp. 233–253.—Stresses the role of co-operatives and labor unions in the transformation of the traditional gild (*pang*) system.

(4) Niida Noboru 仁井田陞, "Chūgoku nōson shakai to kafuchō ken'i" 中國農村社會と家父長權威 (The rural society of China and the patriarchal authority), pp. 255–283,—These same problems are more fully discussed in Niida's **Chūgoku no nōson kazoku**, 8.4.1.

(5) Naoe Hiroji 直江廣治, "Kiu shūzoku" 祈雨習俗 (Customs of praying for rain), pp. 285–309.

(6) Sakai Tadao 酒井忠夫, "Gendai Chūgoku bunka to kyūshūkyō" 現代中國文化と舊宗敎 (Contemporary Chinese culture and the old religions), pp. 311–341.

(7) Nohara Shirō 野原四郎, "Chūgoku gendai bunka no minzokuteki keishiki" 中國現代文化の民族的形式 (National forms of China's contemporary culture), pp. 345–350.

See reviews by Satoi Hikoshichirō 里井彥七郎 in **Shirin** 35.2 (Aug. 1952), 201–204; and by Kanda Nobuo 神田信夫 in **Shigaku zasshi**, 60.8 (Aug. 1951), 745–748.

5. 7. 2 Gaimushō Chōsakyoku Daigoka 外務省調査局第五課 (The Fifth Section of the Bureau of Research and Documentation of the Foreign Office), **Sengo ni okeru Chūgoku seiji** 戰後における中國政治 (Postwar politics of China), printed for limited circulation, Dec. 1948, 340 pp.

Written by members of the Fifth Section of the Bureau of Research and

Documentation, which is in charge of China problems, this handy volume surveys China's post-war politics chronologically at home (to pp. 84) and abroad (to pp. 155) by countries down to Aug. 1948. It also prints translations of 36 documents, with a 5-page table of changes in personnel and 33-page chronology. As distinct from the contemporary F.O. volumes on Mao and the CCP (see 5.6.6 and 5.6.7), this one is primarily devoted to the affairs of the Nationalist Government of China. Since it was completed before the publication (in Aug. 1949) of the monumental documentation in the State Department white paper on United States relations with China, this Japanese appraisal may have some interest for comparative purposes.

5.7.3 Asahi Shimbun Tōaba 朝日新聞東亞部, comp., **Shindankai ni tatsu Chūgoku seiji** 新段階に立つ中國政治 (Chinese politics in their new stage), Getsuyō Shobō 月曜書房, 1947, 331+appendix 54 pp. (Chūgoku sōsho 中國叢書 [Series on China], 3).

This volume represents an attempt by journalists working in the East Asian Section of the **Asahi Shimbun** to give a factual and objective survey of Chinese politics, past and present, for the edification of the general reader. The narrative begins with the late Ch'ing period, and deals with (1) the national revolution and the Kuomintang (to pp. 105), (2) an historical survey of Kuomintang-Communist relations (to pp. 224), and (3) the victory in World War II and the subsequent problem of national unification (down to the Kuomintang-Communist rupture early in 1947). The appendix reprints some post-war documents, but this volume is probably not as useful as the Foreign Office study, **Sengo ni okeru Chūgoku seiji**, 5.7.2.

5.7.4 NIIDA Noboru 仁井田陞, YOSHIKAWA Kōjirō 吉川幸次郎, HIRAOKA Takeo 平岡武夫, MATSUMOTO Yoshimi 松本善海, and KAIZUKA Shigeki 貝塚茂樹, contributors, "Chūgoku no genjo o do miru ka" 中國の現狀をどうみるか (How shall we view the present situation in China?), **Sekai** 世界, pub. by Iwanami Shoten, 44 (Aug. 1949), 12–40.

This symposium in a general magazine includes a main article by Niida Noboru, "Chūgoku kindai kakumei no rekishiteki kadai" 中國近代革命の歷史的課題 (Historical problems of China's modern revolution), pp. 12–20, and rather interesting briefer pieces by the others, all of which provides a panorama of different views.

5.7.5 ANDŌ Hikotarō 安藤彦太郎, **Henkaku to chishikijin** 變革と知識人 (The revolution and the intelligentsia), Tōwasha 東和社, 1952, 218 pp.

Subtitled "thought and learning in China's period of reconstruction", this propaganda work by a member of the Chūgoku Kenkyūjo, assisted by Saitō Akio 齋藤秋男 and Nozawa Yutaka 野澤豐, takes the statements of leading Peking professors and writers, published during or after the remoulding of their thoughts, and demonstrates how these texts (popularly known as *t'an-pai* 坦白 or in Western parlance as "confessions") can be used to paint a party-line picture of new-found happiness. There is of course a good deal more of this type of literature available in Japanese on contemporary China, which we do not attempt to survey.

6. INTELLECTUAL AND CULTURAL HISTORY

Note: Materials under this heading have been the most difficult to arrange in useful compartments, a fact which may indicate the undeveloped nature of this field of study. The literary renaissance, for instance, can be approached through materials in sec. 6.1 and 6.11, but there is no comprehensive and authoritative work on it. The May Fourth Movement (see under sec. 6.10) is dealt with more fully under politics in sec. 5 and also in sec. 6.1. Sec. 2.6 above, on the Reform Movement, is of course an integral part, and one of the most important parts, of the materials on modern China's intellectual history.

6.1 GENERAL WORKS AND ESSAYS

Note: We begin with a thin volume by Ojima Sukema which opens up the unexploited possibilities which lie in the application (to modern Chinese thought) of classical scholarship concerning traditional Chinese thought. Few writers, especially in the West, have combined a classical knowledge with modern interests. Later come several symposia on modern thought and culture. Partly no doubt because of the activity of the new Chinese regime in enlisting and using the intellectuals, as the bearers of China's cultural tradition, the whole field of "culture" and the modern cultural movements of "the new thought", "May Fourth", etc., are receiving increased attention. Our listing is roughly chronological.

6.1.1 OJIMA Sukema 小島祐馬, **Chūgoku no kakumei shisō** 中國の革命思想 (Chinese thought concerning dynastic change and revolution), Kōbundō, 1950, 166 pp.

> This provocative pioneer essay by a leading specialist in traditional Chinese philosophy traces the debt of modern Chinese revolutionary thinkers to their own philosophical heritage. Since the modern term for revolution *ko-ming* 革命, was also the traditional term for the removal of Heaven's mandate to rule the empire from the hands of the reigning dynasty to those of its successor, the modern use of the term in itself involves a semantic problem. The first chapter (pp. 12–68) notes traditional theories of historical change (*yin-yang*, five elements, etc.), the ideas of certain Confucian scholars like Huang Tsung-hsi 黃宗羲 and of Taoist anarchism, and the class and racial basis of ideology. The second chapter traces the rise of popular revolt and the successive phases of political thought centering around 1898, 1911, and Sun's Three Principles.

While this little volume, developed from lectures, is necessarily controversial, it explores an unworked field of research for which few scholars are as yet qualified. See also the author's booklet on Chinese Communism, 5. 6. 12, and his studies thirty years ago of Liao P'ing, 6. 9. 9–10, Liu Shih-p'ei, 6. 9. 11, and other modern Confucian scholars. Having graduated from the Faculty of Law, and then the Faculty of Letters, of Kyōto University, Ojima Sukema (formerly listed in **Shinagaku** as "Y. Ojima") studied under Naitō Torajirō and Kano Naoki, specializing in thought. He also became a close friend of the Marxists, Kawakami Hajime 河上肇 and Nakae Ushikichi 中江丑吉, and through an interest in Marxism as applied to China has become one of the few scholars equipped to see it in the perspective of Chinese intellectual history. For a bibliography of his writings from 1912 to 1941, published to commemorate his sixtieth birthday, see **Shinagaku**, 10, special number (April 1942), 821–826. For a convenient survey of ancient Confucian thought, including the ideas of the other schools also, see his **Shakai keizai shisō** 社會經濟思想 (Social and economic thought), in **Iwanami kōza Tōyō shichō** (9. 8. 4), series 18, 1936, 110 pp.

6. 1. 2 OJIMA Sukema 小島祐馬, "Juka to kakumei shisō" 儒家と革命思想 ("Confucianism and Revolutionary Ideas"), **Shinagaku**, 2. 3 (Nov. ["Dec." sic], 1921), 198–210; 2. 4 (Dec. 1921), 271–280.

A scholarly examination of classical references and traditional ideas concerning (1) the peaceful or violent accession of a new regime and its acknowledgment by the populace, as seen in Mencius, particularly, and also in the concepts of the five elements 五行, of the *kung-yang* school 公羊家, etc., and (2) the veneration of the ruler, as fostered by Hsün-tzu 荀子, the **Spring and Autumn Annals** 春秋, and other classics. This essay is of interest in connection with the author's book of thirty years later with a similar title, 6. 1. 1.

6. 1. 3 TACHIBANA Shiraki 橘樸, **Shina shisō kenkyū** 支那思想研究 (Studies of Chinese thought), Nippon Hyōronsha, 1936 (4th printing, 1939, 519 pp.)

This volume, published in the same year as the author's **Shina shakai kenkyū**, 8. 1. 11, brings together three articles first published in the 1920s and a lecture at Dairen of 1935. Topics treated include a general survey of Chinese thought, religious and ethical concepts, Chinese characteristics, a study of the *ōdō* (*wang-tao* 王道) concept in Japan (pp. 471–519), and a section on the Chinese social revolution which includes a critique of the views of Naitō Torajirō (pp. 360–408). In this book the author seems to be mainly concerned with the social function of the classical ideologies and the actual mentality of the people, especially in rural districts.

6. 1. 4 SANETŌ Keishū 實藤惠秀, ed., **Kindai Shina shisō** 近代支那思想 (Modern Chinese thought), Kōfūkan 光風館, 1942, 366 pp.

This useful volume brings together articles by ten different writers as chapters on aspects of modern Chinese thought. While they appear to vary in quality between research and textbook writing, they present a wide and often repetitive coverage of subject matter, as evidenced by our brief indications of their topical content. On the whole, this volume contains much stimulation for further research.
(1) Kamiya Masao 神谷正男, "Kindai Shina shisō to sono tokushitsu" 近代

支那思想とその特質 (Modern Chinese thought and its special characteristics) 1–51. Traditional, European and modern ideas.

(2) Fujiwara Sadamu 藤原定, "Shimmatsu shisō no chōryū" 清末思想の潮流 (Streams of thought in the late-Ch'ing period), 53–97. The Taipings, Tseng, Li, and Chang Chih-tung. K'ang and Sun.

(3) Ozaki Gorō 尾崎五郎, "Chūkaminkoku go no Shina shisō" 中華民國後の支那思想 (Chinese thought in the Republican period), 99–141. The 1911 revolution and *San-min chu-i*. The May Fourth Movement and after.

(4) Sanetō Keishū 實藤惠秀, "Kindai Shina to gairai shisō" 近代支那と外來思想 (Modern China and ideas from abroad), 143–175. The Jesuit period. *Chung-hsueh wei t'i* 中學爲體, etc., and later theories. Japanese ideas.

(5) Hara Tomio 原富男, "Jukyō shugi no mondai" 儒教主義の問題 (Problems of Confucianism), 177–218. Traditional and modern applications of Confucianism to government.

(6) Takahashi Yūji 高橋勇治, "Sonbunshugi no mondai" 孫文主義の問題 (Problems of Sun Yat-senism), 219–262. As related to Confucianism, socialism, and Japan.

(7) Fujieda Takeo 藤枝丈夫, "Shina shisōkai no genjō" 支那思想界の現狀 (The current situation among Chinese thinkers), 263–284. Modern thought, nationalism, and the *San-min chu-i*.

(8) Yamada Atsushi 山田厚, "Shina tetsugakukai no genjō" 支那哲學界の現狀 (The current situation in Chinese philosophical circles), 285–309. Reform of traditional philosophy and the Chungking government's effort to use it.

(9) Tachibana Shiraki 橘樸, "Shina shisō to sono shōraisei" 支那思想とその將來性 (Chinese thought and its future possibilities), 311–346. Confucianism, Taoism and their relation to society.

(10) Saitō Akio 齋藤秋男, "Kindai Shina shisō bunken" 近代支那思想文獻 (Materials on modern Chinese thought), 347–366. Japanese and Western bibliography.

6.1.5 WADA Sei 和田清, ed. **Kindai Shina bunka** 近代支那文化 (Modern Chinese culture), Kōfūkan 光風館, 1943, 294 pp.

This symposium is a companion volume to **Kindai Shina shakai**, 6.1.6, also edited by Dr. Wada, and to **Kindai Shina shisō**, 6.1.4, edited by Sanetō Keishū. Its chapters, as noted below, touch successively on the traditional, political, Japanese and Western cultural influences on modern China, as well as the successive stages of the new culture movement and the writers and documentation of the period.

(1) Wada Sei, "Rekishisei yori mitaru kindai-Shina bunka no tokushitsu" 歷史性より觀たる近代支那文化の特質 (The characteristics of the culture of modern China, in its historical context), 1–24.

(2) Otake Fumio 小竹文夫, "Seijisei yori mitaru kindai-Shina bunka no tokushitsu" 政治性より觀たる近代支那文化の特質 (The characteristics of the culture of modern China, with special attention to its relationship with politics), 25–52.

(3) Izushi Yoshihiko 出石誠彥, "Kindai Shina bunka shi, Ahen sensō igo no gaikoku kyōiku seido saiyō o chūshin to shite" 近代支那文化史, 阿片戰爭以後の外國敎育制度採用を中心として (The history of modern Chinese culture, with special reference to the introduction of the foreign educational system after the Opium War), 53–84. A posthumous, draft article.

(4) Kamiya Masao 神谷正男, "Kindai Shina bunka undō shi" 近代支那文化運動史 (The history of the cultural movement in modern China), 85–126.

Foreign missionaries; the *Yang-wu yün-tung* 洋務運動; the pre-revolutionary period; the May Fourth movement; socialism; the revival of Confucianism; the New Life movement.

(5) Satō Saburō 佐藤三郎, "Kindai Shina to Nippon bunka" 近代支那と日本文化 (Modern China and Japanese culture), 127–174. Mainly on the Chinese receptivity to things Japanese.

(6) Gotō Sueo 後藤末雄, "Kindai Shina to Seiyō bunka" 近代支那と西洋文化 (Modern China and Western culture), 175–220. The Jesuit period; with a few pages on the late Ch'ing and early Republican period.

(7) Sanetō Keishū 實藤惠秀, "Kindai Shina no shuppan bunka" 近代支那の出版文化 (The printing culture of modern China), 221–258. The introduction of modern printing technology; publications by Westerners; publications by Sino-Japanese collaboration; publication by Chinese alone; the press and magazines; characteristics of modern Chinese publications.

(8) Saitō Akio 齋藤秋男, "Shina Bunkakai no genjō" 支那文化界の現狀 (The current situation in Chinese cultural circles), 259–286.

(9) Henshūbu 編輯部 (The editorial section), "Kindai Shina bunka bunken" 近代支那文化文獻 (Materials on modern Chinese culture), 287–294. A Japanese bibliography, including translations of Chinese works.

6.1.6 WADA Sei 和田清, ed., **Kindai Shina shakai** 近代支那社會 (Modern Chinese society), Kōfūkan 光風館, 1943, 313 pp.

This is a companion volume to **Kindai Shina bunka,** 6.1.5, also edited by Dr. Wada, and to **Kindai Shina shisō,** 6.1.4, edited by Sanetō Keishū. The contributors, as noted below, seek to define the special characteristics of modern Chinese education and religion and provide historical understanding of developments in both these fields, including the Chinese students abroad and also the "New Life Movement" (*Hsin sheng-huo yün-tung* 新生活運動). This symposium for the general reader was of course produced in wartime circumstances and the China in question is sometimes that of Wang Ching-wei rather than Chiang Kai-shek.

(1) Makino Tatsumi 牧野巽, "Kindai Shina shakai no tokushitsu" 近代支那社會の特質 (Characteristics of modern Chinese society), 1–26.

(2) Kaigo Tokiomi 海後宗臣, "Kindai Shina kyōiku no tokushitsu" 近代支那教育の特質 (Characteristics of modern Chinese education), 27–60. Treats primary, middle, higher and social education. Chinese bibliography of 25 items. The author is an outstanding specialist in Japanese educational history.

(3) Yūki Reimon 結城令聞, "Kindai Shina shūkyō no tokushitsu" 近代支那宗教の特質 (Characteristics of modern Chinese religion), 61–84. Treats nature and religion in China; society and religion; modern Chinese religions, their historical background, social activities, and relationships with modern thought.

(4) Masui Tsuneo 增井經夫, "Kindai Shina shakaishi" 近代支那社會史 (History of modern Chinese society), 85–110. Treats the old society characterized by the "eight-legged essay" (*pa-ku wen* 八股文); urban society; the socio-economic transformation.

(5) Kobayashi Sumie 小林澄兄, "Kindai Shina kyōikushi" 近代支那教育史 (History of modern Chinese education), 111–152. Treats the old and new systems, chronologically.

(6) Sanetō Keishū 實藤惠秀, "Kindai Shina ryūgakushi" 近代支那留學史 (History of study abroad in modern China), 153–188. Treats the study of Chinese

141

students in America, in Europe, and in Japan, with some criticism of Japanese attitudes.

(7) Narita Shōshin 成田昌信, "Bukkyō" 佛教 (Buddhism), 189–212. Historical background; the Ch'ing court and the Lama sect; traditional Buddhism; anti-Buddhist attitude of the Taipings; laymen's Buddhism today.

(8) Kubota Ryōon 久保田量遠, "Dōkyō" 道教 (Taoism), 213–234. Historical background; magic; present-day temples and priests; scriptures.

(9) Hiyane Antei 比屋根安定, "Kirisutokyō" キリスト教 (Christianity), 235–252. An historical survey.

(10) Kobayashi Bansei 小林胖生, "Minkan shinkō" 民間信仰 (Popular faiths), 253–274. Relationship with the established religions; various deities; superstition concerning colors; propitious words and propitious designs.

(11) Kinoshita Hanji 木下牛治, "Shinkokumin undō to Shinseikatsu undō" 新國民運動と新生活運動 (The "New National Movement" and the "New Life Movement"), 275–305.

(12) Henshūbu 編輯部 (The editorial section), "Kindai Shina shakai bunken" 近代支那社會文獻 (Materials on modern Chinese society), 307–313. A brief but smartly annotated bibliography.

6. 1. 7 TAKEUCHI Yoshimi 竹内好, YOSHIKAWA Kōjirō 吉川幸次郎, NOHARA Shirō 野原四郎, and NIIDA Noboru 仁井田陞, **Tōyōteki shakai rinri no seikaku** 東洋的社會倫理の性格 (The nature of ethics in the Oriental type of society), Hakujitsu Shoin 白日書院, 1948, 243 pp. (Tōkyō Daigaku Tōyō Bunka Kenkyūjo, **Tōyō bunka kōza,** dai-san kan 東京大學東洋文化研究所——東洋文化講座第3巻 [Studies in Oriental Culture, vol. 3, of the Institute for Oriental Culture, Tōkyō University.])

Of the four contributors to this symposium, Prof. Takeuchi writes on Lu Hsün 魯迅 in the context of European influence on the Far East in an article broadly entitled "Chūgoku no kindai to Nippon no kindai" 中國の近代と日本の近代 (The modern ages of China and Japan), pp. 3–60; this article is a penetrating comparative analysis of Chinese and Japanese intellectual responses to the West, now included in his book, **Gendai Chūgoku ron,** 5. 1. 1. Prof. Yoshikawa's article, "Chūgokujin to shūkyō" 中國人と宗教 (The Chinese and religion), pp. 61–96, stresses the a-religiousness of the Chinese people. Mr. Nohara's article on "Ko Seki shi to jukyō" 胡適氏と儒教 (Hu Shih and Confucianism), pp. 99–140, gives a rather cynical analysis of Hu Shih's view of Confucianism under the impact of Dewey's pragmatism. Dr. Niida's article on "social structure and the legal consciousness", an essay with the same title as the volume, pp. 143–236, deals with the social basis of Oriental authoritarianism, among other things, and is noteworthy as proposing a challenging approach to Chinese history from the point of view of Oriental authoritarianism as he sees it. These highly sophisticated essays cover a very wide range, which defies characterization. See review by Ogura Yoshihiko 小倉芳彦 in **Rekishigaku kenkyū,** 140 (July 1949), 55-57.

6. 1. 8 Chūgoku Kenkyūjo 中國研究所, ed., **Chūgoku no gendai bunka** 中國の現代文化 (The contemporary culture of China), Hakujitsu Shoin 白日書院, 1948, 320 pp.

These essays, as noted below, cover almost all aspects of modern Chinese cultural development in the most recent period and are on the whole well-informed, though often partisan, and undocumented. They seem also less obviously doctrinaire (in 1947–48) than such generally favorable appraisals have later tended to become.

(1) Kaji Wataru 鹿地亙, "Gendai bunka no tokushitsu" 現代文化の特質 (The characterictics of the contemporary culture), pp. 3–31.

(2) Hirano Yoshitarō 平野義太郎, "Shin Chūgoku interi ron" 新中國インテリ論 (On the intellectuals of the new China), pp. 33–50.

(3) Fujiwara Sadamu 藤原定, "Minkoku izen no minshushugi shisō" 民國以前の民主主義思想 (Democratic thought in the pre-Republican period), pp. 51–67.

(4) Iwamura Michio 岩村三千夫, "Samminshugi ron" 三民主義論 (On the San-min chu-i), pp. 69–86.

(5) Saitō Akio 齋藤秋男, "Gakusei undō" 學生運動 (The student movement), pp. 87–117.

(6) Tsurumi Kazuko 鶴見和子, and Nishi Kiyoko 西清子, "Fujin mondai" 婦人問題 (The problem of women), pp. 119–141.

(7) Ono Shinobu 小野忍, "Kokugo mondai" 國語問題 (The national language question), pp. 143–170.

(8) Okazaki Toshio 岡崎俊夫, "Bungaku" 文學 (Literature), pp. 171–195.

(9) Iizuka Akira 飯塚朗, "Engeki" 演劇 (The theatre), pp. 197–219.

(10) Yahara Reizaburō 矢原禮三郎, "Eiga" 映畫 (Motion-pictures), pp. 221–230.

(11) Shimada Masao 島田政雄, "Mokkoku geijutsu" 木刻藝術 (The art of the woodcut), pp. 231–255.

(12) Watanabe Masao 渡邊政男, "Jānarizumu" ジャーナリズム (Journalism), pp. 257–268.

(13) Komiya Yoshitaka 小宮義孝, "Shizen kagaku" 自然科學 (The natural sciences), pp. 269–283.

(14) Iwamura Shinobu 岩村忍, "Shinorojī to America" シノロジーとアメリカ (Sinology and America), pp. 285–302.

(15) Ubukata Naokichi 幼方直吉, "Chūgoku bunka kenkyū no kokusaisei" 中國文化研究の國際性 (The international character of the study of Chinese culture), pp. 303–320.

6.1.9 KUMANO Shōhei 熊野正平, **Gendai Chūgoku shichō kōwa** 現代中國思潮講話 (Lectures on the tide of thought in contemporary China), Ōmon 櫻門 pub. co., 1948, 285 pp.

Drawing on some 30 years of contact with the Chinese intellectual scene, beginning with the May Fourth Movement, the author gives a running account of the ebb and flow of ideas in Republican China: The reformist and revolutionary ideas in the late Ch'ing period, the new culture movement of 1919, the new consciousness of Oriental culture, polemics over philosophy-of-life 人生觀, socialist thought, communism, the Kuomintang movement for ideological unification and criticisms of it, and the rise of nationalistic thought. While far from the last word on the subject, the footnote references and general scope of this volume should make it of value for others in this field. See review by Yamaguchi Ichirō 山口一郎 in **Chūgoku kenkyū**, 6 (Jan. 1949), 72–78.

6.1.10. FUJIWARA Sadamu 藤原定, **Kindai Chūgoku shisō** 近代中國思想 (Modern Chinese thought), Shichōsha 思潮社, 1948, 203 pp.

A rapid survey-essay which sketches in the development of ideas under the Taipings, the subsequent self-strengthening and reform movements, and the Three People's Principles. This is done against a background interpretation (first 70 pp.) of China as a feudal and bureaucratic state in which a popular democratic movement has only with difficulty been able to develop. Writings of Ku Yen-wu 顧炎武 and of Huang Tsung-hsi 黃宗羲 are frequently quoted.

6.1.11 KIMURA Eiichi 木村英一, **Chūgokuteki jitsuzaikan no kenkyū 中國的實在觀の研究** (A study of the Chinese view of reality), Kōbundō, 1948, 382 pp.

This philosophical treatise discusses the problem of the approach to Chinese thought (historically or otherwise) and reprints articles on the uses of *li* 禮 in the old China, the "five elements" 五行 conception, and the ancient *feng-shan* 封禪 ceremony. For modern studies the author's chief contribution is a previously unpublished article on "The thought of the **Ta-t'ung shu 大同書** and its nature" (pp. 191–289), in which he analyzes K'ang Yu-wei's famous book under the following heads: its view of reality, utopias (their position in Chinese thought, the utopia of the **Ta-t'ung shu**), and the view of evolution and of Confucius expressed in the book. Taken with the methodological and metaphysical discussion preceding it, this long chapter may offer useful suggestions. Prof. Kimura is now at Ōsaka University.

6.1.12 TAKAHASHI Yūji 高橋勇治, "Kindai Chūgoku Runesansu—Han-Jukyō undō o chūshin to shite" 近代中國ルネサンス——反儒教運動を中心として ("Renaissance in Modern China"—with special reference to the anti-Confucian movement), **Shakai kagaku kenkyū,** 2.2 (July 1950), 1–40; 2.4 (Jan. 1951), 29–62.

A rather doctrinaire survey of modern China which views nascent capitalism in the late Ch'ing period as having induced three stages of reformism —the *yang-wu* 洋務 movement of the early modernizers, the reform movement of K'ang Yu-wei, and the constitutional and revolutionary movements of Liang Ch'i-ch'ao and Sun Yat-sen. Later developments are similarly schematized in a well-informed manner with citation of a variety of sources including early writings of K. A. Wittfogel. The second half of the article discusses the May Fourth Movement from several angles, and is based on an extensive perusal of the back numbers of **The New Youth (Hsin-ch'ing-nien 新青年)**, the chief organ of the movement. Now in the Institute for Social Sciences of Tōkyō University, the author was a professor at Peking University during the war.

6.1.13 FUKUI Kōjun 福井康順, **Tōyō shisōshi ronkō 東洋思想史論攷** (Studies in the history of Oriental thought), Hōzōkan 法藏館, 1950, 358 pp.

The first of these 8 essays is on "Ethical thought in contemporary China" 現代シナにおける倫理思想 (pp. 2–98), and discusses the changes of the last sixty years in Chinese values as evidenced in the writings of representative men like K'ang Yu-wei, Liang Ch'i-ch'ao, Chang Chih-tung, Chang Ping-lin, Ch'en Tu-hsiu, Hu Shih, et al. This is almost an intellectual history of modern China, in brief form, down to the New Life Movement.

6.1.14 YAMAMOTO Hideo 山本秀夫, "Chūgoku ni okeru itan shisō no tenkai" 中國における異端思想の展開 ("The Development of Non-Authoritative and Unorthodox Thoughts in China"), **Tōyō bunka,** 9 (June 1952), 41–75.

Subsections discuss 1) the use of Taoism and Buddhism as the guiding ideology of peasant rebellions since the Han, particularly in secret societies like the White Lotus, 2) the Taiping ideology, and 3) Sun Yat-senism as the ideology guiding a "bourgeois-democratic" revolution. A broad essay in rather general terms.

6.1.15 TAKEUCHI Yoshimi 竹内好, NOHARA Shirō 野原四郎, SAITŌ Akio 齋藤秋男, and YAMAGUCHI Ichirō 山口一郎, constituting the Gendai Chūgoku Shisō Kenkyūkai, "Gendai Chūgoku shisō no shiteki gaikan" 現代中國思想の史的概觀 (An historical survey of contemporary Chinese thought), **Shisō** 思想, pub. by Iwanami Shoten, 335 (May 1952), 451–459; 337 (July 1952), 670–686; 338 (Aug. 1952), 757–773.

This broad survey is the result of a joint study by specialists who try to take a quick look at the main periods, trends, and individual thinkers in the field of modern Chinese thought, beginning with a critical appraisal (by Takeuchi) of the psychological and scholarly approach made by the Japanese to China problems. Their topical-chronological survey deals with (1) the late Ch'ing period—the shaking of the hold of traditionalistic thought (by Nohara), (2) The May Fourth cultural revolution, as based on a thorough consciousness of the need for renovation (by Takeuchi), (3) the nationalist revolution—the period of reconstruction (by Yamaguchi), (4) the united front period—the maturity of nationalistic consciousness (also by Yamaguchi), and (5) World War II and its aftermath—a period of proposing new values (by Saitō). Biographical and bibliographical notes are given on leading intellectuals, including a number of the leaders of the Democratic League, and the major topics of intellectual concern are briefly summarized. This survey, here presented in a rather sketchy form, has been published in a revised and enlarged form as a separate book in the Iwanami shinsho 岩波新書 series (No. 73): **Chūgoku kakumei no shisō** 中國革命の思想 (The thought of the Chinese revolution), 1953, 222+8 pp.

6.2 TRADITIONAL THOUGHT

Note: Most of the works listed here do not refer primarily to modern China but may be of use for background purposes, intellectual history being so much less bound by chronology than other types of history. Note should also be taken of Itano Chōhachi's articles in **Gendai Chūgoku jiten** (9.5.3.), and of the writings of Ojima Sukema (6.9.8–12). Item 6.2.7 by Shimada Kenji offers a new view of the early modern period.

6.2.1 TAKEUCHI Yoshio 武内義雄, **Shina shisō shi** 支那思想史 (History of Chinese thought), Iwanami Shoten, 1936 (9th printing, 1941, 346+8 pp., Iwanami zensho, 73). Post-war new printing, 1953, under the title of **Chūgoku shisō shi** 中國, with 16-page bibliographical guide.

While this small volume covers the whole range of Chinese thought from ancient times, and in fact devotes only its last ten pages to the Ch'ing period, it can be recommended as a convenient essay-type summary and survey by a competent specialist for purposes of putting early Ch'ing thinkers like Huang Tsung-hsi 黃宗羲 and Ku Yen-wu 顧炎武 into perspective.

6.2.2 NAKAE Ushikichi 中江丑吉, **Chūgoku kodai seiji shisō** 中國古代政治思想 (Ancient Chinese political thought), Iwanami Shoten, 1950, 673 pp.

While this masterly volume by an influential scholar who lived in Peking for 30 years and was deeply versed in Western (especially German) and Chinese learning, does not seek to deal with modern times, its account of the *kung-yang* 公羊 school (pp. 327–446) and other sections on Ch'ing scholarship should be of interest. Mr. Nakae (1889–1942), son of a liberal thinker, Nakae Chōmin 兆民, was the teacher of Suzue Gen'ichi and a friend of Dr. Ojima Sukema. He was a student of German philosophy and of Marxism inter alia. He devotes one chapter to the question of feudalism in China (pp. 233–282). On his life in Peking and his last years there, see Itō Takeo's 伊藤武雄 article in **Chūgoku kenkyū,** 12 (April 1950), 60–71. See the review by Niida Noboru 仁井田陞 in **Chūgoku kenkyū,** 13 (Sept. 1950), 54–57.

6.2.3 NAITŌ Torajirō 內藤虎次郎, **Shina shigaku shi** 支那史學史 (A history of the historical study of China), Kōbundō, 1949, 656 pp.

This scholarly research in Chinese historiography deals with an immense range of subject matter in chronological sequence. The 21 sections under the Ch'ing period do not go far into the final century. There is a concluding essay on the origins of the historical way of thought in China (pp. 586–611). See reviews by Okazaki Fumio 岡崎文夫 in **Shigaku zasshi,** 59.1 (Jan. 1950), 85–91; and by Kaizuka Shigeki 貝塚茂樹 in **Shirin,** 33.1 (Jan. 1950), 100–101.

6.2.4 MORIMOTO Sugio (Chikujō) 森本杉雄 (竹城), **Shinchō Jugakushi gaisetsu** 清朝儒學史概說 (A general survey of the history of Confucian learning in the Ch'ing dynasty), Bunshodō 文書堂, 1931, 460+index 27 pp.

This general survey of 300 years of Chinese thought and scholarship in the great tradition is arranged mainly by individuals under schools (*p'ai* 派), among them being Juan Yuan 阮元 (pp. 216–226), the *kung-yang* 公羊 school (pp. 281–301, including the *chin-ku-wen* 今古文 controversy), Wei Yuan 魏源 and his writings (pp. 314–319), K'ang Yu-wei, T'an Ssu-t'ung 譚嗣同, and Liang Ch'i-ch'ao (pp. 325–374), Tseng Kuo-fan (pp. 430–435), and many others, ending with Hu Shih and Ch'en Tu-hsiu (pp. 443–459). By a scholar who was teaching at the Tōkyō High School 東京高等學校 (now assimilated into the College of General Education of Tōkyō University), this volume, although rather pedestrian, provides a useful perspective on late-Ch'ing scholars.

6.2.5 SHIMIZU Nobuyoshi 清水信良, **Kinsei Chūgoku shisō shi** 近世中國思想史 (History of modern Chinese thought), Meiji Tosho Kabushiki Kaisha 明治圖書株式會社, 1950, 477 pp., pref. by Tezuka Yoshimichi 手塚良道.

This volume continues the author's earlier one on ancient and medieval Chinese thought completed in 1949, and deals mainly with Sung and also Yuan, Ming and Ch'ing. K'ang and Liang are only touched upon (pp. 438–448) and the concluding chapter mentions Sun, Hu Shih, and Mao briefly.

6.2.6 ŌTANI Kōtarō 大谷孝太郎, "Hinraku seikatsu oyobi shisō" 貧樂生活及び思想 (A poor but happy way of life and thought), **Tōa keizai ronsō,** 2.2 (May 1942), 25 pp.

In generalizing about this Chinese view of life, the writer quotes Tseng Kuo-fan's 曾國藩 family letters.

6.2.7 SHIMADA Kenji 島田虔次, **Chūgoku ni okeru kindai shii no zasetsu** 中國に於ける近代思惟の挫折 (The breakdown of modern thought in China), Chikuma Shobō 筑摩書房, 1949, 311 pp.

As a disciple of Naitō Torajirō, the author (now at the Kyōto Institute) means "Sung or later" when speaking of "modern" China and actually presents in this volume an important study of Ming intellectual history. He pays special attention to the left wing of the school of Wang Yang-ming 王陽明 as having achieved valuable goals (in their studies of the subjective consciousness), within the limits of the scholarly Confucian culture; this involves a critical reappraisal of the time-worn theory concerning traditional China's intellectual "stationariness", and so should be noted in any study of modern interpretations.

6.3   SINOCENTRISM

Note: These studies relate to a feature of Chinese traditional thought which was closely integrated with the old political institutions dealt with above in sec. 3, especially sec. 3.3.   Note item 2.6.4 (Itano) and item 2.3.2 (Ueda).

6.3.1 NABA Toshisada, 那波利貞 **Chūka shisō** 中華思想 (Sinocentrism), in **Iwanami kōza Tōyō shichō** (9.8.4), series 17, 1936, 67 pp.

This well known study of "the Middle Kingdom idea" discusses ten different ancient designations used by the Chinese for their country, the origin of this sinocentric or center-of-civilization idea, its characteristics and historical development down to the time of Lord Macartney and the Opium War.   The author gives textual examples of background interest for modern studies.

6.3.2 ABE Takeo 安部健夫, "Shinchō to ka-i shisō" 清朝と華夷思想 ("Ch'ing Dynasty and the Thought of 'Flowery kingdom'"), **Jimbun kagaku,** 1.3 (Dec. 1946), 137–159.

An interesting essay, by a specialist in early Ch'ing institutions, on the Manchu adaptation to the age-old distinction between Chinese and barbarian; this is a suggestive developmental analysis of the ideological aspect of the Manchu-Chinese joint government (or synarchy).

6.3.3 HIRAOKA Takeo 平岡武夫, "Tenkateki sekaikan to kindai kokka" 天下的世界觀と近代國家 (The universal-state or *t'ien-hsia* world view and the modern state), **Tōkō,** 2 (Nov. 1947), 2–23.

A scholar versed in the Chinese classical learning here notes the conflict against, and nevertheless the persistence of, an ancient and traditional Chinese view of the world order. His discussion centers about the relation of the ruler (or government) to the people, touches on various political devices and institutions of the past, like the examination system, and describes the peculiarly traditional concept of sovereignty and territory. Finally the author speculates on the possible future utility of this traditional view of the world.

6.3.4 AKIYAMA Kenzō 秋山謙三, **Shinajin no mitaru Nippon** 支那人の觀たる日本 (Japan in the eyes of the Chinese), in **Iwanami kōza Tōyō shichō** (9.8.4), series 2, 1934, 62 pp., illus.

Beginning with a few pages on sinocentric thought (*chūka shisō* 中華思想) and the Chinese idea of the barbarians, this essay outlines the Chinese view of Japan by periods down to the piratical invasions of the late Ming period, in the context of its basic sinocentric orientation. There is a final section on China's geographical knowledge of Japan. Interesting illustrations.

6.3.5 OTAKE Fumio 小竹文夫, "Shina ni okeru 'bōeki' no kannen" 支那に於ける貿易の觀念 (The concept of trade [*mao-i* 貿易] in China), **Shina kenkyū,** 17 (July 1928), 57–75.

An interesting study of the Chinese (or Ch'ing) attitude concerning trade, as evidenced in tributary relations in Asia and early relations with the West. The author seems to detect a persistence of the traditional idea of foreign trade among modern Chinese who urge boycotts.

6.4 POPULAR RELIGION

Note: These items provide some introduction to this field. For further materials, see sec. 8.6 below. The last two items are on Chinese Moslems.

6.4.1 KUBO Noritada 窪德忠, **Dōkyō to Chūgoku shakai** 道教と中國社會 (Taoism and Chinese society), Heibonsha, 1948, 189 pp.

This small survey for general readers, one of the few available, was written at the Tōkyō Institute. It studies the historical development and present state of Taoism, the Taoist canon, and the esoteric methods for prolonging life (such as 方術). The author, who is one of the rare academic specialists in Taoism, is now an assistant-professor of the Institute for Oriental Culture, Tōkyō University. Reviewed by Sakai Tadao 酒井忠夫 in **Chūgoku kenkyū,** 6 (June 1949), 64–71.

6.4.2 YOSHIOKA Gihō 吉岡義豐, **Dōkyō no kenkyū** 道教の研究 (A study of Taoism), Hōzōkan 法藏館, 1952, 356 pp.

Written partly in Peking and based on field study, this recent volume explores the ramifications of Chinese popular religion as represented in (1) parts of the Taoist literature or canon (the so-called *pao-chüan* 寶卷), including writings ascribed to worthies like Lo Tsu (羅祖 "ancestor Lo"), and other writings in the work called **Kan-ying-p'ien** 感應篇 probably written by Li Ch'ang-ling 李昌齡 of the Sung period and in the so-called **Kung-kuo-ko** 功過格 ("the table of rewards and punishments")—all as a clue to "popular" Taoism; (2) the modern cult (*ch'üan-chen-chiao* 全真教) founded by Wang Chung-yang 王重陽 (b. 1113) and others; and (3) the present (i.e. recent) state of Taoism in Chinese society against the historical background of the "ecclesiastical" Taoism patronized and controlled by the state. Its temples, organization and order, services, pantheon, etc., are dealt with.

6.4.3 KUBO Noritada 窪德忠, "Chūkyō no shūkyō seisaku to minshū dōkyō" 中共の宗教政策と民衆道教 ("An interrelationship between the religious policy and popular Taoist association in New China"), **Tōyō bunka**, 11 (Nov. 1952), 1–26.

A specialist in the history of Taoism discusses the situation under the new regime, using rather scant but laboriously collected information. One section (6 pp.) deals with secret society activity.

6.4.4 SAKAI Tadao 酒井忠夫, "Chūgoku minshū no shūkyō ishiki" 中國民衆の宗教意識 (The Chinese masses' religious consciousness), **Chūgoku kenkyū**, 1 (June 1947), 69–85.

This essay by a specialist in popular religion seeks to distinguish elements of the traditional popular attitude toward religion which may remain valid in the modern period and so, for example, serve as a check on the idea of class warfare. See also 8.6.2. Prof. Sakai is now at Tōkyō Kyōiku Daigaku.

6.4.5 HAYASHI Gemmei 林彥明, "Shina genkon no nembutsu kyōgi" 支那現今の念佛教義 (The Amida [Buddhist] teaching of present-day China), **Ryūkoku gakuhō** 龍谷學報, ed. by the Ryūkoku Gakkai and pub. quarterly by Kōkyō Shoin 興教書院, 315 (June 1936), 309–317.

A brief note (but informative to the initiated) on contemporary developments of the Pure Land sect 淨土教 of Chinese Buddhism, and its connection and comparison with the Japanese counterpart, with brief reference also to individuals like the monk T'ai Hsü 太虛.

6.4.6 SAWASAKI Kenzō 澤崎堅造, "Shina ni okeru Kaikyō no kyōdōsei" 支那に於ける回教の共同性 (The communal character of Mohammedanism in China), **Tōa jimbun gakuhō**, 1.4 (Feb. 1942), 940–956.

A brief survey of the historical and contemporary state of communal life among the Chinese Mohammedans—religious, social, economic and political—with tables and charts.

6.4.7 IWAMURA Shinobu 岩村忍, **Chūgoku kaikyō shakai no kōzō** 中國回教社會の構造 (The structure of the Chinese moslem societies), **Shakai kōseishi taikei** (9.8.7), part one, in series 5, 1949, 134 pp; part two, in series 6, 1950, 90 pp.

This careful account by a leading specialist at the Kyōto Institute, one of the few recent studies of the Chinese moslem community, is based partly on wartime field work in North China and Inner Mongolia. It describes the distribution of communities along the Great Wall borderland, their religious organization (mosques, etc.), religious personnel such as the *ahun* アホン and their interconnections, their religious constituency and its local leadership and sects in specific detail. Throughout, this examplary field report is illuminated by a general command of the literature and history of the field.

6.5   THE JESUITS

Note: While we have generally tried to avoid materials dealing with events before 1800, this section is included as a necessary background for the modern period, just as many of the works in sec. 2.1 on Ch'ing history concern the earlier period of the dynasty.

6.5.1 YAZAWA Toshihiko 矢澤利彦, "Kō-ki-tei to tenrei mondai" 康熙帝と典禮問題 ("A Study of the Rites Controversy"), **Tōyō gakuhō**, 30.1 (Feb. 1943), 1–23; 30.2 (May 1943), 133–162; 30.3 (Aug. 1943), 346–385.

An authoritative analysis, mainly from the emperor's point of view as evidenced in newly published Chinese documents. Presumably this study will be of interest in connection with the more recent volume by Father A. Rosso, **Apostolic Legations to China,** which makes extensive use of Western ecclesiastical sources. Prof. Yazawa is now at Saitama Daigaku 埼玉大學.

6.5.2 YAZAWA Tohihiko 矢澤利彦, **Chūgoku to Seiyō bunka** 中國と西洋文化 (China and Western culture), Nakamura Shoten 中村書店, 1947, 199 pp.

This study centers on the Jesuit activity in China, especially the Rites Controversy (pp. 32–104). Other chapters treat St. Francis Xavier, Adam Schall, Western music, and the economic basis of Catholic evangelism. Most of the book was written at the Tōyō Bunko in the period 1936–41. Footnote references are given in general terms. Chapter 7, on Chinese Catholicism and the question of the nature of woman 支那天主教と女性の問題, is a revision and condensation of an article of the same title in **Rekishigaku kenkyū**, 6.11 (Nov. 1936), 33–46.

6.5.3 ISHIDA Mikinosuke 石田幹之助, **Tōyō bunka no kōryū, Shina bunka to Seihō bunka to no kōryū** 東洋文化の交流，支那文化と西方文化との交流 (Oriental cultural contact, the contact of Chinese and Western culture), in **Iwanami kōza Tōyō shichō** (9.8.4), series 18, 1936, 152 pp.

A preliminary survey of East-West cultural contact from ancient times down to the Jesuits (pp. 1–62) is followed by some attention to their influence on China and major attention to Chinese cultural influence on Europe in the seventeenth and eighteenth centuries (pp. 87–152). A 7-page bibliography leads the reader to more substantial works by Japanese scholars including Gotō Sueo's 後藤末雄 massive volume on the Chinese influence upon 18th century French thought as conveyed through the Jesuits' reports: **Shina shisō no Fu-**

ransu seizen 支那思想のフランス西漸 (The westward movement of Chinese thought into France), Daiichi Shobō 第一書房, 1933, 668 pp., many photos; cheaper edition, without photographs, entitled **Shina bunka to Shinagaku no kigen** 支那文化と支那學の起源 (Chinese culture and the origin of [Western] Sinology), by the same pub., 1933 (3rd printing, 1941, 668 pp.).

6.5.4 GOTŌ Sueo 後藤末雄, **Ken-ryū-tei den** 乾隆帝傳 (A biography of the Ch'ien-lung emperor), Seikatsusha, 1942, 290 pp.

This is actually an account of the Jesuits in Peking during the Ch'ien-lung period, based on their **Lettres édifiantes et curieuses** and a few other Western sources—full of color and detail concerning Yüan-ming-yüan, cartography, astronomy, the Pei-t'ang 北堂, etc., but based on translation rather than research.

6.5.5 TASAKA Okimichi 田坂興道, "Seiyō rekihō no tōzen to kaikai-rekihō no ummei" 西洋曆法の東漸と回回曆法の運命 ("Introduction of European astronomy in China and the Mohammedan Calendarers"), **Tōyō gakuhō,** 31.2 (Oct. 1947), 141–180.

An outstanding specialist on Sino-Islamic contacts discusses here the position in China during the Ming period of the Mohammedan calendar which had been introduced under the Yüan dynasty; the circumstances surrounding the adoption of European astronomy in the late Ming period; and the abortive recovery movement on the part of the Mohammedan calendarers in the early Ch'ing period.

6.5.6 YABUUCHI Kiyoshi 藪内清, "Seiyō temmongaku no tōzen, Shindai no rekihō" 西洋天文學の東漸——清代の曆法 (Introduction of Western astronomy into the East, calculation of the calendar in the Ch'ing period), **Tōyō gakuhō, Kyōto,** 15, part 2 (Jan. 1946). 133–154.

This article, part of the author's project for a general history of Chinese calendars, discusses both the background and the consequences of the introduction of Western astronomy by the Jesuits.

6.5.7 MIKAMI Yoshio 三上義夫, "Shinchō jidai no katsuenjutsu no hattatsu ni kansuru kōsatsu" 清朝時代の割圜術の發達に關する考察 (An investigation of the development of the method of finding the Ludolph's number [pi] in the Ch'ing period), **Tōyō gakuhō,** 18.3 (Aug. 1930), 301–345; 18.4 (Oct. 1930), 439–489.

A study of the development of a phase of Chinese mathematics under the influence of Western mathematicians, especially the Frenchman, Pierre Jartoux (1669–1720) 杜德美, who participated in the Jesuit map making sponsored by Emperor K'ang-hsi and died in China. This article appears to touch upon many aspects of mathematics in the Ch'ing period, including algebra, trigonometry and more abstruse subjects. The author was a leading specialist in the history of mathematics in Japan.

6.6 CHRISTIANITY IN MODERN CHINA

Note: Christianity in 19th century China seems to have been less thoroughly studied, being much more complex, than Christianity in earlier periods, particularly that of the Jesuits. Recent contributions on the early 19th century have been made by Yazawa Toshihiko from the Ch'ing documents, 6.6.4–8, and important studies of the biblical translation process have been made by Shiga Masatoshi, 6.6.16–17.

6.6.1 SAEKI Yoshirō 佐伯好郎, **Shinchō Kirisutokyō no kenkyū** 清朝基督教の研究 (A study of Christianity under the Ch'ing Dynasty), Shunjūsha 春秋社, 1949, 640+24 pp.

Dr. Saeki studied in both America and China and had already produced three solid volumes on Christianity in China before the Ch'ing period entitled **Shina Kirisutokyō no kenkyū** (Shunjūsha, 2 vols., 1943; vol. 3, 1944). The third volume on the Ming period covers also the Rites Controversy under the Ch'ing. The present volume, really the fourth in this series, is necessarily somewhat less thorough; it begins with the Yung-cheng period and traces the fortunes of Catholicism down to the 1920s (pp. 4–353). The Protestant approach to China is then dealt with more briefly. There are long chapters on the Chinese publications of the Roman Catholics (pp. 179–313) and Protestants (pp. 489–637),with examples of translation and transliteration problems, which should be of considerable interest to Western scholars of Chinese thought. Unfortunately, footnote references are inadequate in this work. Dr. Saeki was formerly at Tōkyō Bunrika Daigaku and a member of Tōhō Bunka Gakuin, to which the above studies were originally presented.

6.6.2 MIZOGUCHI Yasuo 溝口靖夫, **Tōyō bunkashijō no Kirisutokyō** 東洋文化史上の基督教 (Christianity in the cultural history of the Orient), Risōsha 理想社, 1941, 450 pp.+index 5 pp., illus.

A study of Christianity in Persia, India (to p. 190) and China, which sets out to trace its cultural influence in the T'ang, Yuan and Jesuit periods (to p. 341) and then recounts the modern Protestant influence: Morrison and his successors, Gutzlaff, the early Americans and their methods of proselytism; Taiping Christianity; Hudson Taylor, etc. The last section on Christianity and the Chinese revolution relates it to the Boxers, the anti-monarchic movement, etc., and winds up with the New Life Movement. Although based on both Western and Japanese literature (e. g. writings of Saeki Yoshirō 佐伯好郎), this survey relies heavily on the former for the modern period and so appears to add little to standard works of Dr. K. S. Latourette and others. It also fails to come to grips with "cultural influence".

6.6.3 KOBAYASHI Matazō 小林又三, "Shina ni okeru Kirisutokyō dendō shi" 支那に於ける基督教傳道史 (History of Christian evangelism in China), **Mantetsu chōsa geppō,** 20.2 (Feb. 1940), 73–99; 20.3 (Mar. 1940), 33–54 (unfinished, no continuation published).

While this survey is mainly on earlier periods, its conclusion touches the 19th century. K. S. Latourette, **A History of Christian Missions in China,** is used extensively and little if anything new seems to have been added by the author.

6.6.4 YAZAWA Toshihiko 矢澤利彦, "Chūgoku Tenshu kyōto to denkyō-sha tono mondai" 中國天主敎徒と傳敎者この問題 ("Chinese Catholics during the Period of Persecution [1723–1843]"), **Shigaku zasshi,** 59.3 (Mar. 1950). 181–199.

Based on the Ch'ing documents (e. g., **Shih-lu 實錄** and **Ch'ing-tai wai-chiao shih-liao 清代外交史料**), this article carefully appraises the attitudes of Chinese Catholics as well as the governmental policies in the period of severe persecution. It also deals with the hierarchical organization of the Chinese clergy in this period.

6.6.5 YAZAWA Toshihiko 矢澤利彦, "Hakugaiki Tenshu kyōto no sei-kaku" 迫害期天主敎徒の性格 ("Characters of the Roman Catholics under the Persecution"), pp. 763–778 in **Wada hakushi kanreki kinen Tōyōshi ronsō,** 9.8.12.

An expert appraisal of the types and motivations of Chinese Christians in the social scene during the mid-Ch'ing century of persecution (ca. 1723–1843), based on Chinese sources such as **Ch'ing shih-lu, Wen-hsien ts'ung-pien 文獻叢編** and **Ch'ing-tai wai-chiao shih-liao 清代外交史料.**

6.6.6 YAZAWA Toshihiko 矢澤利彦, "Ka-kei jūnen (1805) no Tenshukyō kin'atsu" 嘉慶十年 (一八〇五) の天主敎禁壓 (The suppression of Catholicism in 1805), **Tōa ronsō,** 1 (July 1939), 147–188.

6.6.7 YAZAWA Toshihiko 矢澤利彦, "Ka-kei jūrokunen no Tenshukyō kin'atsu" 嘉慶十六年の天主敎禁壓 ("Persecution of Christianism in China in 1811"), **Tōyō gakuhō,** 27.3 (May 1940), 351–392.

Case studies using both Chinese documents **(Ch'ing-tai wai-chiao shih-liao 清代外交史料)** and **Nouvelles lettres édifiantes,** which illustrates Dr. Yazawa's thesis that the Chinese records concerning Christianity have been relatively neglected.

6.6.8 YAZAWA Toshihiko 矢澤利彦, "Kishū Tenshukyōshi ni kansuru ichishiryō" 貴州天主敎史に關する一史料 ("A Document concerning the History of Catholic Missions in Kuei-chou unknown hitherto"), **Tōyō gakuhō,** 26.2 (Feb., 1939), 260–291.

A leading specialist on Christianity under the Ch'ing reproduces, with notes and comment, a memorial of the Kweichow governor, Ch'ing-pao 慶保, concerning the suppression of Christianity in 1814–15, found among his ms. papers in the Tōyō Bunko.

6.6.9 YAZAWA Toshihiko 矢澤利彦, "Kyūhokudō bosshū no jijō ni tsuite" 舊北堂沒收の事情に就いて (On the circumstances of the confiscation of the old Pei-t'ang [Peking Cathedral]), pp. 903–924 in **Katō hakushi kanreki kinen Tōyōshi shūsetsu,** 9.8.5.

An interesting account of the Chinese official taking-over of the old Catholic cathedral in Peking in 1827, later rebuilt.

6.6.10 UEDA Toshio 植田捷雄, "Shina ni okeru Kirisutokyō senkyōshi no hōritsuteki chii" 支那に於ける基督教宣教師の法律的地位 (The legal status of Christian missionaries in China), **Tōyō bunka kenkyūjo kiyō**, 1 (Dec. 1948), 147–194.

Discusses many aspects of the legal status of the foreign missionaries in China—including their rights of evangelism, residence and purchase of property in the interior and the problems connected with Christian converts—down to the 1930s. Dr. Ueda then adds some discussion of the status of Japanese Buddhists in China, and of the Nationalist Government interest in limitation of church land-holding.

6.6.11 YAMAGUCHI Noboru 山口昇, **Ō-Beijin no Shina ni okeru bunka jigyō** 歐米人の支那に於ける文化事業 (Cultural enterprises of Europeans and Americans in China), Shanghai: Nippondō Shoten 日本堂書店, 1921 (3rd printing, 1922, 1371 pp.), tables, illus.

Concentrating on missonary good works in order to observe their methods and results, this volume begins with an historical summary of the Christian approach to China before the 19th century (first 100 pp.), and then deals with each of 33 British mission societies or agencies, not excluding the Baptist Zenana Mission, some 46 American agencies, Germans, Catholics, negotiations over missions, missions in Manchuria, etc., etc. Later sections then deal with 26 educational aspects or institutions, 20 hospitals, and various libraries and eleemosynary works. This big catalogue was put together by a Maritime Customs officer for the Nikka Jitsugyō Kyōkai 日華實業協會. Its voluminous data are undoubtedly not to be found in any one Western work.

6.6.12 ADACHI Ikutsune 安達生恒, "Uchi Mōko chōjō chitai ni okeru nōgyō no tenkai to Kasorikku" 內蒙古長城地帶に於ける農業の展開とカソリック ("Catholicism and the development of agriculture in the Inner Mongolia along the Great Wall"), **Jimbun kagaku,** 1.3 (Dec. 1946), 230–250.

An elaborate analysis, based on the author's field investigation, of the agricultural development in Catholic villages situated in Inner Mongolia along the Great Wall, which date mainly from the late 19th century. This article emphasizes the decisive role played by the Catholic churches on the spot as large-scale buyers of Mongolian banner-estates, and also in generously supplying the means of production for poor peasants, instructing them in agricultural technology, and maintaining order.

6.6.13 SAWASAKI Kenzō 澤崎堅造, "Mōko dendō to Mōkogo seisho" 蒙古傳道と蒙古語聖書 (Christian mission work in Mongolia and the Bible in Mongolian), **Tōa jimbun gakuhō,** 4.2 (Mar. 1945), 391–416.

A brief account of evangelical work in Mongolia, Catholic as well as Protestant, and of the translation of the Bible into Mongolian by I. J. Schmidt and others, noting the meagre results achieved as compared with the painstaking efforts made. Half a dozen different missions (Swedish, American, et al.) are described.

6.6.14 SAWASAKI Kenzō 澤崎堅造, "Nekka Utan ni okeru Katorikku-mura" 熱河烏丹に於けるカトリック村 (Catholic villages in Wu-tan, Jehol), **Tōa jimbun gakuhō,** 3.3 (Jan. 1944), 648-669, maps.

Based on a field investigation report of 1943, this article surveys the current state of the deep-rooted Catholic life found in three villages near Wu-tan, as seen againt the historical background of Catholic penetration into Inner Mongolia.

6.6.15 MIYAGAWA Hisayuki 宮川尚志, "Honshoku kyōkai ni tsuite" 本色教會に就いて (On the Pen-se chiao-hui), **Tōyōshi kenkyū,** 8.1 (Jan.-Feb. 1943), 44-47.

A brief note on the movement for a Chinese Christian church separated from foreign assistance, viewed against the background of the religious history of China.

6.6.16 SHIGA Masatoshi 志賀正年, "Seikyō (Bible) hon'yaku hōhōron kō, Shin'yaku kayaku o chūshin to shite" 聖經 (Bible) 翻譯方法論考――新約華譯を中心として ("The methodological Study of the Bible Translation into Chinese", with special reference to the translation into Chinese of the New Testament), **Tenri daigaku gakuhō,** 1.2–3 (Oct. 1949), 83–115.

A detailed technical study which lists some 50 translations dating from 1800 to 1924 and analyzes the methods used by individuals such as Griffith John (1885 and 1889); by joint efforts such as those of Marshman and Lassar; in the famous "delegates" version of Boone, Stronach and Lowrie (1854); and by committees (e. g. under Edkins in 1866 or Mateer in 1907).

6.6.17 SHIGA Masatoshi 志賀正年, "Kayaku seikyō o tōshite mitaru 'kami' kō" 華譯聖經 (Bible) を通して見たる「神」考 ("The idea of God as seen through the Bible done into Chinese"), **Tenri daigaku gakuhō 天理大學學報** ("Bulletin of Tenri University"), 8 or "vol. IV, no. 1" (July 1952), 79–100.

Distinguishing five periods of Biblical translation since the 13th century, the author conducts an intensive word-count of the use made of different terms for the deity (神, 天主, 上帝 etc.) in various translations, in various books of the Bible, as between the Old and New Testaments, etc., as well as in dialect versions.

6.6.18 OZAWA Saburō 小澤三郎, "Shina zairyū Yasokyō senkyōshi no Nippon bunka ni oyoboseru eikyō" 支那在留耶蘇教宣教師の日本文化に及ぼせる影響 (Influence of Protestant missionaries stationed in China upon Japanese culture), **Shikan,** 28–29 (June 1942), 72–91.

While this study concerns Japan, not China, it casts an interesting sidelight on the work of Bridgman, Edkins, Legge, Muirhead, Wylie and others by indicating how their writings were translated into Japanese.

6.7 EDUCATION

Note: The first five items are on traditional education, especially the examination system. Later works describe the rise of the new education and certain aspects of it. These few studies make it evident that in the context of contemporary events and in the absence of the basic sources for research, the long history of Western educational efforts in China can hardly be expected to emerge from research conducted in Japan.

6.7.1 MIYAZAKI Ichisada 宮崎市定, **Kakyo** 科舉 (The examination system), Osaka: Akitaya 秋田屋, 1946, 288+4 pp., illus.

After a brief summary of its earlier history, with special reference to the interconnection between the examination system and the despotic political structure, Prof. Miyazaki describes the Ch'ing examination system in careful detail, including the arrangements for Manchu bannermen and others, and for Chinese entrance into official life outside the system (pp. 171–201). Final sections comment on the place of the old examination system in Chinese life, and the manner of its collapse. While sources are indicated only very occasionally, this volume provides a convenient summary of the subject.

6.7.2 SUZUKI Torao 鈴木虎雄, "Kobun hihō no zenku" 股文比法の前驅 ("The Development of the Forms of *Pa Ku Wen*") **Shinagaku, 4.1** (July 1926), 27–46.

A study, with textual examples, of the origin of the so-called "Eight Legged Essay" style in the Ming period, with a further look at its antecedents in earlier times.

6.7.3 TANAKA Kenji 田中謙二, "Kyū-Shina ni okeru jidō no gakujuku seikatsu" 舊支那に於ける兒童の學塾生活 (The life of private school pupils in the old China), **Tōhō gakuhō, Kyōto, 15,** part 2 (Jan. 1946), 217–231.

This pioneer study of an unexplored topic uses some of the autobiographical writings of men of letters like Lu Hsün, Hu Shih, Kuo Mo-jo, Shen Ts'ung-wen 沈從文, and Ch'en Tu-hsiu, to throw an interesting light on the actual life of private school children in the late Ch'ing period. The author stresses the formalistic and rigorous character of the training given potential candidates for the examinations (k'o-chü 科舉) in their early school life, with many references to the texts used.

6.7.4 KATSUMATA Kenjirō 勝又憲治郎, "Pekin no kakyo jidai to kōin" 北京の科舉時代と貢院 ("The Period of the Grand Public Examination in Peking and the Examination-Hall"), **Tōhō gakuhō, Tōkyō, 6** "The extra number" (Sept. 1936), 203–240, plates.

A scholarly essay describing the general circumstances of the examination system as administered in Peking under successive dynasties. Only the last dozen pages deal with the Ch'ing. Presumably this account has been superseded by Prof. Miyazaki Ichisada's volume, 6.7.1.

6.7.5 TAUCHI Takatsugu 田内高次, **Shina kyōikugaku shi** 支那教育學

史 (History of Chinese educational learning), Fuzambō, 1942, 545+table of contents 27 pp.

A quoting of practically all the ancient philosophers' remarks on education, ending with a quick 30 pages on the K'ang-Liang school, Chang Chih-tung and the new education which followed.

6.7.6 Kigen Nisenroppyakunen Kinenkai 紀元二千六百年記念會 (Society to commemorate the 2600th year of the imperial era), of the Tōkyō Bunrika Daigaku and Tōkyō Higher Normal School 高等師範學校, **Gendai Shina Manshū kyōiku shiryō** 現代支那滿洲教育資料 (Materials on education in contemporary China and Manchuria), Baifūkan 培風館, 1940, 456 pp.

A history of modern Chinese education by periods, from 1840 to 1937 (pp. 1–96), is followed by translations of a great number of key documents of the late Ch'ing and (principally) the republican period on (1) the aims of education, (2) policies, (3) regulations, (4) the school system, (5) statistics in many tables, plus a chronology (pp. 271–311) and a very extensive, topically arranged bibliography (pp. 312–396); all compiled by Dr. Aritaka Iwao 有高巖 and two other scholars. The same is then done, on a smaller scale, for Manchoukuo.

6.7.7 ŌKUBO Sōtarō 大久保莊太郎, "Shina ni okeru shinkyōiku ni tsuite" 支那に於ける新教育について (On the new education in China), **Tōa jimbun gakuhō**, 1.3 (Dec. 1941), 714–737.

This article provides a quick survey, with useful bibliography and documentation, of the development of the educational system and ideas in China under Western influence from the 1860s down to the 1920s. The author stresses the persistence and revival of classical tradition, the gap between imported foreign ideas and economic and social realities in China, and the decisive influence of politics on education.

6.7.8 HIRATSUKA Masunori 平塚益德, **Kindai Shina kyōiku bunka shi** 近代支那教育文化史 (History of modern Chinese education and culture), Meguro Shoten, 1942, 419+2+index 13 pp.

Subtitled as centering on educational activities of third countries in China, this is a systematic research study of Western missonary educational work in China, from before 1840 to 1900 (to p. 132) and from 1900 to 1912, with later chapters on the educational changes under the early Republic and expansion of Western efforts down through the anti-Christian and rights recovery movements of the 1920s until 1931. The notes include details of bibliography by chapters, pp. 351–419, and make this the most comprehensive Japanese study of the subject. See long review by Ōkubo Sōtarō 大久保莊太郎 in **Tōa jimbun gakuhō**, 3. 1 (Mar. 1943), 213–231.

6.7.9 ONO Shinobu 小野忍 and SAITŌ Akio 齋藤秋男, **Chūgoku no kindai kyōiku** 中國の近代教育 (Modern education in China), Kawade Shobō, 1948, 223 pp., charts. (Kyōiku bunko 教育文庫, 16.)

This survey volume by two of the contributors to the Chūgoku Kenkyūjo volume on contemporary Chinese culture, also of 1948, 6.1.8, summarizes in a similar vein the main phases of the institutional history of modern education in China, the national language (*kuo-yü* 國語) movement (pp. 48–82, by Ono, other chapters being all written by Saitō), the movements for social education and rural reconstruction (T'ao Hsing-chih 陶行知 and Liang Sou-ming 梁漱溟), missionary education (de-emphasized, pp. 106–117), the student movement, CCP education in the Liberated Areas and postwar tendencies. No sources noted but a bibliography of some 60 items.

6.7.10 KAWAI Shingo 河合愼吾, "Minzokushugiteki keikō o chūshin to shite mitaru kyū Chūgoku Kokumintō no kyōiku seisaku no rinkaku" 民族主義的傾向を中心として觀たる舊中國々民黨の教育政策の輪廓 (An outline of the educational policy of the former Kuomintang, with special attention to its nationalistic orientation), **Tōa kenkyū shohō**, 7 (Dec. 1940), 1-49.

A very critical, apparently well informed, but undocumented account of the "partification" (*tōka* 黨化) of education under the Nationalist Government and its essentially anti-foreign spirit, based on general knowledge (周知の如く...). The documentation of this subject is plentifully supplied in the same author's monograph produced by the Tōa Kenkyūjo, **Kokumintō Shina no kyōiku seisaku** 國民黨支那の教育政策 (Kuomintang China's educational policy), 1941, 523 pp., mimeographed, which covers the same ground.

6.7.11 SAITŌ Akio 齋藤秋男, "Chūgoku kyōiku no risō to genjitsu, gimukyōiku mondai wa dō suiishite kitaka" 中國教育の理想と現實——義務教育問題はどう推移してきたか ("Ideals and Realities of Education in China, How the Problems of Compulsory Education have developed"), **Chūgoku kenkyū,** 16 (Sept. 1952), 1–25.

A brief historical review of modern public education programs since the late Ch'ing down to the wartime and recent communist developments.

6.7.12 ŌKUBO Sōtarō 大久保莊太郎, "Kindai Shina no heimin kyōiku undō, 'Teiken Kahoku jikkenku' o chūshin to shite" 近代支那の平民教育運動——「定縣華北實驗區」を中心として (The popular education movement in Modern China, with special reference to "the North China experimental district in Ting-hsien"), **Tōa jimbun gakuhō,** 2.3 (Dec. 1942), 353-400.

In the context of the nationalistic and democratic response to Western penetration in the sphere of education since the time of the May Fourth Movement of 1919, this article recounts in detail the literacy campaign and other educational activities in "the experimental district" of Ting-hsien, Hopei, from the mid-1920s. This study of the well known project headed by Y. C. James Yen and aided by the Rockefeller Foundation is of interest as an independent appraisal which summarizes data and cites a good deal of bibliography.

6.7.13 SAITŌ Akio 齋藤秋男, **Shin Chūgoku kyōshi no chichi, Tō Kō-chi** 新中國教師の父・陶行知 (The father of new China's teachers, T'ao Hsing-chih), Tōkō Shoin, 1951, 218 pp., illus.

T'ao Hsing-chih (1891–1946), one of John Dewey's favorite pupils, was a pioneer of mass education and one of the true fighting liberals of the Kuomintang era. This eulogistic biography gives his life history and translates 8 of his writings and 7 poems. Dr. Tao's memory having been appropriated by the new regime, this history of his work, based on recent Chinese publications, fits him deftly into the revolutionary saga. See review by Yakawa Tokumitsu 矢川德光 in **Chūgoku kenkyū,** 15 (Jan. 1952), 78–80.

6. 8 SINO-JAPANESE CULTURAL RELATIONS

Note: This strikes us as one of the most fertile fields awaiting exploitation by modern scholars. An extensive pioneer reconnaissance has been carried out by Sanetō Keishū (see the first five items), while interesting studies in literary influence are being pursued by Nakamura Tadayuki at Tenri University, 6. 8. 8–12.

6. 8. 1 SANETŌ Keishū 實藤惠秀, **Chūgokujin Nippon ryūgaku shikō** 中國人日本留學史稿 (Draft history of Chinese students in Japan), Nikka Gakkai 日華學會, 1939, 368 pp., illus.

The compiler was a professor in Waseda Kōtō Gakuin 早稻田高等學院 attached to Waseda University, and has long been interested in this subject. His extensive materials are arranged by topics within a chronological sequence of periods from the 1890s to the 1940s and deal with all the various aspects of personnel, educational institutions, publications, political and administrative problems, personal experiences and incidents, involved in this large topic. Together with the bibliography, this volume should provide useful raw material for more analytic study. Prof. Sanetō is now in Waseda University.

6. 8. 2 SANETŌ Keishū 實藤惠秀, **Nippon bunka no Shina e no eikyō** 日本文化の支那への影響 (The influence of Japanese culture on China), Keisetsu Shoin, 1940, 313+3 pp., illus.

As the most comprehensive of Prof. Sanetō's pioneer studies in this virgin field, this volume makes certain wartime assumptions concerning the role of Japan as cultural mentor of modern China. These essays, however, while not in the form of academic research, open up many interesting topics under the following general scheme of arrangement: (1) the Japanization of modern Chinese culture, influence of Japanese literature in China; (2) Chinese students in Japan (various incidents, Japanese reactions); (3) Japanese relations with China's new daily and periodical press, Meiji criticism of Confucianism and its influence in late-Ch'ing China, stories of the period before China's anti-Japanese movement; (4) aspects of future cooperative cultural activity. See review by Ishihara Michihiro 石原道博 and Satō Saburō 佐藤三郎 in **Rekishi-gaku kenkyū** 10. 9, whole number 81 (Sept. 1940), 976–981.

6. 8. 3 SANETŌ Keishū 實藤惠秀, **Kindai Nisshi bunka ron** 近代日支文化論 (On modern Sino-Japanese culture), Daitō Shuppansha 大東出版社, 1941, 269 pp., illus.

The author sought contact with early Japan-returned students like Ts'ao Ju-lin 曹汝霖 and T'ang Pao-e 唐寶鍔. This volume in rather colloquial style deals with Chinese students in Japan, including experiences of the above-named pioneers and of the writer Wang T'ao 王韜, the translation and publication of Japanese works in Chinese, the sending of Chinese students to America as contrasted with Japan, and various intellectual and cultural offshoots of Sino-Japanese contact. Although loosely conceived and organized, the volume gives much information and raises interesting questions.

6.8.4 SANETŌ Keishū 實藤惠秀, **Meiji Nisshi bunka kōshō** 明治日支文化交涉 (Sino-Japanese cultural relations in the Meiiji period), Kōfūkan 光風館, 1943, 394 pp., illus.

This volume is also written in colloquial style and story form, evidently to evoke student interest. Many Chinese and Japanese impressions of the other country are quoted and lists of source materials are occasionally given: e. g., the diary of a Chinese diplomat (Ho Ju-chang 何如璋), stories of the Chinese legation, of the Chinese Christian Ma Hsiang-po 馬相伯, of the Chinese students association in Japan, etc. However, these materials are so thrown together, revised into colloquial style, and treated as bits and pieces that it is difficult to use this volume for research purposes.

6.8.5 SANETŌ Keishū 實藤惠秀, comp., for Kokusai Bunka Shinkōkai 國際文化振興會 (KBS), (Chung-i Jih-wen-shu mu-lu) **Chūyaku Nichibun-sho mokuroku** 中譯日文書目錄 (Bibliography of Japanese books translated into Chinese), pub. by KBS, 1945, 204 pp.

Written in Chinese, this basic research tool for the study of Japanese influence on China has a 29-page introduction characterizing the various modern periods of Japanese-to-Chinese translation work and listing the main translating agencies, bookstores and personnel. Books only are listed, not articles, and arranged in the usual categories. They total about 2500 works in Chinese.

6.8.6 SATŌ Saburō 佐藤三郎, "Rekishigaku kankei hōsho Shinayaku mokuroku" 歷史學關係邦書支那譯目錄 ("Ein Register der ins Chinesiche übersetzten japanisch-geschichtlichen Literatur"), **Rekishigaku kenkyū,** 9.10, whole number 71 (Nov. 1939), 1116–1121.

This list of 152 academic works translated, usually under their original titles, from Japanese into Chinese is arranged by major series (e. g. the items in the **Ti-shih hsiao-ts'ung-shu** 地史小叢書 published by the Commercial Press in Shanghai) or by topics (e. g. economic history, social history, etc.) and thus provides a useful basis for study of Japanese intellectual influence in modern China. Chinese title, Japanese author, Chinese translator and Chinese publisher are listed in each case, but no dates or other data. These are all academic works dealing with Japanese, Chinese, Western or world history.

6.8.7 NAGAI Kazumi 永井算己, "Iwayuru Shinkoku ryūgakusei torishi-mari kisoku jiken no seikaku" 所謂清國留學生取締規則事件の性格 (The nature of the so-called incident concerning the regulations for the control of Chinese students in Japan), **Shinshū daigaku kiyō** 信州大學紀要, 2 (July 1952), 11–34.

A study of the protest of Chinese students in Japan against regulations of the Ministry of Education issued in November 1906 to curb revolutionary, anti-Ch'ing activity among them. See also his "Iwayuru Go Son jiken ni tsuite" 所謂呉孫事件に就て ("On the *Wu Sun* Incident"), **Shigaku zasshi**, 62. 7 (July 1953), 649–667.

6.8.8 NAKAMURA Tadayuki 中村忠行, "Chūgoku bungei ni oyoboseru Nippon bungei no eikyō" 中國文藝に及ぼせる日本文藝の影響 (The influence exerted by Japanese on Chinese literature), **Taidai bungaku 臺大文學**, pub. by the Taidai Bungakukai (Literary studies society of Taihoku Imperial University), Taihoku, Taiwan, 7. 4 (Dec. 1942), 214–243; 7. 6 (April 1943), 362–384; 8.2 (Aug. 1943), 86–152; 8. 4 (June 1944), 27–85; 8. 5 (Nov. 1944), 42–111.

By a specialist in this field now at Tenri Daigaku, this important article surveys Japanese influence on China in the late 19th century during the time of the Reform Movement as reflected in writings of K'ang Yu-wei and Chang Chih-tung, translations of Yen Fu 嚴復 and Liang Ch'i-ch'ao, and political and other writings of the latter published particularly during his residence in Japan in journals such as **Hsin-min ts'ung-pao 新民叢報**. Useful bibliographical references are given, and many examples of Japanese passages translated or condensed by Liang into Chinese. Just as the father of the Kuomintang got vital Japanese support, such research demonstrates how the father of Chinese liberalism made use of Japanese materials.

6.8.9 NAKAMURA Tadayuki 中村忠行, "'Shin Chūgoku miraiki' kōsetsu, Chūgoku bungei ni oyoboseru Nippon bungei no eikyō no ichirei" 「新中國未來記」攷說——中國文藝に及ぼせる日本文藝の影響の一例 ("A study of **Hsin Chungkuo Weilai Chi 新中國未來記**, An Instance of the Influence of Japanese Literature upon that of China"), **Tenri daigaku gakuhō, 1. 1** (May 1949), 65–93.

A study in literary history, using the methods of comparative literature and tracing Japanese influence from the turn of the century through a "political novel" of Liang Ch'i-ch'ao, on "New China's future" (1902). The writer, one of the few specialists on Sino-Japanese literary relations, first gives a synopsis of the five instalments of Liang's novel, then goes into the circumstances surrounding its composition (Liang was then 30), notes its Japanese prototypes, and analyzes its literary structure and general significance. The author stresses the historical significance of the literary movement inspired by Liang and others as the preliminary stage in the literary revolution led by Ch'en Tu-hsiu, Hu Shih and Lu Hsün. A model study in a neglected field.

6.8.10 NAKAMURA Tadayuki 中村忠行, "Tokutomi Roka to gendai Chūgoku bungaku" 德富蘆花と現代中國文學 ("Roka Tokutomi and modern Chinese Literature"), **Tenri daigaku gakuhō, 1. 2–3** (Oct. 1949), 1–28; 2. 1–2 (Nov. 1950), 55–84.

This is a detailed and textual study of the influence exercised upon Chinese literature by the writings of the Japanese humanistic novelist Tokutomi Kenjirō 德富健次郎 (pen-name Roka 蘆花), younger brother and critic of the famous journalist-historian Tokutomi Iichirō 德富猪一郎 (pen-name Sohō 蘇峰). The author first tells how the name of the novelist was introduced to the Chinese intellectual scene through the translation, in 1902 and after, by Liang Ch'i-ch'ao and others of his political or detective novels (which were actually Tokutomi's translations of Western works). Then he dwells at length and factually upon the remarkable repercussions caused by Tokutomi's romantic and psychological novel of family life, entitled **Hototogisu 不如歸** (1898) introduced into China through a retranslation by Lin Ch'in-nan 林琴南 in 1908 from an English version, and through a play adapted from the original novel. Finally this article discusses why this famous novel was so enthusiastically received by the Chinese in contrast to another famous novel of the Meiji period of a quite different type, **Konjiki yasha 金色夜叉** ("The golden demon" or "The money-grubber") by Ozaki Kōyō 尾崎紅葉.

6.8.11 NAKAMURA Tadayuki 中村忠行, "'Shinrōma denki' sakki"「新羅馬傳奇」札記 (Notes on the **Hsin Lo-ma chuan-ch'i**), **Gakugei 學藝,** pub. by Akitaya 秋田屋 (in Osaka), 34 (Dec. 1947), 15–22.

Analysis of this unfinished drama written by Liang Ch'i-ch'ao suggests that its technique of story-telling was taken from Japanese models cited by the author.

6.8.12 NAKAMURA Tadayuki 中村忠行, "Ban-Shin ni okeru engeki kairyō undō" 晚清に於ける演劇改良運動 ("Play-Reform-Movement in the Manchu Dynasty"), **Tenri daigaku gakuhō,** 3.3 (Mar. 1952), 37–62; 4.1 (July 1952), 51–78.

As a first instalment centering on the contact between the old Chinese drama and the Japanese theatre of the Meiji period, this article describes the influence exerted through contact with Chinese like Huang Tsun-hsien 黃遵憲 and Wang Tzu-ch'üan 王紫詮, a poet; through their writings and subsequently through those of K'ang Yu-wei on drama reform; and by Wang Hsiao-nung 汪笑儂, a Manchu actor, Liang Ch'i-ch'ao and others in the early 20th century. The movement was stimulated by the late Ch'ing reform program in general and by patrons like the Manchu governor-general at Nanking, Tuan-fang 端方, as well as the opening of modern theaters in Shanghai about 1908. This study explores a new field. See also his article "Ban-Shin ni okeru bungaku kairyō undō" 晚清に於ける文學改良運動 (The literary reform movement in the late Ch'ing period), **Kokugo kokubun 國語・國文,** pub. by the Kokubun Gakkai at Kyōto University, 21.1 (Dec. 1951), not seen by us.

6.8.13 OKAZAKI Seirō 岡崎精郎, "Chūgoku kenkyū ni okeru jitsugaku-teki senku" 中國研究における實學的先驅 ("A Modern Japanese Forerunner in Chinese Studies"), **Tōyōshi kenkyū,** 12.1 (Oct. 1952), 51–67.

Subtitled "With special reference to a letter of Kawashima Jun 河島醇", this interesting article discusses the emergence, in the Meiji period, of a realistic and pragmatic approach to Chinese problems, detached from the earlier Confucian stereotypes. The writings of Fukuzawa Yukichi 福澤諭吉, Taguchi Ukichi 田口卯吉, Tokutomi Iichiro 德富猪一郎 and others are quoted in this connection, and the letter in question, written in 1887 by Kawashima Jun, a treasury official then in Germany, is analyzed against this background.

6.8.14 NAKAMURA Kyūshirō 中村久四郎, "Kinsei Shina no Nippon bunka ni oyoboshitaru seiryoku eikyō" 近世支那の日本文化に及ぼしたる 勢力影響 (The power and influence of modern China over Japanese culture), **Shigaku zasshi,** 25.2 (Feb. 1914), 125–154; 3, 265–296; 4, 448–481; 7, 1011–1021; 10, 1205–1236; 26.2 (Feb. 1915), 131–172.

This long and prolix article loses some of its appeal when discovered to be taking as "modern" anything reaching Japan from the Sung period down to the mid-19th century. But as it is heavily documented, this department-store-like article may be of reference value for further study of any specific topic—academic, political, literary, cultural, religious, linguistic, animal, vegetable, mineral, or miscellaneous.

6.8.15 SATŌ Saburō 佐藤三郎, "Meiji ishin igo Nisshin sensō izen ni okeru Shinajin no Nippon kenkyū" 明治維新以後日清戦争以前に於ける 支那人の日本研究 ("Japanforschung der Chinesen in der Periode nach der Meiji-Restoration und vor dem chinesisch-japanischen Krieg"), **Rekishi-gaku kenkyū,** 10.11, whole number 83 (Nov. 1940), 1147–1187.

This valuable monograph gives a summary of Sino-Japanese contact before the modern establishment of formal relations; traces some of the early Chinese impressions of mid-century Japan, quoting contemporary Chinese accounts; and then takes up the various writings of Chinese who visited or discussed Japan in the last decades of the century (with useful bibliography). The last section then examines several Chinese compendia on Japan.

6.8.16 MIURA Hiroyuki 三浦周行, **Meiji ishin to gendai Shina** 明治維 新と現代支那 (The Meiji Restoration and present-day China), Tōkō Shoin, 1931, 350 pp., illus.

This posthumous volume by a Kyōto professor prints eight lectures on Meiji Japan given in major Chinese universities in 1931 (before Sept. 18), together with several more given on his return to Japan, about contemporary China. The latter give scholarly first hand impressions. This book, as a whole, provides a useful case-study in Sino-Japanese cultural relations.

6.8.17 KŌ Jun-chō 洪淳昶, comp., "Chūgokujin chosaku hōyakusho mokuroku" 中國人著作邦譯書目錄 (Bibliography of works by Chinese translated into Japanese), **Shin Chūgoku,** 2 (April 1946), 61–63; 3 (May 1946), 94–96; 4 (June 1946), 125–128; 5, 157–160; 6, 192–193; 7, 223–225; 8 (Oct. 1946), 253–257; 14 (July 1947), 271–272; 15 (Aug. 1947), 308; 17 (Oct. 1947), 392–396; 18 (Nov. 1947), 445–448.

A list of 641 Japanese translations of modern Chinese works, with a chronological chart (in no. 18) indicating that there was a high point of such work in 1940 (160 items) and hardly a dozen items a year before 1937.

6.9 STUDIES CONCERNING CERTAIN SCHOLARS

Note: Items in this section relate to the ideas of certain outstanding Chinese scholars of the modern period. There is no particular unity among these items, except that each concerns an individual and his work. Arrangement is roughly chronological by persons. Liang Sou-ming is included here (6.9.17–20). But Lu Hsün is in sec. 6.11 on literature (6. 11.7–9).

6.9.1 SUZUKI Torao 鈴木虎雄, "'Jinkyō roshisō' o yomu"「人境蘆詩草」を讀む ("On Huang Tsun Hsien's 'Jen Ching Lu Shih Ts'ao'"), **Shinagaku,** 9.1 (July 1937), 25–48.

A biographical and literary appraisal of the reformer-official Huang Tsun-hsien 黃遵憲 (1848–1905), including his time in Japan.

6.9.2 KAMIYA Masao 神谷正男, "Ryō Kei-chō no rekishigaku" 梁啓超の歷史學 (The historical studies of Liang Ch'i-ch'ao), **Rekishigaku kenkyū,** no. 105 (Dec. 1942), 1069–1096.

A scholarly survey of Liang's writings between the ages of 26 and 57 (also shown in a table), an analysis of the chief tenets of his view of history, and a critique of them.

6.9.3 KUWABARA Jitsuzō 桑原隲藏, "Ryō Kei-chō shi no 'Chūgoku reki-shi kenkyūhō' o yomu" 梁啓超氏の中國歷史研究法を讀む ("On Reading Mr. *Liang Ch'i-ch'ao's* 'A Method of Studying Chinese History'"), **Shinagaku,** 2.12 (Aug. 1922), 883–900.

A critical review of a famous work, by one of the leading Japanese sinologists.

6.9.4 KURAISHI Takeshirō 倉石武四郎, "Ō Shō to Rō Nai-sen" 王照と勞乃宣 (Wang Chao and Lao Nai-hsüan), **Kangakukai zasshi** 漢學會雜誌 (Journal of the Chinese literature society), pub. by the research seminar of the department of Chinese philosophy and literature 支那哲文學研究室 of the Faculty of Letters, Tōkyō Imperial University, 12.1–2 (Dec. 1944), 7–27.

An account of (1) the career of the reformist Hanlin scholar, Wang Chao (1859–1933), who spent two years in Japan at the end of the century and returned to became a pioneer in the movement for literary reform; and (2) the similar efforts of Lao Nai-hsüan, an intransigently anti-Boxer official and the author of a well-known essay on the Boxers.

6.9.5 OKAZAKI Fumio 岡崎文夫, "Shisōka to shite no Shō Hei-rin" 思想家としての章炳麟 (Chang Ping-lin as a thinker), **Geimon,** 7.11 (Nov. 1916), 1090–1102.

An early appraisal of the way in which Chang Ping-lin, as a scholar of the revolutionary era, related himself to the Confucian tradition—one of several studies of him in Japanese.

6.9.6 HONDA Shigeyuki 本田成之, "Shō Hei-rin no bunshū o yomu" 章炳麟の文集を讀む (On reading the collected writings of Chang Ping-lin), **Tōa kenkyū,** pub. by the Tōa Gakujutsu Kenkyūkai 東亞學術研究會, 6.8 (Sept. 1916), 11–20.

Noting Chang Ping-lin's capacity to keep up with Japanese scholarly writings as well as his knowledge, gained through Japanese, of Buddhism and Western culture, the writer reviews his **Chang-shih ts'ung-shu 章氏叢書** and briefly summarizes Chang's views on several philosophical topics.

6.9.7 HONDA Shigeyuki 本田成之, "Shō Hei-rin no gakusetsu" 章炳麟の學説 (The theories of Chang Ping-lin), **Geimon,** 8.2 (Feb. 1916), 122–130; 3 (March), 187–198.

A brief, informed discussion of Chang's approach to the Confucian tradition, his view of human nature, and his theory of nihilism 虚無論.

6.9.8 OJIMA Sukema 小島祐馬, "Shō Hei-rin no 'Hikō' o yomu" 章炳麟の非黄を讀む ("On Reading Chang Ping Lin's Confutation of Huang Tsung Hsi"), **Shinagaku,** 1.3 (Nov. 1920), 230–236.

A note on Chang's criticism of Huang's famous political work, **Ming-i tai-fang lu 明夷待訪録**, which concludes that Chang's views are irrelevant.

6.9.9 OJIMA Sukema 小島祐馬, "Ryō Hei no gaku" 廖平の學 (The learning of Liao P'ing), **Geimon,** 8.5 (May 1917), 426–446.

An interesting intellectual biography of an old-style, productive and representative Szechwan scholar (b. 1852), recounting the development of his views on various classics, his response to K'ang Yu-wei and to modern republicanism, etc.

6.9.10 OJIMA Sukema 小島祐馬, "Roppen seru Ryō Hei no gakusetsu" 六變せる廖平の學説 ("Six Stages in the Development of Mr. Liao P'ing's Opinions"), **Shinagaku,** 2.9 (May 1922), 707–714.

Further analysis of the intellectual biography of the late Ch'ing Confucian scholar, Liao Chi-p'ing 廖季平. See the writer's earlier study of 1917, 6.9.9.

6.9.11 OJIMA Sukema 小島祐馬, "Ryū Shi-bai no gaku" 劉師培の學 (The learning of Liu Shih-p'ei), **Geimon,** 11.5 (May 1920), 355–365; 7 (July), 566–582.

An appraisal of the work of a leading Confucian scholar of the early 20th century. Liu Shih-p'ei (*Tzu*, Shen-hsü 申叙, pen-name Liu Kuang-han 劉光漢) was at first revolutionary and then became conservative, contributed to 82 issues of **Kuo-ts'ui hsueh-pao 國苹學報** between 1905 and 1911, and before his death in 1919 at the age of 35 had produced much further significant work. This is a sympathetic account of his life and ideas.

6.9.12 OJIMA Sukema 小島祐馬, "Kyō Ji-chin no nōsōsetsu" 龔自珍の農宗說 (The theory of *nung-tsung* of Kung Tzu-chen), **Keizai ronsō, 10.6** (June 1920), 765–780.

Kung Tzu-chen (Ting-an 定庵, 1792–1841) is studied here as a scholar interested in statecraft 經世 and political economy who put forward a program to solve the agrarian problem in his theory of *nung-tsung* 農宗. Taking as his theoretical basis the traditional idea of the family based on the ancestor-cult (宗法), he presents a rather moderate and practical scheme of land-equalization.

6.9.13 ANDŌ Hikotarō 安藤彦太郎, "Sai Gen-bai no shōgai to sono hyōka ni tsuite, Sai Shō-shi cho 'Sai Gen-bai gakujutsu shisō denki' oboegaki" 蔡元培の生涯とその評價について──蔡尚思著「蔡元培學術思想傳記」おぼえがき ("The Life of Tsai Yüan-pei and its Estimation, Note on Tsai Shang-ssu, 'Biographical Study of Ts'ai Yüan-pei's Scientific Thought") **Chūgoku kenkyū,** 16 (Sept. 1952), 26–34.

A book report on the above biographical volume (published in Shanghai by T'ang-ti 棠棣 Pub. Co., Oct. 1950, 464 pp.), which summarizes the phases of Ts'ai's career as recounted in it. The volume incidentally, by its selection of subject matter, ties this academic liberal leader firmly into the proto-leftist movement of protest against the KMT at the end of his career in the 1930s.

6.9.14 AOKI Masaru 靑木正兒, "Ko Seki o chūshin ni uzumaite iru bungaku kakumei" 胡適を中心に渦いてゐる文學革命 ("A Literary Revolution in China with Mr. Hu Shih as its central figure"), **Shinagaku, 1.1** (Sept. 1920), 11–26; 1.2 (Oct. 1920), 112–130; 1.3 (Nov. 1920), 199–219.

A rather full account of Dr. Hu's early espousal in 1917 of *pai-hua* literature and the ensuing discussion, including some note of writings of Fu Ssu-nien 傅斯年, Ch'ien Hsüan-tung 錢玄洞 and others in **Hsin-ch'ing-nien.**

6.9.15 NAITŌ Torajirō 內藤虎次郎, "Ko Seki-shi kun no shincho **Shō Jitsu-sai nempu** o yomu" 胡適之君の新著章實齋年譜を讀む ("On Reading Mr. Hu Shih's Latest Work 'A Chronological Sketch of Chang Shih Chai's Life'"), **Shinagaku,** 2.9 (May 1922), 638–654.

A critical review of Dr. Hu's work on the great 18th century scholar, which incidentally indicates something of the relationship between modern Japanese and Chinese sinology.

6.9.16 NOHARA Shirō 野原四郎, "Hitori no kindai Shinashika" 一人の近代支那史家 (A modern Chinese historian), **Rekishigaku kenkyū,** no. 105 (Dec. 1942), 1057–1068.

An appraisal of the historical criticism and views of Ku Chieh-kang 顧頡剛 (known in the West through Dr. A. W. Hummel's **Autobiography of a Chinese Historian,** which translates his **Ku-shih pien 古史辨**). Ku is studied here in the context of his generation, with reference to K'ang Yu-wei, Chang Ping-lin 章炳麟, and others.

6.9.17 OKAZAKI Fumio 岡崎文夫, "Ryō Sō-mei cho 'Tōzai bunka oyobi sono tetsugaku'" 梁漱溟著「東西文化及其哲學」 ("Mr. Liang Sou Ming's View on Oriental and Occidental Culture"), **Shinagaku,** 2.9 (May 1922), 697–701.

An appreciative review of a book which played a role in modern Chinese intellectual history.

6.9.18 KIKUTA Tarō 菊田太郎, "Ryō Sō-mei no sonchi ron" 梁漱溟の 村治論 (Liang Sou-ming's view of village government), **Keizai ronsō,** 52.4 (April 1941), 501–508.

A summary, with useful bibliographical references, of Liang's leading ideas in his famous work at Tsou-p'ing 鄒平, Shantung, and the contemporary debate over them in China. On Liang's Shantung project, see the brief note in **Mantetsu chōsa geppō,** 17.4 (April 1937), 228–230.

6.9.19 KIMURA Eiichi 木村英一, "Ryō Sō-mei no shisō, **Tōzai bunka oyobi sono tetsugaku** ni tsuite" 梁漱溟の思想──「東西文化及其哲學」 について (The thought of Liang Sou-ming, on his book **Culture and philosophy in the East and the West**), **Tōa jimbun gakuhō,** 3.3 (Jan. 1944), 496–542.

As a case study of China's response to the West, this article analyses the thought of Liang Sou-ming (1893—) in his early years as condensed in his influential book **Tung-Hsi wen-hua chi ch'i che-hsüeh** (first ed., 1921). This analysis (pp. 505–542) is preceded by a useful biographical study, one of the few yet made of this significant figure.

6.9.20 ONOGAWA Hidemi 小野川秀美, "Ryō Sō-mei ni okeru kyōson kensetsuron no seiritsu" 梁漱溟に於ける郷村建設論の成立 ("Establishment of the Theory of the rural Reconstruction by Liang Sou-Ming"), **Jimbun kagaku,** 2.2 (Mar. 1948), 86–123.

A valuable analysis of one of the most important modern Chinese thinkers, with useful references. The author takes special note of Liang's inclination to return to the ancient age, as a cultural ideal.

6.9.21 IMAHORI Seiji 今堀誠二, "Chūgoku rekishi gakkai no tembō" 中國歷史學界の展望 (A survey of the historians in China), **Tōa ronsō,** 6 (April 1948), 98–106.

This brief note, completed in May 1946, reflects the false optimism of that period as to the postwar opportunity for Chinese scholarship but comments informatively on wartime trends of thought among Chinese historians, left, right and backward.

6.9.22 TAKIGAWA Masajirō 瀧川政次郎, "En Ko yūki" 燕滬遊記 ("Peking, Tientsin and Shanghai—impressions of Chinese Scholars"), **Shakai keizai shigaku,** 3.10 (Feb. 1934), 1401–1414; 3.11 (Mar. 1934), 1533–1552; 4.1 (April 1934), 103–119; 4.2 (May 1934), 232–241, photos.

The author made his third trip to China in 1933 and here records current information, based on interviews, on the life and work of a dozen individuals, including Ch'en Pao-chen 陳寶琛 at the age of 86, K'o Shao-min 柯劭忞 at 83, Hu Shih, Ch'en Yüan 陳垣, Tung K'ang 董康, and Paul Pelliot.

6.10 THE NEW THOUGHT MOVEMENT

Note: The principal studies of this key topic will be found among the surveys and symposia listed in sec. 6.1 above. What follows here is miscellaneous and secondary. Note also item 1.1.8 (Matsui).

6.10.1 KAMIYA Masao 神谷正男, "Shisō kakumei to sono rekishiteki igi" 思想革命とその歴史的意義 ("Die Ideenrevolution und ihre geschichtliche Bedeutung"), **Rekishigaku kenkyū,** 9.10, whole number 71 (Nov. 1939), 1026–1055.

Discussing the anti-Confucianism of the May Fourth Movement (necessarily on the basis of **Hsin-ch'ing-nien 新青年**), the author deals with (1) the rise of the anti-Confucianist movement as a counter-attack against the mobilization and sponsorship of traditionalistic thinkers by the Yüan Shih-k'ai regime, (2) the main arguments of the anti-Confucianists in their negative criticism of the old ideas, (3) the new view of the world as a positive propounding of new ideals, (4) the influence of the revolution in thought, mainly on the literary revolution, and (5) the historical significance of the new thought. The author attacks the effort of Chinese Marxists to minimize this and stresses the role of the revolution in ideas as the origin and fountainhead of the Renaissance in China.

6.10.2 SANETŌ Keishū さねとうけいしゆう [實藤惠秀], for the Chūgoku Kenkyūjo, ed., **Shin Chūgoku no Jukyō hihan 新中國の儒教批判** (Criticism of Confucianism in the new China), Jitsugyō no Nipponsha, 1948, 180 pp.

This elementary survey has chapters on Confucius in the old China and in modern China, and the relationship of Confucianism to the introduction of modern science, K'ang Yu-wei, Liang Ch'i-ch'ao, **Hsin-ch'ing-nien,** Sun Yatsenism and KMT neo-Confucianism (*pen-wei wen-hua* 本位文化). This appears to be a rather "anti-traditionalistic" interpretation aimed at a middle school audience.

6.10.3 KUMANO Shōhei 熊野正平, "Go-shi fūchō" 五四風潮 ("The Wu Ssu Feng Chao"), **Hitotsubashi ronsō,** 28.6 (Dec. 1952), 595–627.

Subtitled as an investigation into what actually occurred, this is a study not of the "May Fourth Movement" of 1919 in its broader aspects, but rather of the anti-Japanese student demonstration of that period in its factual detail and causal context. The author is a specialist in the Chinese language, also interested in the intellectual history of China. He graduated from the T'ungwen College (Tōa Dōbun Shoin) in Shanghai, and was for many years a professor of that school. Prof. Kumano is now at Hitotsubashi University.

6.10.4 KA Kō (Hua Kang) 華崗, trans. by AMANO Motonosuke 天野元之助譯, **Go-shi undō shi** 五・四運動史 (History of the May Fourth Movement), Osaka: Sōgensha, 1952, 264 pp.

> This translation includes as appendices a useful who's who (25 pp.), glossary of terms and topics (19 pp.), and chronology (8 pp.), added to the original work.

6.10.5 MIKAMI Taichō 三上諦聽, "Minkoku shoki no fujo kaihō undō" 民國初期の婦女解放運動 ("The Woman Suffrage Movement in the Nationalist Regime of China"), **Ryūkoku shidan** 龍谷史壇 ("The Journal of History of Ryūkoku University"), Kyōto: pub. by the Historical Society of Ryūkoku University, 32 (Dec. 1949), 40–53; 34 (Feb. 1951), 48–55.

> Dividing the movement for the "emancipation of women" into periods before, during and after the May Fourth Movement of 1919, the author quotes **Hsin-ch'ing-nien** 新青年 and a few other Chinese periodical sources to trace the main trends of thought about it.

6.10.6 SAKAI Tadao 酒井忠夫, **Shina chishiki kaikyū no minzokushugi shisō** 支那智識階級の民族主義思想 (Nationalistic thought among the Chinese intelligentsia), Tōa Kenkyūjo, vol. 1, 1941, 338 pp., mimeo. (no further volume seen).

> This wartime analysis is concerned with the anti-Japanese movement in China as a first draft volume of a projected larger study. It deals with the period up to the Revolution of 1911 (to p. 209) and then up to May 4, 1919, surveying China's educational development (with tables) and quoting writings of Liang Ch'i-ch'ao, Sun, **Hsin-ch'ing-nien,** and the like, without much inspiration of new ideas. The author distinguishes among the Chinese intellingentsia three main elements—sinocentric, liberal and socialist.

6.11 ON CHINESE LITERATURE

Note: This section barely touches the large field of Chinese modern literature as seen from Japan, since we do not include translations of Chinese literature into Japanese (see item 6.8.17 for a bibliography). Thus far, such translations do not seem to have been studied comprehensively. Of the studies listed below, the first group relate to problems of the language and translation, to which item 6.11.1 provides an introduction. Japanese linguistic and philological work on the Chinese language is of course abundant and often highly competent; we have not tried to include it. The remaining entries deal mainly with specific works or authors—see especially Takeuchi Yoshimi's studies of Lu Hsün (item 6.11.7–9). Many interesting and critical, though brief, studies and comments on modern Chinese literature will be found in the journal **Chūgoku bungaku** (see in sec. 9.9).

6.11.1 KURAISHI Takeshirō 倉石武四郎, **Kanji no ummei** 漢字の運命 (The future fate of the Chinese written characters), Iwanami Shoten, 1952, 200+index 2 pp., illus. (Iwanami shinsho, 93).

This excellent small volume by an outstanding specialist in the Chinese language, both classical and colloquial, gives the Japanese public an interesting analysis of the problem of the Chinese written language in the context of Chinese modern history and society. It has chapters on the nature of *kanji*, the origin and development of phonetic romanization and other transliteration systems in the course of Western contact, a survey of 30 years of the national language (*kuo-yü* 國語) movement, and the CCP *la-ting-hua* or *latinxua* system (pp. 111–152), with many technical examples and illustrations. The last chapter is on the Japanese *rōmaji* movement. Chronology of events, 1839–1951 (pp. 192–200), index of Chinese leaders of the *kuo-yü* movement (2 pp.). Prof. Kuraishi is now in Tōkyō University.

6.11.2 OJIMA Sukema 小島祐馬, "Utility no yakugo ni tsuite" Utility ノ譯語ニ就イテ (On the translation of "Utility"), **Keizai ronsō,** 4.6 (June 1917), 900–909.

A scholarly analysis of the implications of the translations *riyō* (*li-yung* 利用) and *kōyō* (*hsiao-yung* 效用) for the key English term "utility", as judged by uses of the terms in various Chinese classical texts. This appears to be an excellent example of the kind of etymological study needed for the proper understanding of the acceptance of Western ideas into Chinese thought.

6.11.3 YOSHIKAWA Kōjirō 吉川幸次郎, "Hon'yakuron no mondai" 飜譯論の問題 (Problems in the argument about translation), **Chūgoku bungaku,** 72 (May 1941), 84–93. "Hon'yaku jihyō" 飜譯時評 (Current reviews concerning translation), ibid., 76 (Sept. 1941), 256–261; 78 (Nov. 1941), 411–417; 79 (Dec. 1941), 459–467.

These are percipient discussions of the problem of translation from Chinese into Japanese, by a leading scholar of Chinese literature, formerly at the Kyōto Institute and now at the University. The first item is a dialogue carried on by correspondence with the editor, Takeuchi Yoshimi 竹内好, who began the discussion by criticizing Yoshikawa in ibid., 70 (Mar. 1941), 645–649. The argument concerns questions such as word-for-word translation as against semantic or idiomatic equivalence, etc. This should have some parallel relevance to Chinese-English translation problems.

6.11.4 SHIGA Masatoshi 志賀正年, "Chūgoku ni okeru hon'yaku bungaku josetsu" 中國における飜譯文學序説 ("Introduction to Translated Literature in China"), **Tenri daigaku gakuhō,** 3.2 (Jan. 1952), 27–44.

As part 1, "An Etymological Study" 語原考, in a work in progress, this article distinguishes various periods of translation into Chinese and then, with an elaborate series of charts and tables, discusses the etymology and derivative variations of the word "translation" (*fan-i* 飜譯). In the following parts to be published, the author proposes to deal with (1) its development, (2) method, (3) style and (4) translation itself.

6.11.5 KONDŌ Haruo 近藤春雄 **Gendai Chūgoku no sakka to sakuhin**

現代中國の作家と作品 (Contemporary Chinese writers and writings), Shinsen Shobō 新泉書房, 1949, 374+6 pp.

This textbook for Japanese students, completed in 1948, contains much useful data for Western students of modern Chinese literature: biographies of writers (115 pp.), an historical sketch, descriptive bibliography of translations from Chinese into Japanese (52 pp.), a bibliography of translations from Japanese into Chinese (90 pp.), many of them by leading Chinese writers; and brief essays on Hu Shih, Japanese and American influences, Lu Hsün, Yü Ta-fu, Kuo Mo-jo, women writers, and similar topics. Such materials open the door upon the whole untouched question of modern Japanese influence on Chinese literature.

6.11.6 OGAWA Tamaki 小川環樹, "Shōsetsu to shite no Jurin gaishi no keishiki to naiyō" 小説としての儒林外史の形成と内容 ("The Form and Substance of the *Ju Lin Wai Shih* as a Novel"), **Shinagaku,** 7. 1 (May 1933), 49–70.

A careful textual study of this novel of Ch'ing literati, which pursues questions concerning its provenance and author (Wu Ching-tzu 吳敬梓), without, however, going far into its implications as a social document.

6.11.7 TAKEUCHI Yoshimi 竹内好, **Ro Jin** 魯迅 (Lu Hsün), Sekai Hyō-ronsha 1948, 338+index of names 6 pp., photo.

A comprehensive study of Lu Hsün (Lu Sin, Lu Hsin, Chou Shu-jen) by the leading specialist on his works, with a somewhat different approach from the same author's earlier volume of the same title of 1944, 6.11.8.—(1) biography (103 pp.), (2) historical environment, with special reference to the literary revolution (42 pp.), (3) development of his literary activities (91 pp.), (4) the so-called cult of Lu Hsün or Lu Hsün as a symbol of national culture (5 pp.), (5) a biographical chronology (33 pp.), (6) an annotated table of his writings, chronologically arranged (44 pp.), and (7) a pertinent bibliographical guide (16 pp.) to the Japanese translations of his writings and to Japanese and Chinese studies. This is the fullest work on the subject. A new edition of the book, with a slightly revised text and greatly simplified chronology and bibliography, has also been published under another title and by another publisher: **Ro Jin nyūmon** 魯迅入門 (A guide to Lu Hsün), Tōyō Shokan 東洋書館, 1953, 239 pp. Prof. Takeuchi is now at the Tōkyō Metropolitan University (Tōkyō Toritsu Daigaku 東京都立大學).

6.11.8 TAKEUCHI Yoshimi 竹内好, **Ro Jin** 魯迅 (Lu Hsün), Nippon Hyō-ronsha, 1944, 192 pp.

A scholarly appraisal of Lu Hsün with chapters on biographical problems, his works, his main ideas, his view of the relation of politics and literature, etc. The author detects something like a religious consciousness of sin as the moral basis of Lu Hsün's literature. A post-war new edition, slightly revised, in a pocket-book form, has been published by Sōgensha, 1952, 194 pp. (Sōgen bunko 創元文庫, A–108).

6.11.9 TAKEUCHI Yoshimi 竹内好, **Ro Jin zakki** 魯迅雑記 (Miscellaneous notes on Lu Hsün), Sekai Hyōronsha, 1949, 237 pp.

This volume brings together a score of essays on Chinese modern literature. Roughly half deal with Lu Hsün, in his relations with other individuals or in his writings. Other essays concern Lin Yü-t'ang 林語堂, Mao Tun 茅盾, and Chinese and Japanese literature. The author also expresses his own sense of struggle and protest.

6.11.10 MATSUEDA Shigeo 松枝茂夫, "Shū Saku-jin" 周作人 (Chou Tso-jen), **Chūgoku bungaku,** 60 (May 1940), 1–22.

A biographical memoir and appreciation of the brother of Lu Hsün 魯迅 (Chou Shu-jen 周樹人).

6.11.11 HATTORI Takazō 服部隆造, "Kō-Nichi senji no Chūgoku bungaku" 抗日戰時の中國文學 ("Chinese Literature during War against Japan"), **Tenri daigaku gakuhō,** 1.4 (May 1950), 85–116.

A widely informed survey which takes up the development of "reportage" 報告文學 and popular literature, the new drama, the novel, wartime poetry and literary and aesthetic theory, giving numerous examples.

6.11.12 SHIGA Masatoshi 志賀正年, "Chūgoku shimbungei no dansō" 中國新文藝の斷相 ("Glimpses of New Chinese Literature"), **Tenri daigaku gakuhō,** 3.1 (Sept. 1951), 105–111.

A brief schematic effort to note trends in the use of popular speech, new terms, and the like, mentioning briefly the Chinese Christian mentality in its current transition.

6.11.13 HATANO Tarō 波多野太郎, "Chūgoku no gendai bungaku to tochi kaikaku" 中國の現代文學と土地改革 ("The Present-Day Chinese Literature and The Land Reallotment"), **Yokohama daigaku ronsō,** 橫濱大學論叢 ("The Bulletin of the Yokohama City University Society"), pub. by Hyōronsha 評論社, 3.5–6 [whole number, 12–13] (Jan. 1952 [see cover]), 1–26.

An enthusiastic account of the new literature and drama (including **Pai-mao nü** 白毛女, "The white-haired girl"), which accompanied the land reform program.

7. ECONOMIC HISTORY AND INSTITUTIONS

Note: This is the largest section in this volume, as might be expected, and falls rather easily into concrete categories. In particular, the compendia in sec. 7.2 would seem to hold a good deal of raw material for the historian. The post-war spate of studies of North China, some of which are under sec. 7.6 and others under sec. 8.2, are often hard to classify as either "economic" or "social", being in most cases a little of each. A similar ambiguity is evident in our putting "gilds" under sec. 8.5 but "commercial organization" (excepting gilds) under sec. 7.16. In spite of this overlapping, we have avoided the category "socio-economic" because it would have to be called "socio-economical-historical".

7.1 SURVEYS OF ECONOMIC HISTORY

Note: These survey volumes are few in number partly because so many of the general historical surveys, especially of the Marxist type, are heavily economic in subject matter, although not labelled as such. See under sec. 1.1.

7.1.1 KATŌ Shigeshi 加藤繁, **Shina keizaishi gaisetsu** 支那經濟史概說 (A general survey of Chinese economic history), Kōbundō, 1944, 163 pp., illus.

> A completely revised and much enlarged volume, developed from the author's earlier Chinese economic history of 1927, this book is a condensed survey of traditional Chinese economic institutions (population, land system, food production, clothing production, handicrafts, trade, communications, foreign trade, currency, and prices), studied more or less over the whole sweep of Chinese history. Although written by a leading specialist, its scope is perhaps a bit broad for research purposes on modern China. Reviewed by Wada Sei in **Shakai keizai shigaku,** 14.4 (July 1944), 53–54. Note also the same author's "Shina no hōken seido ni tsuite" 支那の封建制度に就いて ("Some remarks on the feudal system in China"), **Shakai keizai shigaku,** 7.9 (Dec. 1937), 991–1000. The recent volume of Dr. Katō's republished articles, **Shina keizaishi kōshō,** 9.8.8, contains many items noted by us elsewhere.

7.1.2 HIRASE Minokichi 平瀬巳之吉, **Kindai Shina keizaishi** 近代支那經濟史 (Modern Chinese economic history), Chūō Kōronsha, 1942, 388 +index 13 pp.

The book opens with an analysis of the Ch'ing dynasty fiscal system with special reference to its military motives and needs. Sections deal with Ch'ing economic policy toward mining, tax policy (salt and customs especially), status of the merchant class (Canton hongs and compradores), types of political intrigues and economic activities of officials, and the problem of industrial enterprise in the late Ch'ing period. On this score the author considers the tea, cotton, silk and pottery industries, especially, and also rural handicraft and spare-time industry and officially sponsored heavy industries. This well informed and scholarly study is based on skillfully selected and extensive sources, and explores many interesting problems. The result is a bold attempt at historical problem-analysis by a scholar versed in economics, who seeks to get a comprehensive grasp of the economic, social and political structure of the Ch'ing period as an approach to understanding the present problem of Chinese native capital formation. Reviewed by Matsumoto Yoshimi 松本善海 in **Shakai keizai shigaku,** 12.8 (Nov. 1942), 996-1001; by Kobayashi Hidesaburō 小林英三郎 in **Tōagaku,** 8 (Sept. 1943), 283-289.

7.1.3 OTAKE Fumio 小竹文夫, **Kinsei Shina keizaishi kenkyū** 近世支那經濟史研究 (Researches in modern Chinese economic history), Kōbundō, 1942, 293 pp.

Professor Otake was for some three decades engaged in research and teaching at the Tōa Dōbun Shoin in Shanghai and this well-known volume of esssays compresses into small compass an insight into the Chinese traditional economy which had been developed over many years. They concern (1) the nature of the old Chinese society (with six reasons why it cannot easily be called "feudal", pp. 3-36); (2) the inflow of foreign silver into China in the Ming and Ch'ing periods (from the Philippines, Japan, England and America, pp. 39-73); (3) fluctuations in the comparative value of gold and silver in the Ch'ing period (compared over the whole length of the dynasty, pp. 77-139); (4) levy of taxes in goods and in money in modern China (as an index to the growth of a money economy, pp. 143-168); (5) opening of new land to cultivation in the Ch'ing (with an estimate of cultivated area, pp. 171-239); (6) population in the Ch'ing period (reviewing the whole field of estimates, pp. 243-293). Each of these key topics in economic history is dealt with authoritatively. Reviewed by Majima Toshio 眞島利雄 in **Shirin,** 28.2 (April 1943) 209-210: by Kobayashi Hidesaburō 小林英三郎 in **Tōagaku,** 8 (Sept. 1943), 283-289 (joint review with 7.1.2). On the author's experience at the Tōa Dōbun Shoin see his memoir, **Shanhai sanjū-nen** 上海三十年 (Shanghai during thirty years), Kōbundō, 1948; rev. by Ikeda Makoto 池田誠 in **Tōyōshi kenkyū,** 11.1 (Sept. 1950), 73-4. Prof. Otake is now at Tōkyō Kyōiku Daigaku.

7.1.4 SHIMIZU 清水泰次, **Chūgoku kinsei shakai keizai shi** 中國近世社會經濟史 (Social and economic history of modern China), Nishino Shoten 西野書店, 1950, 211 pp.

Belying its title, this is a factual study of Ming dynasty economic history, particularly the tax system as the connective element among changing phenomena of landholding and currency. Successive chapters treat the tax system, commutation of tax payments (e. g. paying currency instead of grain), the "single-whip" 一條鞭法 reform of the late-Ming period (systematic commutation of tax-in-kind and labor-service into silver payments), and revenue collections for military purposes. Sources are extensively quoted within the text. Though lacking page references, bibliography or index, the volume presents a painstaking account of a very complex subject, using the Ming History, gazetteers and similar sources and attempting to indicate the relationship between state taxation and social classes. Its content (as well as the anachronistic title) makes this volume of possible interest for late-Ch'ing studies. See review by Yamane Yukio 山根幸夫 in **Shigaku zasshi,** 59. 8 (Aug. 1950), 763–769; by Iwami Hiroshi 岩見宏 in **Tōyōshi kenkyū,** 11. 2 (Mar. 1951), 181–184; by Nakayama Hachirō 中山八郎 in **Shakai keizai shigaku,** 17. 1 (Jan. 1951), 75–82; and by Tanaka Masatoshi 田中正俊 in **Rekishigaku kenkyū,** 147 (Sept. 1950), 55–56.

7. 1. 5 OKADA Takumi 岡田巧, **Kinsei Shina shakai keizai shi** 近世支那社會經濟史 (Social and economic history of modern China), Kyōto: Kyōiku Tosho Kabushiki Kaisha 敎育圖書株式會社, 1942, 365 pp.

A conventional summary of a widely accepted interpretation of Asian history, describing China's feudal society before the inroads of capitalism, the process of capitalist aggression and its effects, including changes in the Chinese family system, population increase, overseas migration, etc. Chinese and Japanese sources are used. This is at best a well-organized textbook, heavily based on secondary writings.

7. 2 GENERAL ECONOMIC COMPILATIONS

Note: The first nine items are general gazetteer-type compilations of data which, while mainly economic, are also politico-socio-geographic, touching on almost everything but the Chinese mind. Similar works on provinces and cities then follow. This section is closely related to sec. 1.2 above, to which it forms a continuation.

7. 2. 1 Tōa Dōbunkai 東亞同文會, **Shina keizai zensho** 支那經濟全書 (China economic series), vols. 1–4, Ōsaka: Maruzen, 1907; vols. 5–12, Tōkyō: Tōa Dōbunkai, 1908; each volume ca. 700–1000 pp., illus.

This series produced by the famous Japanese center of Chinese studies at Shanghai, Tōa Dōbun Shoin, is one of the monuments of Japanese research on modern China comparable to the **Shinkoku gyōseihō,** 1.2.1, or the works of the South Manchurian Railway Co. The first four volumes went through four printings in five months, and even today for the historian the series still contains much of firsthand value. Practically every conceivable economic topic is dealt with and in lieu of attempting to summarize the table of contents, we only call attention to the following sections which would seem in many cases to have been neglected by both Chinese and Western scholars: vol. 1 (pp. 127–174) laborers, (pp. 175–211) capitalists, (pp. 327–436) standard of living of the common people; vol. 2 (pp. 149–326) monopoly merchants (salt and tea), (pp. 327–538) compradors; vol. 3 (pp. 275–440) water transport, (pp. 553–637) the Shansi banks; vol. 5 (914 pp.) railroads; vol. 6 (pp. 567–720) Chinese banks; vol. 7 (pp. 143–210) merchant associations, (pp. 211–260) brokers (*ya-hang* 牙行); and vol. 10 (pp. 367–974) mining. Under the direction of Professor Negishi Tadashi 根岸佶 at the Shanghai center, materials for these volumes were secured through the field trips and graduation theses of some 120 students (listed in vols. 1 and 5). Information on the process of compiling this big work is given in **Zoku tai-Shi kaikoroku,** 4.2.10, vol. 2, pp. 740–741, and in a memoir by Negishi in **Hitotsubashi ronsō,** 23.5 (May 1950), 472–476.

7.2.2 Nisshin Bōeki Kenkyūjo 日清貿易研究所, **Shinkoku tsūshō sōran** 清國通商綜覽 (General survey of the foreign trade of China), vols. 1 and 2 pub. by Maruzen Shōsha Shoten; vol. 3 by Ōkura Shoten 大倉書店, 1892, 1059+666+600 pp., maps, illus.

This ancient compendium has the merit of providing a large collection of data from the relatively unstudied period of the 1880s. In addition to a general geographic survey and vital statistics, volume 1 describes the 25 treaty ports, customs, education, religion, the governmental structure, finance in all aspects, communications, industry, argicultural production, etc., etc., and winds up with what the foreigner needs to know for all aspects of life in China, plus itineraries for all main routes of travel. The second volume is mainly devoted to (necessarily unreliable) history, e. g. of Chinese agriculture, industry and commerce through the ages. Volume 3 then describes in detail a vast number of artisan's and industrial products, as well as products of the Chinese land and sea. This industrious compilation of data on every conceivable aspect of the Chinese scene as of ca. 1890, including the kitchen sinks then available, should provide useful historical evidence on many subjects. The compiling agency, then known as Shinkoku Shanhai Nisshin Bōeki Kenkyūjo 清國上海日清貿易研究所, was the forerunner of the Tōa Dōbun Shoin.

7.2.3 Rinji Taiwan Kyūkan Chōsakai 臨時臺灣舊慣調查會 (Temporary commission of the Taiwan Government-general for the study of old Chinese customs), **Dai-nibu, Chōsa keizai shiryō hōkoku** 第二部調查經濟資料報告 (Report and materials of the economic investigation by the Second Section), 2 vols., 1905, 758+840 pp.

This report of the Second Section of the above commission (as distinct from the First Section, on law, and the Third, on preparation of legislation) presents detailed data under the following major sections: 1) production of crops such as rice, tea, sugar, tobacco, etc., and of gold; 2) production by regions: Keelung, Taipei, etc.; 3) communications: post, telegraph, harbors, navigation, shipping; 4) general economic materials on wages, cost of living, insurance, commercial treaties, and the salt administration. Although out of date in many aspects of form and content, these reports include a wealth of data on contemporary conditions and institutions which should make them of value for historical research.

7.2.4 NEGISHI Tadashi 根岸佶, ed., for Tōa Dōbunkai 東亞同文會, **Shinkoku shōgyō sōran** 清國商業綜覽 (General survey of Chinese commerce and industry), Maruzen, vols. 1–3, 1906, 230+550+401 pp.; vol. 4, 1907, 530 pp.; vol. 5, 1908, 761 pp.

These compilations seek to provide an intimate operational picture of the conduct of economic activity in the late-Ch'ing period: the various types of mercantile enterprise and organization, the system of accounting, with illustrations (vol. 1), the geographical distribution of water-way trade by regions (vol. 2), the railroad system, line by line (vol. 3), the currency and banking system, described for the major centers and for the minor ports and big domestic markets (vol. 4), and the major products and commodities, including opium (96 pp.), manufactured goods, and minerals, discussed geographically. Properly used, these volumes should yield important historical data, including things not known to Western observers of the early twentieth century.

7.2.5 Gaimushō Tsūshōkyoku 外務省通商局 (Foreign Office, Bureau of Commercial Affairs), comp., **Shinkoku jijō** 淸國事情 (Conditions in the Ch'ing empire), 1907, 2 vols.. 1037 and 986 pp.

This compilation of consular reports presents data on current conditions in all the usual meteorological-commercial-industrial-fiscal-educational and other conceivable aspects of life in each of the Japanese consular districts in China—Tientsin, Chefoo, Shanghai, Hankow (vol. 1), Soochow, Hangchow, Nanking, Changsha, Shasi, Foochow, Amoy, Swatow, Canton, and Chungking, all of which may have interest as reflecting the pre-revolutionary situation of a few years before 1911.

7.2.6 Tōa Dōbunkai 東亞同文會, comp., **Shina shōbetsu zenshi** 支那省別全誌 (Comprehensive gazetteer of the various provinces of China), pub. by the Tōa Dōbunkai, 18 vols., 1917–1920, ea. vol. ca. 900–1200 pp., maps and illus.

These comprehensive volumes compiled at the Tōa Dōbun Shoin in Shanghai, with Ōmura Kin'ichi 大村欣一 (to vol. 5) and Yamazaki Chōkichi 山崎長吉 (from vol. 6) as the chief editors, constituted the logical and almost inevitable end product of the effort to get a grasp on China through the amassing of data. The series was based on a systematic collection of information and presentation of it in reports by students of the college, who made research trips along the trade routes deep into the rural districts according to a plan carefully prepared by Prof. Negishi Tadashi and aided by governmental subsidies. As used in this compilation, they begin with the fifth graduating class in 1907, and end with the labors of the sixteenth class in 1918. Each province is treated of course in all the aspects deemed important before World War I, including such matters as commercial centers, trade, capital city and various hsien, communications of all kinds, production, commercial customs and organization, industry and mining, currency (in the days of the variable tael), and the like. Numerous maps were also compiled as part of the project, from which it is evident that many other compilations, such as those of Negishi Tadashi, were also put together. Superficially one gets the impression that the coverage of geography and communications—routes, transport facilities, currency, and other matters within the observation of itinerant note-takers—is one of the stronger features of these volumes.

7.2.7 Shina Shōbetsu Zenshi Kankōkai 刊行會, ed., **Shinshū Shina shōbetsu zenshi** 新修支那省別全誌 (New and revised edition, comprehensive gazetteer of the various provinces of China), pub. by Tōa Dōbunkai (and sold by Maruzen), 8 vols., 1941–1944, ea. vol. ca. 600–1200 pp.

With a foreword by Premier Konoe Fumimaro 近衞文麿, who was also the president of the Tōa Dōbunkai, this revised edition has Yonaiyama Tsuneo

米內山庸夫 (to vol. 3), an old China hand and ex-foreign service official, and Baba Kuwatarō 馬場鍬太郎 (from vol. 4) as its chief editors. It makes use of more than two decades of studies accumulated at the Tōa Dōbun Shoin and so constitutes an entirely new work. Although planned in 22 volumes, to be published with the help of the Asia Development Board 興亞院 and the Foreign Office, this project was cut short by events in 1944. Volumes completed are 1 and 2 for Szechwan: 3, Yunnan; 4 and 5, Kweichow; 6, Shensi; 7, Kansu and Ninghsia; 8, Sinkiang—in other words, the provinces of Free China which lay beyond Japan's wartime contact. Nevertheless they are based largely on the students' field reports, from the pre-war years, supplemented by library research. Contents are similar to the earlier edition, emphasizing communications, products, industry and economy, but with the addition of some interesting sections on governmental organization and administration (e. g., on the revival of the *pao-chia* system as an anti-communist measure in Kweichow and Shensi provinces). Indices are also added. We are informed, though we have not been able to confirm it, that one or two more volumes have been published since the end of the war.

7.2.8 NISHIKAWA Kiichi 西川喜一, **Shina keizai sōran** 支那經濟綜攬 (Economic handbook for China), vols. 1 and 2, Shanghai: Shanhai Keizai Nippōsha 上海經濟日報社, 1922, 600+587 pp.; vols. 3 and 4, Shanghai: Nippondō 日本堂, 1924, 575+645 pp., vol. 5, the same, 1925, 611 pp.

The author worked for 6 years at the Japanese chamber of commerce in Hankow and industriously if somewhat indiscriminately compiled a comprehensive work on China, Mongolia, Manchuria and Sinkiang: climate, ecology (as it would now be called), recent history, the interests and rivalries of the powers and the general system of extraterritoriality (vol. 1), loans and railroads, the Chinese people, their education, the customs service, banking and currency (vol. 2), the cotton industry (vol. 3), mining, raw cotton (vol. 4), Yangtze River trade and the economy of Hankow (vol. 5). The latter volumes contain extensive and presumably useful information on the cotton industry and the economic activities centering upon Hankow in particular.

7.2.9 Nikka Jitsugyō Kyōkai 日華實業協會 (Sino-Japanese industrial association), comp., **Shina kindai no seiji keizai** 支那近代の政治經濟 (Politics and economics of modern China), Gaikō Jihōsha 外交時報社, 1931, 1042+14 pp., many tables, charts.

This association was headed by Viscount Shibusawa Eiichi 澁澤榮一 and had been founded in 1920 to represent Japanese business, banking and industrial circles in seeking to improve economic relations with China and combat the anti-Japanese boycott there. This volume commemorates its tenth anniversary by describing in detail the current state of China's (1) finance and economy and (2) politics and foreign relations. In so doing it compiles extensive accounts of contemporary institutions and events (e. g. the Nanking Government and its finances and policies, the Kemmerer Commission recommendations of 1930, posts and communications, foreign investments, etc., etc.; and the history of the Kuomintang and its opponents, the rights recovery movement, and the like).

7.2.10 KANDA Masao 神田正雄, **Shisenshō sōran** 四川省綜覽 (A general survey of Szechwan province), Kaigaisha 海外社, 1936, 1019 pp., illus., map.

The author spent three years in Szechwan about 1906 and traveled there again in 1935 to collect materials for this comprehensive compilation. Like his subsequent volume on Hunan, 7.2.11, it covers all the easily imaginable aspects of the terrain, cities, and economic products and activities of all kinds, with final attention to Mt. Omei 峨眉 (pp. 932–988). Matters of local government are taken up hsien by hsien.

7.2.11 KANDA Masao 神田正雄, **Konanshō sōran** 湖南省綜覽 (A general survey of Hunan province), Kaigaisha 海外社, 1937, 1052 pp., illus., map.

Based on materials collected by the author on four field trips to the province, with help from the local scholars and officials, this volume records a useful bibliography of these works (pp. 8–9) and condenses the available data (perhaps a bit uncritically) into the usual sections on all aspects of Chinese life, particularly the economy and local customs. The author was a journalist with the **Asahi Shimbun** and later became a Diet member. This volume, based partly on mss. of Chinese scholars, is considered superior to the companion work on Szechwan (7.2.10).

7.2.12 Shinkoku Chūtongun Shireibu 清國駐屯軍司令部 (Headquarters of the army stationed in China), comp., **Tenshin shi** 天津誌 (A gazetteer of Tientsin), Hakubunkan, 1909, 662 pp., maps, many photos.

Taking the similar gazetteer of Peking, 1.2.4, as its prototype, this smaller volume required some five years to be produced, under military auspices, by some three dozen researchers. It similarly deals with all aspects of the ·city, but principally with its economic life, including, for instance, aspects like the K'ai-p'ing 開平 mines (pp. 591–6).

7.2.13 ICHIKI Yoshimichi 市來義道, comp., for the Nankin Nippon Shōkōkaigisho 南京日本商工會議所 (Japanese Chamber of Commerce and Industry of Nanking), **Nankin** 南京 (Nanking), Shanghai: pub. by the chamber, 1941, 688+32+22 pp., many tables and illus.

This gazetteer of Nanking under the Wang Ching-wei regime, produced during the Japanese wartime ascendancy there, describes the many aspects of that place: geography, history, communications of all kinds, government both central and local, economy in detail, foreign relations briefly, Japanese residents, etc., plus a hopeful tourist's guide. Many photos. Sections on local economic activity probably provide as much data as is available anywhere.

7.2.14 IDE Kiwata 井出季和太, **Minami Shina no sangyō to keizai** 南支那の産業と經濟 (The production and economy of South China), Ōsakayagō Shoten, 1939, 675 pp., maps, tables, many photos.

This industrious compilation presents a great many facts on the various economic aspects of Fukien, Kwangtung (pp. 207–462), and Kwangsi, with general chapters on South China trade (591–628), and relations with the overseas Chinese. Table of contents 28 pp.

7.3 ECONOMIC METHODOLOGY AND STATISTICS

Note: The study of population, on which we have found rather little, is included here, as well as accounting. On accounting note also 7.16.2 (Negishi).

7.3.1 ADACHI Ikutsune 安達生恒, **Shōgyō shihon to Chūgoku keizai** 商業資本と中國經濟 (Commercial capital and the Chinese economy), Yūhikaku, 1953, 212 pp.

Subtitled "A research concerning the modernization of the Chinese economy", this study was developed from a paper delivered at the Kyōto Institute in 1948 (following study in China in 1944–45), and deals necessarily with precommunist China. The author divides this sophisticated study into (1) "actual conditions" (i. e., those of permeating control by commercial capital), (2) "basic factors" (making the control by commercial capital possible), and (3) "functions" (of commercial capital in the Chinese economy under the aggressive impact of foreign capital). In this framework the author criticizes, on the one hand, the approach of those like Magyar who assert the decisive importance of the imperialistic penetration of foreign capital as the essential factor in breaking down the stationary and "feudal" economy of old China, and, on the other hand, the approach of those like Wittfogel who, he says, in the last analysis posit a sort of naturalistic determinism (e. g. their emphasis upon "irrigation" and "water control"). The author then tries to view the structure of China's economy and its transformation (i. e., "modernization") as promoted mainly from within by the native commercial capital.

7.3.2 KASHIWA Sukekata 柏祐賢, **Keizai chitsujo kosei ron, Chūgoku keizai no kenkyū** 經濟秩序個性論——中國經濟の研究 (On the individual character of the economic order, a study of the Chinese economy), Jimbun Shorin 人文書林, vol. 1, 1947; vols. 2 and 3, 1948; total 930 pp.

This is a methodological work in economic theory by a Kyōto professor interested particularly in attempting to establish a frame of reference useful for the understanding of the individuality of any specific economic order. He therefore devotes his first volume entirely to theory, his second to testing certain hypotheses against the Chinese social scene (e. g. economic ethics based on the idea of *pao* 包), and the third to a more factual study of agriculture, especially in North China. This big work seems to be a comprehensively rearranged and refined version of the author's earlier production entitled "Hoku–Shi no nōson keizai shakai", 7.5.3.

7.3.3 TANAKA Tadao 田中忠夫, **Shina keizai no hōkai katei to hōhō ron** 支那經濟の崩壊過程と方法論 (A treatise on the progressive breakdown of the Chinese economy and the methodology [for analysis of it]), Gakugeisha 學藝社, 1936 (reprinted 1937, 782 pp.).

With a 21-page table of contents, this massive tome seeks to settle the China question once and for all and accordingly begins with several chapters on methodology inasmuch as research on China is as disorganized as China itself. Subsequent sections deal with the economy, agriculture, and currency and fiscal matters (pp. 599–782). While somewhat disorganized by undigested Marxist problem-analysis, this volume presents materials on a miscellany of subjects, e. g., the *piao-chü* 鏢局 or forwarding firms (pp. 683–694), the problem of opium cultivation (pp. 295–333), or the overseas Chinese of Southeast Asia in the great depression (pp. 232–240). The author also wrote **Shina bukken kanshū ron 支那物權慣習論** (On Chinese customs concerning real rights), Shanghai: Jih-pen t'ang shu-tien (Nippondō Shoten) 日本堂書店, 1925, 434 pp., among other works. See also 8.2.21.

7.3.4 KAMIMURA Shizui 上村鎮威, "Chūgoku no shūkakudaka tōkei" 中國の收穫高統計 ("Yield statistics in China"), **Jimbun kagaku,** 2.1 (Dec. 1947), 22–40.

This brief article appraises the nature and value of crop yield statistics in China, through an analysis of the actual process of statistics-making as carried out by the government agencies. It concludes that the appalling unreliability of Chinese official statistics is due to deep-rooted evils inherent in the Chinese bureaucratic system.

7.3.5 FUJII Hiroshi 藤井宏, "Mindai dendo tōkei ni kansuru ichi-kōsatsu" 明代田土統計に關する一考察 ("A Critical Examination of the Statistics of the Cultivated Fields in the Ming Dynasty"), **Tōyō gakuhō,** 30.3 (Aug. 1943), 386–419; 30.4 (Aug. 1944), 506–533; 31.1 (Feb. 1947), 97–134.

This important article, by an outstanding specialist in the economic history of the Ming period, is noted here because it provides a basis for a similar approach to the Ch'ing period, which has not yet been made. The author examines the figures submitted by the officials of Hukuang, Honan, Kwangtung, and Kiangsi in particular, and the figures given in the **Ta-Ming hui-tien 大明會典** and other sources, for various periods, and finds serious discrepancies among them.

7.3.6 KAMIMURA Shizui ("Shidzui") 上村鎮威, "Chūgoku no jinkō tōkei" 中國の人口統計 ("The Statistics of Population in China"), **Kōchi daigaku kenkyū hōkoku, Jimbun kagaku 高知大學研究報告──人文科學** ("Reports of the Kōchi University, Cultural Science"), 2 (Mar. 1952), 19–30.

This attempt to estimate the degree of error in the modern Chinese population figures goes over a considerable body of literature on the recent period concerning farm families and suggests, through an analysis of the process of making statistics, a possible range of error of 8 percentage points (+3 to −5%) in the estimates of farm population. The author thus infers a lower degree of error than is commonly assumed, for certain stated reasons.

7.3.7 HORIE Yoshihiro 堀江義廣, "Shina ni okeru kaikeishi seido ni tsuite" 支那に於ける會計師制度に就いて (On the accountant system in China), **Shina kenkyū,** 55 (May 1940), 187–282, tables.

A report on the modern development of the accountant's profession in China, the regulations established by successive government ministries, the accountants' associations and the nature, scope, and qualifications of the profession.

7.3.8 ARIMOTO Kunizō 有本邦造, "Shina koyū no bokihō gaisetsu" 支那固有の簿記法概説 (An outline of the peculiarly Chinese system of book-keeping), **Shina kenkyū,** 18 (special Shanghai research number, Feb. 1930), 591–653; "Shina kaikeigaku no kompon mondai" 支那會計學の根本問題 (The fundamental problems of the Chinese science of accounting), ibid., 22 (May 1930), 483–536; "Shina koyū no kaikei to sono kairyō mondai" 支那固有の會計と其改良問題 (The peculiarly Chinese method of accounting and the problem of its reform), ibid., 23 (July 1930), 91–112; "Shina koyū no boki ni kankei aru chōtan gaisetsu" 支那固有の簿記に關係ある賬單概説 (A general treatise on the bills [*chang-tan* 賬單] relating to the peculiarly Chinese system of book-keeping), ibid., 23 (July 1930), 219–262, illus.

These long articles discuss the currency factors which predetermine accounting systems (such as the necessity of transfers from taels to dollars) and go into the ramifications of contemporary Chinese book-keeping, with a wealth of illustrations, as well as tracing the historical origin of the traditional Chinese system and analyzing its defects. The last article describes 16 kinds of bills, checks, receipts and similar documents. The author also published a book, **Shina kaikeigaku kenkyū** 支那會計學研究 (Researches in the Chinese science of accounting), Daidō Shoin 大同書院, 390 pp., which we have not seen but which apparently includes these and other articles.

7.4 GEOGRAPHICAL STUDIES

Note: Geography being both a natural and a social science, we have barely touched upon it and have not gone into the major geographical journals. For Japanese studies of Chinese geography, see the pages of **Chirigaku hyōron** 地理学評論 ("The geographical review of Japan"), ed. and pub. by Nippon Chiri Gakkai 日本地理学会 ("The association of Japanese geographers"), with offices in the Department of Geography, Faculty of Science, Tōkyō University, monthly, 1.1 (March 1925)—26.6 (June 1953); together with **Chirigaku hyōron sōsakuin** 総索引 (A general index for [vols. 11–20 of] the **Chirigaku hyōron**), comp. and pub. by the same, 1951, 151 pp. A survey book by Kobori Iwao 小堀巖 is expected to be published in 1954 and will be useful as a bibliographical guide: **Chūgoku** 中国 (China), in 2 or 3 vols, to be included in a forthcoming series of geographical studies, published by Kokon Shoin 古今書院. For recent bibliography, see Jimbun Chiri Gakkai 人文地理学会, comp., **Chirigaku bunken mokuroku 1945–1951** 地理学文献目録 1945–1951 (Bibliography of geographical studies for the years 1945–1951), Yanagihara Shoten 柳原書店, 1953, 148 pp. This is a list of 4923 items, together with indices indicating about 120 items on China, and many more on

Asia. Note that extensive data of a geographic nature is included in the gazetteers listed in sec. 7.2 above, especially 7.2.6.

7.4.1 BABA Kuwatarō 馬場鍬太郎, **Shina keizai no chiriteki haikei** 支那經濟の地理的背景 (The geographical background of the Chinese economy), Shanghai: Tōa Dōbun Shoin, 1936, 299 pp.

Building on his previous experience in compiling geographic data, and influenced also by the more recent example of G. B. Cressey and other Western geographers, the author here takes up the range of topics of modern geographical interest: expansion of China's boundaries, geographic regions, factors of soil, climate, irrigation, and natural resources. The volume is well arranged, though the author seems to have lacked basic scientific training, and gives useful bibliographic references, many of them to works in English.

7.4.2 FUJITA Motoharu 藤田元春, **Kōga kadō hensen no chibungakuteki kōsatsu** 黃河河道變遷の地文學的考察 (A geomorphological study of changes in the course of the Yellow River), **Shirin,** 7.2 (1922), 189–218, maps.

A survey which is informed by both historical and geographical knowledge of the subject. See the author's comparable study "Shina daiunga no chirigakuteki kōsatsu" 支那大運河の地理學的考察 (A geographical study of China's Grand Canal), ibid., 11.3 (1926), 375–399. Both articles touch on the late-Ch'ing but are mainly devoted to a general survey of earlier times.

7.4.3 IKEDA Shizuo 池田靜夫, "Kōnan kurīku bunkashiron" 江南クリーク文化史論 ("Signification des 'creeks' dans l'histoire de la civilisation de Kiangnan"), **Tōa keizai kenkyū,** 22.1 (Jan. 1938), 77–108.

Discusses the geographic origin, omnipresent multiplicity, and relation of these Yangtze delta waterways to the farm economy and to the cities.

7.4.4 YONEKURA Jirō 米倉二郎, "Yōsukō sankakusu heiya no kaihatsu to kurīku no tenkai" 揚子江三角洲平野の開發とクリークの展開 ("Exploitation of the delta in Yang-tze-kiang and development of the creeks"), **Shirin,** 23.2 (April 1938), 375–384, map.

With reference to writings of J. Sion and G. B. Cressey the author quotes Chinese sources indicating the historical development of these waterways.

7.4.5 MIKAMI Masatoshi 三上正利, "Shinkyō shūhen no kōtsū" 新疆周邊の交通 ("The Marginal Traffic of Sin-Kiang"), **Shirin,** 27.3 (July 1942), 198–232, maps.

This geographical study notes the routes of communication from Chinese Turkestan to India, Russia, and China and appraises the feasibility of rail and motor connections with Russian and Chinese railheads, citing a good deal of bibliography.

7.4.6 MOMOSE Hiromu 百瀨弘, "Shimmatsu Chokureishō no sonzu sanshu ni tsuite" 清末直隸省の村圖三種について (On three village maps of Chihli province in the late Ch'ing), pp. 841–860 in **Katō hakushi kanreki kinen Tōyōshi shūsetsu**, 9. 8. 5.

A technical examination of certain maps mainly preserved at the Tōhō Bunka Gakuin, Tōkyō, noting the vital statistics and other data recorded on them, as an indication of their use in administration, and evaluating their usefulness as materials for socio-economic history.

7. 4. 7 KIKUTA Tarō 菊田太郎, "Hoku-Shi keizai shakai ni taisuru shizen no seiyaku" 北支經濟社會に對する自然の制約 (Natural limitations on the economy and society of North China), **Tōa jimbun gakuhō**, 1.3 (Dec. 1941), 599–626.

This article discusses the influence of geographical factors such as movements of the earth's crust, climate, soil erosion and silting.

7. 5 FARM ECONOMY

Note: These materials mainly concern agricultural production and technology, but inevitably interpenetrate those on landholding in the next section. The work of Amano Motonosuke is outstanding in this field. Note also 7. 10. 1 (Nishijima).

7. 5. 1 AMANO Motonosuke 天野元之助, **Shina nōgyō keizai ron** 支那農業經濟論 (A treatise on the Chinese farm economy), Kaizōsha 改造社, vol. 1, 1940, 734+bibliography and index 27 pp.; vol. 2, 1942, 736+bibliography and index 37 pp.

The author spent 22 years (1926–1948) in research under the SMR research bureau at Dairen, traveling widely in China and publishing extensively in this field (see preface for listed items). These volumes put together a great body of data and observation on the system of land use, rural classes, tenancy, hired labor, subsidiary farm industries (all in vol. 1), agrarian taxation and finance, and production of commercial goods. The projected third volume of this trilogy was published under the title **Chūgoku nōgyō no shomondai**, 7. 5. 2. This work makes very wide use of Chinese sources (periodicals and reports) not available in Japan, and is useful as a tremendous compilation of factual data. Dr. Amano is now in the Jimbunkagaku Kenkyūjo, Kyōto University.

7. 5. 2 AMANO Motonosuke 天野元之助, **Chūgoku nōgyō no shomondai** 中國農業の諸問題 (Problems of Chinese agriculture), Gihōdō 技報堂, vol. 1, 1952, 374+21 pp.; vol. 2, 1953, 302+16 pp.

This is the originally projected third volume of the author's **Shina nōgyō keizai ron**, 7. 5. 1. Vol. 1 deals principally with (1) agricultural production (grains, beans, roots, fibres, tea, tobacco, oils, etc., and cotton and tobacco for market) and (2) the various aspects of peasant life, mainly material (food, clothing, etc.) but also annual customs, beliefs and the like. A supplementary section (pp. 335–366) summarizes Chinese reports of rural changes since the "Liberation". Vol. 2 deals with (1) agricultural products as transformed into merchandise, (2) primitive markets in rural districts, such as *shih-chi* 市集 and *miao-hui* 廟會, (3) trade in rural markets, (4) sources of mercantile profit, and (5) rice circulation, its mechanisms and costs. Pages 253–299 discuss agricultural problems since the "Liberation". Bibliographical essays and lists in both volumes total 37 pp.

7.5.3 KASHIWA Sukekata 柏祐賢, **Hoku-Shi no nōson keizai shakai, sono kōzō to tenkai** 北支の農村經濟社會——その構造と展開 (The rural economy and society of North China, its structure and development), Kōbundō Shobō, 1944, 458 pp.

Developed from a series of some twenty articles, this volume uses the Western and Asian publications in this field to present an analysis of (1) the traditional, self-sufficient and under-capitalized farm economy of North China, which was, however, part of a money economy; (2) the factors of cooperation, finance, technology, etc., which play a part in its modern agrarian development; and (3) problems concerning population pressure. The literature of the subject, as of the time of writing, appears to have been rather fully used, through study at the Institute of Humanistic Sciences (Jimbunkagaku Kenkyūjo) at Kyōto, which also published the volume under a different title, 7.5.4. This wartime study was aimed at providing a basis for future development of North China in the Japanese sphere.

7.5.4 KASHIWA Sukekata 柏祐賢, **Hoku-Shi nōson keizai shakai no kōzō to sono tenkai** 北支農村經濟社會の構造とその展開 (The structure of the rural economy and society of North China and its development), Kyōto Teikoku Daigaku, Jimbunkagaku Kenkyūjo 京都帝國大學人文科學研究所 (Institute of Humanistic Sciences, Kyōto Imperial University), 1944, 437 pp. (Research reports of the Institute, no. 1.)

This is textually the same book as the volume of similar title published in Tōkyō six weeks later by Kōbundō, 7.5.3, omitting only a few pages of back matter appearing in the latter work.

7.5.5 Minami Manshū Tetsudō Kabushiki Kaisha Chōsabu, ed., **Kita-Shina no nōgyō to keizai** 北支那の農業と經濟 (The agriculture and economy of North China), Nippon Hyōronsha, 1942, 2 vols., total 903 + indices of tables 13 + bibliog. 14 pp., maps, many tables.

This factual compilation is made up of the contributions of 8 members of a research team at the SMR research institute on the North China economy, and is accordingly of rather uneven quality. Vol. 1 deals with the natural and technological bases of North China agriculture, vol. 2 with its social and economic characteristics (pp. 327–556), farm management and farm economy, and the whole question of livestock and a livestock economy (pp. 769–903).

7.5.6 Mantetsu Chōsabu 滿鐵調査部, **Shina tochi mondai ni kansuru chōsa shiryō** 支那土地問題に關する調査資料 (Research materials concerning the Chinese land problem), printed by the SMR, 1937, 982 pp.

A weighty collection of local hsien government and other documents, investigator's field reports and translations, picked up mainly in Peking and North China and compiled in larger reports by Amagai Kenzaburō 天海謙三郎 and two colleagues. This activity seems to have been, in a sense, a forerunner of the wartime SMR field investigation and brings together many details of the rural scene.

7.5.7 FUKUSHIMA Yōichi 福島要一, "Kahoku nōgyō no gijutsu suijun" 華北農業の技術水準 ("The level of agricultural technology in North China"), **Tōyō bunka**, 4 (Nov. 1950), 61–93, illus., maps and tables.

This is a condensed version of the unpublished report of the author's thorough investigation, in June 1940, of a single farmstead in a village near Tsinan (Chi-nan 濟南). This farm is analyzed as a typical case against the background of general agricultural conditions in North China, which are briefly summarized. Vital statistics, conditions of dwelling and of land use, crops and farm technology are all examined and the importance of irrigation is repeatedly stressed. The final section mentions also the possible future development, in North China, of the ammonium sulphate industry, which was projected by the Japanese during the war.

7.5.8 Minami Manshū Tetsudō Kabushiki Kaisha Shanhai Jimusho Chōsashitsu 南滿洲鐵道株式會社上海事務所調查室, comp., **Kōsoshō Shōkōken nōson jittai chōsa hōkokusho** 江蘇省松江縣農村實態調查報告書 (A report of agrarian field investigation in Sungkiang hsien, Kiangsu province), Shanghai: pub. by the Shanghai branch office, 1941, 233 pp., illus., maps, many tables; **Kōsoshō Mushakuken** 無錫縣 (Wusih hsien) **nōson jittai chōsa hōkokusho,** pub. by the same, 1941, 156 pp., illus., maps, many tables; **Kōsoshō Nantsūken** 南通縣 (Nan-t'ung hsien) **nōson jittai chōsa hōkokusho,** pub. by the same, 1941, 186 pp., illus., maps, many tables.

These detailed field surveys were led by Amano Motonosuke, who had many years experience as a membr of the SMR research staff in Dairen. While each study presents details concerning local population, land use, tenancy, cultivation, labor, living standards and similar matters, the main value of each lies in its treatment of the special production of each locality: in Sungkiang, rice; in Wusih, silk; and in Nan-t'ung, cotton.

7.5.9 NEGISHI Benji 根岸勉治, **Minami Shina nōgyō keizai ron** 南支那農業經濟論 (On the farm economy of South China), Taihoku: Noda Shobō 野田書房, and Tōkyō: Maruzen, 1940, 598 pp., maps, tables.

The author traveled as an official researcher under the army in South China and was assisted by the Taihoku Imperial University. His book outlines the semi-colonial status of South China under foreign (Western) influence in a brief historical survey, studies the overseas Chinese connection with the region, and then expatiates on the agrarian production and resources of Kwangtung at considerable length—touching most aspects of land use, credit and class structure—and then doing the same for Hainan Island (pp. 333–598). Much data is presented.

7.5.10 KASHIWA Sukekata 柏祐賢, "Hoku-Shi nōgyō ni okeru shōhinseisan no kichō" 北支農業に於ける商品生產の基調 (The underlying circumstances of commercial goods production in North China agriculture), **Tōa jimbun gakuhō,** 1.4 (Feb. 1942), 920–939.

An analysis of certain peculiar aspects of the highly money-economized agriculture in North China. The author points out that the predominantly numerous small-scale petty farmers are producing commercial goods and yet can support themselves only with the help of side-jobs, particulary wage-earning ones, that their markets are locally limited and closed notwithstanding the connections between North China and foreign markets, and that they are ruthlessly controlled by the merchants.

7.5.11 Anonymous, "Teiken ni okeru nōson keizai" 定縣に於ける農村經濟 (Agrarian economy in Ting-hsien), **Mantetsu chōsa geppō,** 16.1 (Jan. 1936), 107–145.

This is a report of a field investigation by members of Daitō Kōshi (*Ta-tung kung-ssu* 大東公司), evidently a Japanese firm, in Tientsin, and may have interest as a somewhat independent appraisal of a Sino-American project. It presents data on the Ting-hsien terrain, population, history, area, division of the land, cultivation, various aspects of rent arrangements and payments, land-lord-tenant relations, hired labor, subsidiary farm industries, taxes, communications, irrigation, public order, banks, cooperatives, and peasant life in general.

7.5.12 (Tōa Kenkyūjo) Daigo Chōsa Iinkai (東亞研究所) 第五調查委員會 (The fifth research committee [of the Tōa Kenkyūjo]), "Shina inasaku no gijutsu suijun" 支那稻作の技術水準 (The technological level of Chinese rice cultivation), **Tōa kenkyū shohō,** 11 (Aug. 1941), 533–576, map, many tables.

This article, certainly written by Yamada Moritarō 山田盛太郎, is a businesslike attempt to calculate the area of rice cultivation, the rice yield from it and hence the product per hectare; then the production by regions and per cultivator. Stated simply, this is no mean task. The studies of Dr. J. L. Buck are used and a map constructed (on the basis of the Herrmann atlas). Final conclusion is that Chinese rice culture is about 65% as efficient as Japanese. This appears to have been a very thorough effort. The author is now a professor at the Faculty of Economics of Tōkyō University.

7.5.13 (Tōa Kenkyūjo) Daigo Chōsa Iinkai (東亞研究所) 第五調查委員會 (The fifth research committee [of the Tōa Kenkyūjo]), "Shina inasaku nōka keizai no kichō" 支那稻作農家經濟の基調 (Basic trends in the economy of rice-cultivating farmers in China), **Tōa kenkyū shohō,** 14 (Feb. 1942), 1–34.

In continuation of the same group's research on the technical level of Chinese rice culture, 7.5.12, this article, also written by Yamada Moritarō 山田盛太郎, goes back over the subject to make an appraisal of its statistical basis, with many further tables and charts and discussions of problems involved.

7.5.14 SATŌ Haruo 佐藤晴生, "Shina keizaishi ni okeru kangai" 支那

經濟史に於ける灌漑 (Irrigation in China's economic history), **Mantetsu chōsa geppō**, 20.4 (April 1940), 63–122.

A detailed scientific and technical study, widely based both on Western technical works and on Chinese classical sources. The author notes references to the development of dry-farming and draws some general conclusions as to the social effects of irrigation.

7.5.15 NAGAO Gyōsuke 長尾行介, "Hoku-Man ni okeru kikai nōgyō" 北滿に於ける機械農業 (Mechanized farming in North Manchuria), **Mantetsu chōsa geppō**, 18.7 (July 1938), 49–65.

This appraisal of current conditions and future possibilities for modern agriculture opens with a useful seven-page review of the growth of mechanized farming in North Manchuria, introduced by Russians and then by Anglo-Saxons, since 1904.

7.6 LANDHOLDING

Note: These studies are, for the most part, products of the South Manchurian Railway Company's long-continued and fruitful research program (see the last entry below for a bibliography), which reached a climax, after three decades, in the wartime project for field investigation of North China villages (on this project see sec. 8.2, and especially item 8.1.1). Materials in this section concern primarily the question of land relations (landlordism, tenancy, rent) but necessarily overlap materials in sec. 7.5 above and 8.2 below, on village studies, as well as sec. 3.6 above, on law. Finally, the post-war communist land reform is touched upon in sec. 7.19 below. Note item 9.6.4, a dictionary of terminology relating to the land, and items 1.4.7 and 1.4.9, on the Sung period.

7.6.1 HATADA Takashi 旗田巍, "Chūgoku tochi kaikaku no rekishiteki seikaku" 中國土地改革の歷史的性格 ("The historical traits of land reform in China"), **Tōyō bunka**, 4 (Nov. 1950), 33–60.

This boldly interpretative article attempts to define the historical types of landownership in North China, as a basis for understanding the Chinese Communist land reforms. The author believes the so-called "feudal" or "semi-feudal" character of Chinese landownership can be clarified only by comprehensive study of both the landlords and the rich owner-farmers. In North China, as compared with other regions, he enumerates: (1) the numerical predominance of owner-farmers, (2) the striking unevenness of land distribution, (3) the class-differentiation among owner-farmers, (4) the underdevelopment of tenant-farmer relations, and (5) as to the types of farm-rent, the predominance of *métayage* and money-rent and the lack of labor-rent. Prof. Hatada then distinguishes three types of current landownership, and tries to find historical and social interconnections among them: (1) "Landownership of the antique family type", in which labor relations are essentially slavery disguised by intrafamilial collaboration. Although now vestigial, this type is historically important as preceding the other two types. (2) "Landownership of the rich owner-farmer type", notwithstanding its seemingly "modern" aspects, is seen to retain pre-modern traits in its labor relations; the political importance of the overwhelmingly great number of poor owner-farmers and employee-farmers is indicated. (3) "Landownership of the landlord type" also retains pre-modern and even pre-feudal characteristics which are analyzed carefully through the various types of farm-rent. The conclusion predicts future difficulties for the Chinese revolution because of the persistence of these very old elements in China's social structure. The author began as a specialist in Korean and early Manchu history and later participated in the famous field-survey of North China sponsored by the South Manchurian Railway Co. and the Tōa Kenkyūjo. He seems to have been deeply influenced by the legal and sociological approaches of such scholars as Suginohara Shun'ichi 杉之原舜一. He is now a professor of Oriental history in the Tōkyō Toritsu Daigaku 東京都立大學 (Tōkyō Metropolitan University).

7.6.2 Isoda Susumu 磯田進, "Hoku-Shi no kosaku, sono seikaku to sono hōritsu kankei" 北支の小作——その性格とその法律關係 (North China tenancy, its nature and its legal relationships), **Hōgaku kyōkai zasshi 法學協會雜誌**, 60.7 (July 1942), 1089–1117; 12 (Dec. 1942), 1999–2040; 61.3 (Mar. 1943), 375–409; 5 (May 1943), 635–672; 7 (July 1943), 939–980. (Unfinished but no continuation has appeared.)

This excellent and clear-cut analysis is one of the most valuable of the extensive and detailed reports based on the field study of Chinese "rural practices" conducted under the wartime auspices of the SMR in North China. It provides through a socio-legal approach a useful basis for comparing the rather loose, impersonal and therefore often harsh North China landlord-tenant relationships with the more personal relationship generally prevailing in Japan—as evidenced, e.g., in the greater use of middlemen 中人 in interpersonal relations in the Chinese village. The author, who is now an assistant professor at the Institute for Social Sciences 社會科學研究所 of Tōkyō University, was trained by the late Dr. Suehiro Itsutarō 末弘嚴太郎, and is now one of the most competent legal scientists in the field of Japanese labor problems.

7.6.3 Yagi Yoshinosuke 八木芳之助, for the Tōa Kenkyūjo, **Keizai ni kansuru Shina kankō chōsa hōkokusho, tokuni Hoku-Shi ni okeru kosaku seido 經濟に關する支那慣行調査報告書——特に北支に於ける小作制度** (A report of investigation of economic customs in China, with special attention to the tenancy system in North China), Tōa Kenkyūjo, 1943, 401 pp.

Based on SMR field investigation reports from China, this volume by a team of Kyōto University economists studies the various types of tenancy, on the part of tenant-laborers, sharecroppers, or tenants who pay in kind or in money; also perpetual tenancy, contracts, deposit money, and various other ramifications.

7.6.4 Kashiwa Sukekata 柏祐賢, "Kita-Shina ni okeru nōgyōsha no seikaku" 北支那に於ける農業者の性格 (The characteristics of farmers in North China), **Tōa jimbun gakuhō**, 1.1 (Mar. 1941), 1–39.

This article analyses the well known situation in which some 70% of the farmers in North China are so-called owner-farmers. Under this single category, the author distiguishes (1) large-scale self-managing farmers who employ many wage-laborers but whose economic motivations are not capitalistic but rather self-supporting and autarchic; (2) small-scale petty farmers who are highly involved in a money-economy and produce for the market, but support themselves only by doing side-jobs such as wage-labor and family industry controlled by the putting-out system; and (3) medium-size genuine owner-famers who are supported by family labor and without side-jobs, but are quantitatively negligible. The author then discusses the status of the farmers under the first two sub-categories, in the context of the Chinese social structure.

7.6.5 KAWANO Shigetō 川野重作, "Kosaku kankei yori mitaru Hoku-Shi nōson no tokushitsu" 小作關係より見たる北支農村の特質 (Characteristics of North China villages with special reference to tenancy), pp. 285–329 in Tōa Kenkyūjo, ed., **Shina nōson kankō chōsa hōkokusho** 支那農村慣行調査報告書 (Reports of investigations of rural customs in China), pub. by the same, series 1 第一輯, 1943, 329 pp.

Based on reports of the SMR field investigation, this article by a specialist in the modern economic approach to the study of rural economy, analyzes several aspects of tenancy in a village in Hopei province,—(1) the predominance of money-rent and its non-modern character as a short term and unstable arrangement, dependent on crop prices, (2) the independence of tenants from landlords in the management of agricultural production, and (3) the resulting stationariness of production. Prof. Kawano is now at the Tōyō Bunka Kenkyūjo, Tōkyō University.

7.6.6 YAGI Yoshinosuke 八木芳之助, and YAMAZAKI Takeo 山崎武雄, for the Tōa Kenkyūjo, **Keizai ni kansuru Shina kankō chōsa hōkokusho, Hoku-Shi ni okeru tochishoyū no idō to bumpu narabini tochi no kaikon** 經濟に關する支那慣行調査報告書——北支に於ける土地所有の移動と分布並に土地の開墾 (A report of investigation of economic customs in China, on the transfer and distribution of land-ownership and opening to cultivation of land in North China), Tōa Kenkyūjo, 1944, 293 pp.

This volume by Kyōto University economists uses SMR field data to describe the transfer or dividing up of landholdings (1) through inheritance, (2) transfer by sale, (3) by pawning (pp. 94–147), or (4) by pledging as security 抵押, as well as the problem of opening new land to cultivation.

7.6.7 KONUMA Tadashi 小沼正, "Kahokushō Jungiken ni oite kankisan seiri no Yōwakyū kōtōchi ni oyoboseru eikyō ni tsuite 河北省順義縣に於て官旗産清理の雍和宮香燈地に及ぼせる影響について (Concerning the effect brought upon the estates of the [Lama temple] Yung-ho-kung by the liquidation of the official and bannermens' estates in Shun-i hsien, Hopei), pp. 263–288 in **Katō hakushi kanreki kinen Tōyōshi shūsetsu**, 9.8.5.

This article, by a participant in the SMR field survey, analyses the complexities of the land system on the estates of a wealthy Lama temple at Peking, formerly patronized by the Ch'ing court, with special attention to the two modes of taxation—collections through the temple itself and through government agencies.

7.6.8 NIIDA Noboru 仁井田陞, "Shina kinsei no ichiden ryōshu kankō to sono seiritsu" 支那近世の一田兩主慣行と其の成立 (The modern Chinese custom of having two owners for one field, and its establishment), **Hōgaku kyōkai zasshi** 法學協會雜誌, pub. by the office of the Hōgaku Kyōkai in the Faculty of Law of Tōkyō Imperial University, 64.3 (Mar. 1946), 129–154; 64.4 (April 1946), 241–261.

This important article describes how in Kiangsu, Kiangsi and Fukien one landowner may have rights to the usufruct of a piece of land while the other landowner is entitled to receive rent from it. This study was later summarized on pp. 290–297 of Dr. Niida's **Chūgoku hōseishi**, 3.6.1.

7.6.9 KEN Yō (Hsien Jung) 憲容, a lecture given by, "Kahokushō no tochi seido ni tsuite" 河北省の土地制度に就いて (On the land system of Hopei province), **Tōa ronsō,** 1 (July 1939), 189–207, illus.

A brief account, with facsimiles of documents, of the origin and development of the special tax status of lands originally belonging to the emperor, the Manchu banners and Buddhist temples. The author was an imperial clansman of the Manchu dynasty and his article reflects a rather pathetic nostalgia for the good old days and bitter feelings against Yuan Shih-k'ai and the warlords.

7.6.10 OZAKI Shōtarō 尾崎庄太郎, "Chūgoku ni okeru jinushiteki tochishoyū no rekishiteki tokushitsu" 中國における地主的土地所有の歷史的特質 (Historical characteristics of landlordism in China), **Chūgoku kenkyū,** 3 (Mar. 1948), 1–31.

A theoretical and conventional discussion of the Chinese type of "feudal" landlordism in the Chinese "manor" (*shōen* 莊園) system.

7.6.11 NAGANO Akira 長野朗, **Shina tochi seido kenkyū** 支那土地制度研究 (A study of the Chinese land system), Tōkō Shoin, 1930, 424 pp.

While this volume discusses the origin of China's land relations (including the legendary *ching-t'ien* 井田 system), questions of land use and ownership, administration and taxation, tenancy and the like, it is now so out of date as to have little value in this rapidly developing field of empirical study. The author, an ex-officer and graduate of the Japanese military academy, studied in Peking and for a time was attached to the SMR research bureau. His books on China (such as **Shina no shakai soshiki** 支那の社會組織 [The social structure of China], 1926, and **Shina no rōdō undō** 支那の勞働運動 [The labor movement of China], 1927), although somewhat journalistic, constituted pioneer research on a number of problems. See review by Nohara Shirō 野原四郎, in **Shigaku zasshi,** 42.2 (Feb. 1931), 253–255.

7.6.12 Mantetsu Chōsabu, comp., **Manshū tochi mondai kankei bunken mokuroku** 滿洲土地問題關係文獻目錄 (Bibliography of materials relating to the land problem in Manchuria), Dairen: pub. by the SMR, 1943, 233+70 pp.

Compiled in the SMR and other libraries in Dairen, Mukden and Hsinking in the period 1940-1943, this bibliography reproduces the complete tables of contents, insofar as they are pertinent, of a large body of Japanese books and articles relating to the land system and attendant problems in Manchoukuo, with reference also to Mongol lands but not to China. Amagai Kenzaburō 天海謙三郎 supplies an article of 70 pages as a "key to the understanding of the Manchoukuo land system."

7.7 MONGOL AND MANCHU LANDS

Note: Mainly on the opening of grazing lands to Chinese colonization. On Mongol-Chinese land relations, see also sec. 3.5. Note also 6.6.12.

7.7.1 Minami Manshū Tetsudō Kabushiki Kaisha 南滿洲鐵道株式會社 (South Manchurian Railway Co.), **Manshū kyūkan chōsa hōkoku** 滿洲舊慣調查報告 (Reports on investigations into the old customs of Manchuria), 9 vols., 1913-1915 (3rd printing, Shinkyō [Hsinking] 新京: Daidō Inshokan 大同印書館, 1935, each vol. 100 to 400 pp., tables and maps).

These volumes of officially-sponsored research, comparable to other large compilations like **Shinkoku gyōseihō, 1.2.1,** are similarly based on the official Ch'ing compendia (e. g. **Ta-Ch'ing hui-tien**) and published documents. The series is divided into two parts, the first six volumes dealing with real estate 不動產 and the last three with rights affecting it. Each volume treats its topic historically and in comprehensive and annotated detail, and reproduces key documents in an appendix. The topics of the volumes are: (1) Kamenofuchi Ryūchō 龜淵龍長, compiler, **Ippan minchi** 一般民地 (A general survey of the lands of the people), including the types of land and land holding in all their variety, land taxation, hunting and grazing land and opening of lands to cultivation. (2) Kamenofuchi Ryūchō, comp., **Mō chi** 蒙地 (Lands of the Mongols), divided by banners, with attention also to tenancy (*tien-hu* 佃戶). (3) Amagai Kenzaburō 天海謙三郎, comp., **Naimufu kansō** 內務府官莊 (Official villages under the Imperial Household Department), their origins, types, distribution, administration, etc. (4) Amagai Kenzaburō, comp., **Kōsan** 皇產 (Imperial properties), including official villages under the various Boards (in Mukden) of Revenue (*Sheng-ching hu-pu kuan-chuang* 盛京戶部官莊), etc. (5) Miyauchi Kishi 宮內季子, comp., **Ten no kanshū** 典の慣習 (Practices concerning mortgages). (6) Miyauchi Kishi, comp., **Ō no kanshū** 押の慣習 (Practices concerning pawning). (7) Sugata Kumaemon 姶田熊右衞門, comp., **Soken** 租權 (Rights of rent). These last three volumes are thinner and more strictly legal in nature.

7.7.2 KAWAKUBO Teirō 川久保悌郎, "Shindai Manshū no ijō" 清代滿洲の圍場 (The official hunting-grounds in Manchuria in the Ch'ing period), **Shigaku zasshi,** 50.9 (Sept. 1939), 1123-1171; 10 (Oct. 1939), 1262-1277; 11 (Nov. 1939), 1350-1375.

Mainly based on Ch'ing documents and **Manshū kyūkan chōsa hōkoku,** 7. 7. 1, this article deals with (1) the original meaning of the term *wei-ch'ang* 圍場, (2) group hunting as practiced among the Manchu bannermen, (3) the development of the hunting system (*hsing-wei* 行圍) as a military training method for bannermen, (4) the institutional establishment of the hunting grounds (*wei-ch'ang*) in Manchuria, and (5) the eventual decline or devastation of the hunting grounds. The final breakdown of the system and the opening of the lands to Chinese peasant colonists in the late 19th century are reserved for another study.

7. 7. 3 HIROTA Gōsuke 廣田豪佐, "Manshū ni okeru shiteki tochi shoyū no hatten" 滿洲に於ける私的土地所有の發展 (Development of private landownership in Manchuria), **Mantetsu chōsa geppō,** 21.8 (Aug. 1941), 58–115.

In the context of the crumbling of the old feudal land system, this study analyzes the original Manchu system of landholding and then traces the process of dismemberment which overtook the official-village (*kuan-chuang* 官莊) system, including the breaking up of the old peasant society, transfers of land, the breakup of family joint ownership, and the change in types of rent.

7. 7. 4 ANZAI Kuraji 安齋庫治, "Shimmatsu ni okeru Suien no kaikon" 清末に於ける綏遠の開墾 (The spread of cultivation in Suiyuan in the late Ch'ing), **Mantetsu chōsa geppō,** 18.12 (Dec. 1938), 1–43; 19.1 (Jan. 1939), 14–62; 2 (Feb. 1939), 36–68.

This historical study traces the development of Ch'ing policy up to the decision of 1903 to promote colonization, and the subsequent process of land opening in four different regions. The author notes the historical background of Ch'ing policy in each case, and some of the problems faced and methods used.

7. 7. 5 ANZAI Kuraji 安齋庫治, "Shimmatsu ni okeru Domokutoku no tochi seiri" 清末に於ける土默特の土地整理 (Revision of the land system in the Tumet area in the late Ch'ing), **Mantetsu chōsa geppō,** 19.12 (Dec. 1939), 29–82.

In this study, a survey of land relationships and the expansion of cultivation by Chinese immigrants in the Tumet area around Kweihua 歸化 during the Ming and Ch'ing periods (pp. 29-52) is followed by an account of the active policy and evils which developed in the late Ch'ing. These reflected a persistent clash between the interests of Mongol landowners and those of Chinese peasants, as well as opposition to any increased burden imposed by the Manchu government, on the part of both Mongols and Chinese on the spot. The author received assistance from the land administration of the Mongolian authorities.

7. 7. 6 ADACHI Ikutsune 安達生恒, "Hoku-Man Mōchi no kaihō katei" 北滿蒙地の開放過程 (The process of opening of Mongol lands in northern Manchuria), **Tōa jimbun gakuhō,** 4.2 (Mar. 1945), 352–390.

A theoretical analysis of the process by which, in the late 19th century, the estates of Mongolian nobles attached to the Cherim League 哲里木盟 were opened to Chinese colonists. The article deals with (1) the historical background of the problem, (2) de facto and then legal openings, and (3) large-scale buying of perpetual lease-holds by the so-called *lan-t'ou* 攬頭, their types and economic functions. Finally the author notes, as the results of the opening process, the decline of the traditional land system and the increase of soybean exports from northern Manchuria.

7.7.7 SHIBA Sakuo 柴三九男, "Shimmatsu ni okeru Kita Manshū Kairin Haisen chihō no tochi kaihatsu" 清末に於ける北滿洲海倫拜泉地方の土地開發 (Opening of land to cultivation in the region of Pai-ch'üan in Hai-lun in northern Manchuria in the late Ch'ing), **Shikan,** 4 (June 1933), 80–115.

On the issues, problems, regulations, procedures and results (in 1904) in this case of colonization.

7.8 SALT TRADE AND ADMINISTRATION

Note: A few general studies, which bring out the importance of salt-smuggling, followed by monographs mainly on the Liang-huai or North Kiangsu producing area. The institutional role of China's traditional salt gabelle seems to have been generally neglected by Western scholars. Note items 8.5.9 on salt merchant gilds, 3.1.8 on salt merchant families.

7.8.1 SUMIYOSHI Shingo 住吉信吾 and KATŌ Tetsutarō 加藤哲太郎, **Chūka engyō jijō** 中華鹽業事情 (Conditions in the Chinese salt industry), Ryūshukusambō 龍宿山房, 1941, 455 pp.

With 63 tables and a 10-page bibliography, this wartime volume surveys factually the Chinese salt industry both in general and by major regions of production (Ch'ang-lu 長蘆, Shantung, Liang-huai 兩淮, Fukien, Shansi, Shensi, Hupei and nearby areas; Szechwan being out of range). The authors were both business employees handling salt in North China. Their modern survey may be compared with the brief research of Katō Shigeshi, based on the salt administration gazetteers of the major producing areas, "Shindai no empō ni tsuite" 清代の鹽法に就いて ("The Salt laws of the Ch'ing period"), **Shichō,** 7.1 (Feb. 1937), reprinted in his **Shina keizaishi kōshō** (9.8.8), 2.493–504, which outlines the Ch'ing system and the evils which led to its reform in the middle of the 19th century.

7.8.2 KUTSUMIZU Isamu 沓水勇, "Hoku-Shi ni okeru shien no ryūtsū, Shina ensei kenkyū no ichikadai" 北支における私鹽の流通——支那鹽政研究の一課題 (The circulation of private [i.e. smuggled] salt in North China, a problem in the study of the salt administration in China), **Tōa kenkyū shohō,** 27 (April 1944), 179–228; 28 (June 1944), 308–352, maps.

Estimating that roughly one-half the salt distributed in North China has been illegally produced and sold in recent times—a significant fact affecting government revenue, official corruption, and banditry—, this article assesses historically the effect of the China Incident on the salt administration; the

North China production and its legal and illegal distribution, by regions; its taxation and the proportions handled outside official channels.

7.8.3 ISHII Masayasu 石井正泰, "Chōro enden chōsa" 長蘆鹽田調査 (An investigation of the Ch'ang-lu salt-fields), **Mantetsu chōsa geppō,** 16.2 (Feb. 1936), 101–118.

This field report on the various aspects of salt production, preservation, distribution and administration as observed by a competent investigator in 1936 may be noted as of interest in connection with the earlier history of this important producing area.

7.8.4 AMANO Motonosuke 天野元之助, "Kōhoku no enkon kōshi kō" 江北の鹽墾公司考 (A study of salt-producing firms in North Kiangsu), **Tōa keizai ronsō,** 2.3 (Sept. 1942), 24 pp., maps and tables.

A leading reporter on the Chinese agrarian economy here describes the scope, capital sources, and personnel of a dozen or more of the salt firms in this important region.

7.8.5 SHIMA Yasuhiko 島恭彦, **Chūgoku okuchi shakai no gijutsu to rōdō** 中國奧地社會の技術と勞働 (Technology and labor in the society of the Chinese interior), Kōtō Shoin 高桐書院, 1946, 165 pp., maps, illus.

Becoming interested in the salt industry of Tzu-liu-ching 自流井 in Szechwan and the question of its technological modernization, the author made this study at the Kyōto Institute in 1944–46, using Chinese sources like **T'ien-kung k'ai-wu 天工開物** and **Ssu-ch'uan yen-fa chih 四川鹽法志** to describe the complex traditional technology, equipment and processes used and the type of labor force employed (chapters 1 and 2 to p. 106). He then explores the broad basic question of the social factors and repercussions involved in technological modernization and industrialization of this traditional salt production system. See the review by Ubukata Naokichi 幼方直吉 in **Chūgoku kenkyū,** 2 (Nov. 1947), 179-82.

7.8.6 SAEKI Tomi 佐伯富, "Shindai ni okeru engyō shihon ni tsuite" 清代における鹽業資本について ("The Capital of Salt Trade under the Ch'ing Dynasty"), **Tōyōshi kenkyū,** 11.1 (Sept. 1950), 51–65; 11.2 (March 1951), 128–140.

This survey article explores the heavily documented and extremely complex Ch'ing salt gabelle by concentrating on the operations of the big salt transporting merchants of Yangchow 揚州 in the Huai-nan 淮南 (now North Kiangsu) producing area, from which came about 40 per cent of the Ch'ing salt revenue (the latter constituting about a quarter of the central government's total revenue). Tracing the rise of the Yangchow merchants from the Ming period to their height under Ch'ien-lung, the author summarizes their administrative vicissitudes in the 19th century, compares Huai-nan with other areas, and notes the influence of the merchant class. The problem as to why the huge capital accumulated by salt merchants could not be transformed into modern industrial capital is also mentioned.

7.8.7 HATANO Yoshihiro 波多野善大, "Shindai Ryōwai seien ni okeru seisan soshiki" 清代兩淮製鹽における生産組織 ("The Producing System of Salt in the Liang-huai 兩淮 Province under the Ch'ing 淸 Dynasty"), **Tōyōshi kenkyū**, 11.1 (Sept. 1950), 17–31.

A summary, based on more extensive work in progress, of the complex salt-producing system inherited and reorganized by the Ch'ing in this important area, indicating something of the technology, division of labor and production relationships involved. The predominant control of production by merchant-capital through a form of "putting-out system" (*toiyasei* 問屋制) is stressed. See also the author's article on bureaucracy, 3.1.8.

7.8.8 SUZUKI Tadashi 鈴木正, "Shinsho Ryōwai enshō ni kansuru ichikōsatsu" 清初兩淮鹽商に關する一考察 (A study concerning the Liang-huai salt merchants of the early Ch'ing), **Shien** 史淵, 35 (Mar. 1946), 101–134; 36–37 (Mar. 1947), 125–158.

A rather extensive background study of the salt control system, merchant organization, the basis of the merchants' activity (their relations with official-dom) and their methods of capital accumulation (through unofficial sale or smuggling of salt), in the period before the reform of the early 19th century. The author uses the **Liang-huai yen-fa chih** and other contemporary sources and builds up an impressive picture of the official-merchant-smuggler relationship which milked the profits from the salt gabelle.

7.8.9 YAMAMURA Jirō 山村治郎, "Shindai Ryōwai no sōko ippan" 清代兩淮の竈戶一斑 (A general survey of the *tsao-hu* [salt producers] of the Liang-huai area in the Ch'ing period), **Shigaku zasshi**, 53.7 (June 1942), 827–859; 11 (Nov. 1942), 1330–1359.

A general description of the salt-production process in this major producing area, and of the organization and functioning of the licensed producers, as they existed down into the 18th century.

7.9 MINING

Note: These articles are supplementary to the data presented in the compilations listed in sec. 7.2. Note also the bibliography on mining, item 9.2.7. Nevertheless, the subject, like mining itself, seems to have been neglected in the case of China.

7.9.1 SATOI Hikoshichirō 里井彥七郎, "Shindai kōgyō shihon ni tsuite" 清代鑛業資本に就いて ("The Mining Capital under the Ch'ing Dynasty"), **Tōyōshi kenkyū**, 11.1 (Sept. 1950), 32–50 (to be continued; continuation not yet published).

A description reconstructed from edicts, memorials, gazetteers and similar records, of types of private mine operators and smelters and their supply of capital, with special attention to the use of capital from merchants and officials. Some 68 references are given but relatively few statistics.

7.9.2 SUGAYA Kazuo 菅家一雄, "Kairan kōmukyoku no enkaku" 開滦礦務局の沿革 (History of the Kailan Mining Administration), **Mantetsu chōsa geppō**, 14.6 (June 1934), 71–90; 7 (July), 131–162; 8 (Aug.), 149–162.

Derived from a report of the same title by the Bureau of Industry (Shih-yeh-t'ing 實業廳) of Hopei province, this detailed source material presents translations of the major contracts, regulations, memorials and other documents connected with the Kaiping 開平 mines and their development into the KMA, together with its growth down to the 1920s, thus providing raw material (in Japanese translation) for the historian.

7.9.3 TEZUKA Masao 手塚正夫, "Shina tankō no dohō keitai" 支那炭礦の土法形態 (Native methods in the Chinese coal mines), **Tōa kenkyū shohō**, 20 (Feb. 1943), 117–176, illus.

Distinguishing new methods 新法 used in modern mines from the old local practices 土法 used in the small traditional mining operations in many parts of China, the author describes in detail the old style mines' technology, forms of enterprise and management, extractive methods and capacities, and conditions of labor, drawing copiously on the extensive foreign and Chinese bibliography on this subject.

7.9.4 TEZUKA Masao 手塚正夫, "Shina ni okeru tangusuten-kō no seisan to ryūdō" 支那に於けるタングステン礦の生産と流動 (The production and outflow from tungsten mines in China), **Tōa kenkyū shohō**, 6 (Aug. 1940), 1–49.

By a specialist in Chinese industrialization, this is a detailed statistical study of the Chinese tungsten mining industry—its distribution, production figures, forms of enterprise, market prices, foreign sales, smuggling, etc., with details also on the history and conditions of mining in Kiangsi, Hunan, Kwangtung and elsewhere.

7.9.5 KIKUTA Tarō 菊田太郎, "Minami Shina no suzu tangusuten anchimonī kōgyō no kankyō" 南支那の錫・タングステン・アンチモニー礦業の環境 (The environmental situation of the tin-, tungsten-, and antimony-mining industry in South China), **Tōa jimbun gakuhō**, 3.3 (Jan. 1944), 612–647.

Using mainly Chinese publications, this is an analysis of the natural, social and economic circumstances surrounding the world famous production of tin, tungsten and antimony in Yunnan, Hunan and Kiangsi provinces. The persistence of the traditional way of production (t'u-fa 土法) is appraised as an adequate adaptation to the environment, much as its harsh or inefficient features may be deplored abroad.

7.10 COTTON, TEA AND SILK

Note: These three rural products have in common the fact that they are created by subsidiary farm industries in which farm manpower can produce for a commercial market. The varying fortunes of these industries have therefore been connected, by some historians, with the incidence of foreign trade and the growth of a domestic money economy. On cotton and silk production note the SMR studies in item 7.5.9.

7.10.1 NISHIJIMA Sadao 西嶋定生, "16・17 seiki o chūshin to suru Chūgoku nōson kōgyō no kōsatsu" 16・17世紀を中心とする中國農村工業の考察 (An inquiry into Chinese rural industry, with specific reference to the 16th and 17th centuries), **Rekishigaku kenkyū,** 137 (Jan. 1949), 15-24.

On the basis of four earlier monographic articles on the early development of the cotton industry in China, especially in Sungkiangfu 松江府, the author attempts to extrapolate the phases in the development of rural industry for commercial production before the modern period, and so puts forward a significant thesis as to China's domestic economic development before the impact of Western trade. See review by Hatano Yoshihiro 波多野善大 in **Rekishigaku kenkyū,** 139 (May 1949), 51-53. Prof. Nishijima is now in the Faculty of Letters, Tōkyō University.

7.10.2 NISHIJIMA Sadao 西嶋定生, "Shina shoki mengyō shijō no kōsatsu" 支那初期棉業市場の考察 ("Inland trade of cotton cloth in China at its beginning stage"), **Tōyō gakuhō,** 31.2 (Oct. 1947), 262-288.

This scholarly analysis deals with the vicissitudes of the trade in raw cotton and cotton cloth, centering in Sungkiang 松江, Kiangsu, before the import of Indian raw cotton was begun by the British East India Co. The author first analyzes the trade in raw cotton produced in the cotton planting area near Sungkiang, secondly the trade in raw cotton produced in Shantung and other northern provinces and transported to Sungkiang, and thirdly the trade in raw cotton produced near Sungkiang and transported to Canton and Fukien. He then discusses the trade in cotton cloth produced near Sungkiang and the local market for it and traces its distribution, through Shansi and Shensi merchants, to the nothern provinces where, though raw cotton was produced in large quantity, no considerable cotton cloth industry had yet developed. This complementary relationship between Sungkiang cotton cloth and North China raw cotton disappeared when a cotton cloth industry was developed in the north around the beginning of the Ch'ing period. Thereupon, the cotton cloth industry near Sungkiang began to weave another type of cloth and sought new markets in Central and South China, whither cloth was transported by the so-called Hsin-an 新安 merchants. All this provides an informative background for China's early modern trade with India.

7.10.3 NISHIJIMA Sadao 西嶋定生, "Shina shoki mengyō no seiritsu to sono kōzō" 支那初期棉業の成立とその構造 ("L'établissement et la structure des manufactures cotonnières en Chine dans les 16 et 17 siècles"), **Orientalica,** 2 (1949), 88-141.

A study of the growth of cotton production in Sungkiangfu 松江府 near Shanghai, with reference to cotton growing, textile-making methods, management, labor and other local factors, viewed in a context of China's political and social structure. This article makes some use of gazetteers as well as dispersed materials of other kinds, and continues the author's earlier article, "Shina shoki mengyō shijō no kōsatsu", 7.10.2, and also one on cotton in the Ming period, "Mindai ni okeru momen no fukyū ni tsuite" 明代に於ける木綿の普及に就いて (On the distribution of cotton in the Ming period), **Shigaku zasshi**, 57.4 (Aug. 1948), 193–214.

7.10.4 UBUKATA Naokichi 幼方直吉, "Nankin momen kōbōshi" 南京木棉興亡史 (History of the rise and fall of "Nankeen" cotton), **Tōa ronsō,** 1 (July 1939), 257–285.

A survey article of interpretation rather than original research, which traces the growth in the 18th and early 19th centuries of China's raw cotton imports from India and exports of "Nankeens" from Canton, and attempts to relate this to the growth of capitalism in China.

7.10.5 MURAKAMI Sutemi 村上捨巳, **Hoku-Shi nōgyō keizai ron** 北支農業經濟論 (On the North China farm economy), Nikkō Shoin, 1942, 361 pp.

Subtitled "with special reference to cotton production and the problem of cooperatives", this study pursues a number of basic questions: the place of cotton in the North China economy, how to increase agricultural production generally, the commercialization of agriculture in Shantung (a case study), problems of crop management in Shansi, irrigation and cotton production in Hopei, and the general question of cooperatives. The author was well informed and widely traveled.

7.10.6 HATANO Yoshihiro 波多野善大, "Chūgoku yushutsucha no seisan kōzō" 中國輸出茶の生產構造 ("Manufacturing Structure of the Chinese Tea for export"), **Nagoya daigaku bungakubu kenkyū ronshū** 名古屋大學文學部研究論集, II, **Shigaku** 史學, 1 (1952), 183–210, illus.

Subtitled "before the Opium war", this scholarly monograph makes good use of the extensive Western literature (by Robert Fortune and others) on Chinese tea so as to work out the types of product and the kinds of producers, middlemen and merchants engaged on the Chinese side of the export trade. The author pays attention also to the financial control exercised over the Hong merchants by the E.I.C., and over the tea-collecting merchants (ch'a-chuang 茶莊) in turn by the Hong merchants. The latter at the same time controlled some tea production directly.

7.10.7 Sanshigyō Dōgyōkumiai Chūōkai 蠶絲業同業組合中央會 (Central Association of the Co-operative Unions for the Sericulture and Silk-spinning Industry), comp., **Shina sanshigyō taikan** 支那蠶絲業大觀 (A general survey of the sericulture and silk-spinning industry in China), Okada Nichieidō 岡田日榮堂, 1929, 1102 pp.

The product of an extensive five-year field study by Uehara Shigemi 上原重美, a T'ung-wen College graduate and son of a Japanese silk-industrialist, begun in 1923 for the Japanese silk industry, this large compilation surveys practically every conceivable aspect of Chinese silk production, manufacture, consumption and trade. Major sections are devoted to Central China, North

China, Szechwan, and South China, with attention also to relations with the Japanese and Western industries. This is one of the most informative and detailed surveys of the industry.

7.10.8 Tōa Kenkyūjo, **Shina sanshigyō kenkyū** 支那蠶絲業研究 (A study of the sericulture and silk-spinning industry in China), Ōsakayagō Shoten, 1943, 622 pp.

This wartime research study, written by Fujimoto Jitsuya 藤本實也, Dr. of Agriculture, goes into the basic factors of Chinese production and trade in silk, its relation to the farm economy and general organization, and then discusses its relation to the silk industry of Japan and how the two might be fitted together in future.

7.10.9 HORIE Eiichi 堀江英一, for the Tōa Kenkyūjo, **Keizai ni kansuru Shina kankō chōsa hōkokusho, Shina sanshigyō ni okeru torihiki kankō** 經濟に關する支那慣行調査報告書——支那蠶絲業における取引慣行 (A report of investigation of economic customs in China, on the trade customs in the sericulture and silk-spinning industry in China), Tōa Kenkyūjo, 1944, 205 pp.

Produced at Kyōto on the basis of SMR field data and the published literature on the subject, this volume of theoretical analysis describes the system for production and distribution of (1) silk cocoons and (2) silk thread in various regions of China. The author's article entitled "Shina seishigyō no seisan keitai", 7.10.10, is reprinted in this volume.

7.10.10 HORIE Eiichi 堀江英一, "Shina seishigyō no seisan keitai" 支那製絲業の生產形態 (The forms of production in the Chinese silk-spinning industry), **Tōa keizai ronsō**, 2.4 (Dec. 1942); 3.1 (Feb. 1943), 16+17 pp.

An interpretative study of the early history of the machine silk-spinning industry in the 19th century, development of factories and subsidiary farmstead production, and the native Chinese capitalization of the new industry, as an aspect of the inroads of capitalism in China. The author stresses the slight development of machine silk industry in China (1) as affected from outside by the largescale persistence of the silk industry based on native methods and also by the increasing competition of the Japanese modern silk industry; and (2) as checked from within by the pre-modern character of its native capitalization (i.e. controlled by commercial capital).

7.11   INDUSTRY AND LABOR

Note: These items begin with industry and then turn to its counterpart, labor. Note should be taken of the work of Toda Yoshirō, who was a leading specialist at the Tōa Dōbun Shoin and conducted many careful on-the-spot studies. The SMR study of Wusih, 7.11.16, is also noteworthy. See also 7.10.1 (Nishijima).

7.11.1 TEZUKA Masao 手塚正夫, **Shina jūkōgyō hattatsushi** 支那重工業發達史 (History of the development of Chinese heavy industry), Taigadō 1944, 548+14 pp., maps, tables.

This solid survey takes account of wartime developments in appraising the basic conditions for, and the actual growth of, modern-type industries in coal-mining and iron and steel production, and the modern-type administration of such enterprises. The last section is a history of foreign capital investment in Chinese mines—the Kailan 開灤 Mining Administration (pp. 410-448), the Peking Syndicate, Ltd. (*Fu kung-ssu* 福公司) in Shansi and Honan (pp. 449-493), etc. There are 152 tables, a 14-page index and a full bibliography (pp. 540-547). On the whole this volume combines the merits of economic history with those of a contemporary survey and would appear to be an outstanding work of its kind, even though somewhat pedestrian and over-factual in style.

7.11.2 ASOBE Kyūzō 遊部久藏, "Chūgoku kindai kōgyō no hatten" 中國近代工業の發展 (Development of industry in modern China), pp. 167–203 in Niida Noboru, ed., **Kindai Chūgoku no shakai to keizai**, 5.7.1.

A useful factual and statistical summary of China's industrial growth, based on a highly sophisticated interpretation. It is divided into three periods: (1) 1895–1913 (the establishment of foreign factories inside China and the emergence of native modern industry); (2) 1914–22–31 (the golden age of native industry followed by the post-war depression); and (3) 1932–37 (the repercussions of the great depression and the outbreak of the Sino-Japanese war).

7.11.3 ASOBE Kyūzō 遊部久藏, "Shina kindai kōgyō kōsei, Shina kōgyō kihon tōkei no ichiseijo" 支那近代工業構成──支那工業基本統計の一整序 (The structure of Chinese modern industry, an arrangement of the basic statistics of Chinese industry), **Tōa kenkyū shohō**, 17 (Aug. 1942), 432–547.

With 60 pages of statistics as an appendix and many charts and tables, this article summarizes the data compiled in a big report on Chinese industry **Chung-kuo kung-yeh tiao-ch'a pao-kao** 中國工業調查報告, published by the Military Affairs Commission of the Chinese Government in 1937. The main headings are (1) statistical materials, (2) structure of industry (types, distribution), (3) structure of expenditures (by industries), (4) forms of enterprise and (5) geographical location.

7.11.4 OZAKI Gorō (Shōtarō) 尾崎五郎 (庄太郎), **Shina no kōgyō kikō** 支那の工業機構 (China's industrial structure), Hakuyōsha, 1939, 453 pp.

These essays, of which several were published earlier in **Mantetsu chōsa geppō**, center around the problems posed by the impact of the depression on both the modern and the old-style sectors of China's industrial production. Main sections deal with (1) the process by which China's modern industry has developed (including the depression and after), (2) an analysis of the inner components of Chinese industry (types of rural industries, organic composition of native capital, labor, etc.), (3) features of native industries in North China, (4) aspects of North China rural industry as observable in Ting-hsien, (5) characteristics of Chinese labor in the cities and in rural areas. This is one of the more competent studies of the subject. The author, Ozaki Shōtarō, who formerly used pen-names such as Ozaki Gorō and Tamaki Hideo 玉木英夫, is now a leading member of the Chūgoku Kenkyūjo.

7.11.5 HASHIMOTO Hideichi 橋本秀一, "Chūgoku no kōgyōka mondai" 中國の工業化問題 ("Industrialisation of China"), **Tōyō bunka kenkyū,** 10 (Feb. 1949), 1–17.

A brief *aperçu*, quoting secondary works, of the historical growth of Chinese industries, their types and locations, etc., with an analysis of factors necessary for their future development, and its possibilities. Prof. Hashimoto is now at the Tōyō Bunka Kenkyūjo, Tōkyō University.

7.11.6 MIYAZAKI Ichisada 宮崎市定, "Min-Shin jidai no So-shū to kei-kōgyō no hattatsu" 明清時代の蘇州と輕工業の發達 ("The development of the light industries in Su-Chou under the Ming and Ch'ing dynasties"), **Tōhōgaku,** 2 (Aug. 1951), 64–73.

A brief but interesting survey which puts together various evidences on its topic, as a by-product of a joint study, now in progress, of terminology in Chinese economic history.

7.11.7 HOSHI Ayao 星斌夫, "Mindai ni okeru sōun gunshi no shūeki jōtai" 明代に於ける漕運軍士の就役狀態 ("Life of the official labourers at the grand canal during the Ming 明"), **Tōyō gakuhō,** 32.2 (Jan. 1949), 103–119; 32.3 (April 1950), 85–106.

Subsections discuss (1) group 幫 organization of laborers and its functioning, (2) terms of payment and economic maintenance, (3) private trade on canal vessels and the laborers' livelihood, (4) hardships of the laborers at work, (5) the escape of laborers and its causes. This is a factual study, mainly based on the **Ming shih-lu 明實錄** and **Ta-Ming hui-tien 大明會典**. No comparable work seems to have been attempted for the Ch'ing period. Prof. Hoshi is now at the Yamagata Daigaku 山形大學.

7.11.8 OKABE Toshinaga 岡部利良, "Chūgoku no kigyō keiei ni okeru ningen kankei" 中國の企業經營に於ける人間關係 ("Human Relations in the Business Enterprises in China"), **Jimbun kagaku,** 1.3 (Dec. 1946), 180–210.

Subtitled "with special reference to the system of management in the native cotton-spinning industry in China", this article discusses the interpersonal relations in seemingly modern business enterprises in China, stressing their backwardness and irrationality. Subsections deal with topics such as capitalists, engineers, foremen and nepotism.

7.11.9 ASOBE Kyūzō 遊部久藏, **Chūgoku rōdōsha kaikyū no jōtai** 中國勞働者階級の狀態 (Conditions among the Chinese labor class), Kōgakusha 好學社, 1948, 228 pp., tables.

Based on field study in Shanghai and Wusih, this sober and sophisticated volume describes conditions of labor, the history of labor policy, the historical development of China's modern industry, and new and old elements in company partnerships (合股).

7.11.10 TODA Yoshirō 戶田義郎, "Shina kōgyō rōdōsha no nōritsu to shotoku, Shina kōgyō rōdō kenkyū joron" 支那工業勞働者の能率と所得 ——支那工業勞働研究序論 (Efficiency and income of Chinese industrial laborers, an introduction to the study of Chinese industrial labor), **Shina kenkyū,** 47 (July 1938), 95–127.

A study in low efficiency and low income, using statistics and the international literature on the subject.

7.11.11 TODA Yoshirō 戶田義郎, "Shina ni okeru chingin hōsoku" 支那における賃銀法則 (Wage principles in China), **Shina kenkyū,** 48 (Dec. 1938), 91–122; 52 (Nov. 1939), 113–149; 56 (Nov. 1940), 153–182.

Professor Toda first notes the close connection in the West between the wage theory and economic system and then looks at theories as to the economic order prevailing in China (feudal, capitalist, early and late liberal, communist), before noting and criticizing the various theories of wages applicable to conditions in China. This article is, in the main, a detailed and critical review of Paul Arnt and others, **Der Arbeitslohn in China** (Leipzig, 1937), and uses the German literature and terminology.

7.11.12 TODA Yoshirō 戶田義郎, "Shina kōgyō rōdōsha to nōson tono ketsugō ni kansuru jisshōteki kenkyū" 支那工業勞働者と農村との結合に關する實證的研究 (A factual study concerning the close connection between the Chinese industrial laborers and the rural districts), **Shina kenkyū,** 50 (March 1939), 255–300.

As in his other studies, Professor Toda here marshals the evidence of Western and other literature on this subject and combines it with his own study to produce a valuable monograph—in this case, on labor migration and the peasant origin of factory labor, bringing together a number of statistical case-studies showing various aspects of the connection between factory and village.

7.11.13 TODA Yoshirō 戶田義郎, "Shina bōseki kaisha no keiei ni tsuite" 支那紡績會社の經營について (On the management of the Chinese cotton-spinning firms), **Shina kenkyū,** 36 (March 1935), 203–240; 37 (June 1935), 129–159; 38 (Nov. 1935), 143–163; 40 (March 1936), 187–220.

An extensive study of efforts and factors making for increased profit in production (increased productive capacity, longer hours, greater intensity of labor) and in distribution (in the forcing down of wages, the by-passing of middlemen, and the cutting of costs), based on Chinese, Japanese and English-language materials.

7.11.14 TODA Yoshirō 戶田義郎, "Shina bōseki rōdō no gimmi" 支那紡績勞働の吟味 (An appraisal of labor in the Chinese cotton-spinning industry), **Shina kenkyū,** 45 (June 1937), 67–151.

The author describes conditions of management, absentee ownership, native capitalization, material plant, etc., affecting the labor force in the cotton-spinning and textile industry, and seeks to appraise their position technologically and economically.

7.11.15 TODA Yoshirō 戸田義郎, "Zoku Shina bōseki rōdō no gimmi" 續支那紡績勞働の吟味 (A further appraisal of labor in the Chinese cotton-spinning industry), **Shina kenkyū,** 46 (March 1938), 89–123.

This continuation of the author's previous study takes up such problems as those of low wages, assistants and apprentices, the labor-contracting system, and the comprador system. Prof. Toda is now in the Faculty of Economics of Kōbe University.

7.11.16 Kōain Seimubu 興亞院政務部, **Mushaku kōgyō jittai chōsa hōkokusho** 無錫工業實態調査報告書 (Report of an investigation of conditions of industry in Wusih), printed by the above agency and classified confidential 祕, 1940, 877 pp., maps, illus. (Investigation materials, no. 18).

Produced through two weeks of field study in 1939 by the Shanghai research staff of the SMR, this bulky volume studies the industries of the manufacturing center of Wusih, Kiangsu, from all angles: silk filatures, textiles, dyeing and weaving, manufacturing of goods such as socks, gloves and shirts, foodstuffs, ironworks, chemicals, etc., and also power supply and labor conditions, together with the political and economic environment of the district and its market conditions. Throughout, this highly factual and comprehensive survey notes the historical transition from pre- to post-occupation circumstances.

7.11.17 NAKAMURA Takatoshi 中村孝俊, **Hatō seido no kenkyū** 把頭制度の研究 (A study of the labor contractor system), Ōsakayagō Shoten, 1944, 143 pp.+15 tables. (Reports of the Rōdōkagaku 勞働科學 Kenkyūjo [Research institute on labor science], fifth section, on labor management in Greater East Asia, vol. 4).

This interesting study was based largely on a field survey conducted by the author at the Lung-yen iron mine 龍煙鐵鑛 near Hankow in 1941–42, as a field research member of the above institute. It first describes the general conditions of a half-peasant, half-worker labor force and then studies the activities, institutional setting, status, methods, origin, background, types, etc. of two main kinds of labor contractors — *tsu-chang* 組長 (i.e., a contracter, properly speaking) and *pan-chang* 班長 (i.e., a foreman under the contractor), with tables of data on some two dozen cases.

7.11.18 MIYAMOTO Michiharu 宮本通治, "Saikin Shina ni okeru rōdō undō no hatten keikō" 最近支那に於ける勞働運動の發展傾向 (Developmental tendencies in the labor movement in modern China), **Mantetsu Shina gesshi** ("The Shanghai S.M.R. Co's Monthly"), 7.5 (May 1930), 79–94.

This brief article surveys the forms of labor organization in the old China, the growth of the modern labor movement, the KMT and government unions, labor contractors (*kung-t'ou* 工頭), yellow unions and their difference from government unions, etc., on the basis of an article by the CCP leader Ch'ü Ch'iu-pai 瞿秋白 in **Bolshevik** 布爾塞維克, 3.2-3.

7.11.19 SHIOWAKI Kōshirō 鹽脇幸四郎, **Chūgoku rōdō undō shi** 中國勞働運動史 (A history of the Chinese labor movement), Hakuyōsha 白揚社, 1949, 2 vols., total 452 pp.

> With a preface by Hirano Yoshitarō, this orthodox success-story by a member of the Chūgoku Kenkyūjo is based, the author says, on facts whenever possible. No references are given. See review by Hosoi Shōji 細井昌治 in **Rekishigaku kenkyū**, 144 (Mar. 1950), 54–57.

7.11.20 TAKATA Gensei 高田源清, **Manshū oyobi Shina no kumiai seido** 滿洲及支那の組合制度 (The trade union system of Manchuria and China), Sōbunkaku 叢文閣, 1941, 275+app. 348 pp.

> A legal study devoted mainly to trade union organization in Manchuria, and Kwantung in particular, with a section (pp. 147–244) on China, and an appendix of 139 laws and regulations. Under China the author deals with the traditional gilds 同業公會, the modern cooperative societies 合作社, and the unions in transport, commerce and industry newly introduced by Japanese initiative through the puppet regime.

7.11.21 KISHIMOTO Eitarō 岸本英太郎, "Manshū ni okeru kindai shihon =rōdōryoku keiseishi josetsu" 滿洲に於ける近代資本＝勞働力形成史序說 (An introduction to the history of the formation of modern capitalism and labor power in Manchuria), **Rekishigaku kenkyū,** 124 (Oct. 1946), 32–47; 125 (Jan. 1947), 18–37.

> An academic, documented study, from a Marxist point of view, of the rise and collapse of Manchurian agriculture under the Ch'ing, and the growth of industry and labor under the stimuli of trade through Newchwang, the Sino-Japanese War, and subsequent developments.

7.11.22 MUKŌYAMA Hiroo 向山寬夫, "Chūka Jimmin Kyōwakoku ni okeru rōdōsha no keiei sanka" 中華人民共和國における勞働者の經營參加 ("'Workers' enterprise participation in the People's Republic of China'"), **(Aichi Daigaku) Hōkei ronshū** (愛知大學) 法經論集 ("The journal of the association of legal, political and economic sciences"), pub. by the association at Aichi University, 6 (April 1953), 19–42.

> A sober legal discussion which takes up, in its historical and social background, the workers' participation in management in public and private enterprises under the new regime. Because of the scarcity of materials, this article is inevitably reduced to a textual analysis of the relevant regulations, rather than a study of what actually transpires.

7.12 COMMUNICATIONS AND TRANSPORT

Note: Mainly on shipping and water transport, and the official post routes of the Ch'ing period. Communications are described at length in the gazetteers in sec. 7. 2, especially the series **Shina shōbetsu zenshi** (7. 2. 6). Note also 7. 11. 7 (Hoshi).

7. 12. 1 Mantetsu Chōsabu 滿鐵調查部 (Research Department of the SMR), ed., **Chū-Shi no minsengyō** 中支の民船業 (The native shipping industry [junk trade] of Central China), Hakubunkan, 1943, 363 pp., maps and illus.

This "report of a factual investigation of native shipping at Soochow" was based on 15 days investigation in 1941 by a 10-man team from the SMR Shanghai research section led by Amano Motonosuke in the Soochow region, where the Grand Canal running to Hangchow intersects the delta traffic between Shanghai and Wusih. Information was secured from 40 junks and 9 shipping and transport firms. The report deals with (1) junk construction and capacity, reasons for the same, types, (2) the junkmen (labor force) including those on grain junks, wages and living standards, gild organization, (3) ownership relations, (4) business management, (5) junk distribution of commercial goods, (6) boat firms and transport firms, their finances. An exemplary job of field investigation.

7. 12. 2 BABA Kuwatarō 馬場鍬太郎, **Shina sui'un ron, fu Manshūkoku sui'un** 支那水運論附滿洲國水運 (On Chinese water transport, supplement on Manchoukuo water transport), Shanghai: Tōa Dōbun Shoin, 1936, 395 pp.

Although the author at first planned a third volume on "Production and trade" to follow his first two volumes of **Shina keizai chiri shi**, 1. 2. 6, he found history shifting under his feet beyond the capacity of compilers to stabilize and so published this lighter and narrower work. It is a factual and somewhat superficial study of the value and position of waterways in Chinese life; the problem of their rehabilitation, with many specific examples; steam traffic and water transport, with many more case studies which cover most of the known rivers. Pages 305 ff. do the same for Manchoukuo.

7. 12. 3 SUDŌ Yoshiyuki 周藤吉之, "Shindai no Manshū ni okeru ryōmai no sōun ni tsuite" 清代の滿洲に於ける糧米の漕運に就いて (On the water transport of government grain in Manchuria under the Ch'ing), **Tōa ronsō,** 3 (Sept. 1940), 141–165.

Using the **Shih-lu** and other Chinese records, this article traces the vicissitudes of the grain-transport administration, and its relation to the officially sponsored and later suspended Chinese immigration into Manchuria.

7. 12. 4 KATŌ Shigeshi 加藤繁, "Shindai Fukken Kōso no senkō ni tsuite" 清代福建江蘇の船行に就いて (On the boat firms of Fukien and Kiangsu in the Ch'ing period), **Shirin,** 14. 4 (1929), 529–537.

A note describing, from gazetteers and official memorials, how these boat transport firms took responsibility for transporting merchants' goods, viewed mainly as a case-study of transportation insurance. Also in the same author's **Shina keizaishi kōshō** (9. 8. 8), 2. 585–594.

7. 12. 5 MIYAZAKI Ichisada 宮崎市定, "Shōshōkyoku no ryakushi, Chū-goku no dokusenteki kisen kaisha" 招商局の略史, 中國の獨占的汽船會社 ("A Short History of Chao-shang-chü," China's monopolistic steamship company), **Tōyōshi kenkyū,** 11. 2 (Mar. 1951), 63–69.

Professor Miyazaki here makes available the substance of a statement which appeared in **Shen-pao** on Nov. 1, 1930, quoting Li Chung-kung 李仲公, manager of the famous but little-studied China Merchants Steam Navigation Co., initiated in 1872. He states mainly the terms on which the Company was set up.

7. 12. 6 KŌNO ("Kohno") Michihiro 河野通傳, "Shindai no ba'ekiro" 清代の馬驛路 ("Ma-I-Lu [Official Overland Stage Route] in Chin Dynasty of China"), **Jimbun chiri 人文地理** ("The Human Geography"), 2. 1 (Jan. 1950), 13–24, maps and tables.

As the first part of a study of domestic communications in the Ming and Ch'ing periods, this article distinguishes the various types of foot, water, horse, etc., stage services, works out in overall terms the main horse routes as recorded in such sources as the **Ta-Ch'ing hui-tien,** and locates them on a sketch map with some further analytic comments.

7. 12. 7 KŌNO Michihiro 河野通博, "Shindai Santōshō no kansei rikujō kōtsuro" 清代山東省の官制陸上交通路 ("The Official Overland Communication Routes in Ch'ing Dynasty"), **Shirin,** 33. 3 (May 1950), 317–336, maps, tables.

An important dissection of the official post routes in Shantung as recorded in the early nineteenth century (mainly in the Chia-ch'ing edition of the **Ta-Ch'ing hui-tien,** 1818). The author distinguishes the horse-post routes (*ma-i lu* 馬驛路) and their extensive transport establishment from the slower foot-post routes (*p'u-lu* 鋪路) and other contemporary communication routes. Shantung had some 1054 foot-post stations at intervals of roughly 10 *li*, compared to some 150 horse-post stations. Changes in the system are noted. This is one of the few such studies as yet available. See also the author's "Shindai no ba'ekiro," 7. 12. 6.

7. 12. 8 HIBINO Takeo 日比野丈夫, "Arutai gundai ni tsuite" 阿爾泰軍臺について ("On the public post road leading to Altai in the Ch'ing dynasty"), **Tōhō gakuhō, Kyōto,** 16 (Aug. 1948), 141–160, illus., maps.

A factual summary of the history and recent condition (described by sections) of the public post road from Kalgan (Chang-chia-k'ou 張家口) to Kobdo (K'o-pu-to 科布多), which was the key strategical route for the Ch'ing control of Mongolia.

7. 12. 9 Tetsudō Daijin Kambō Gaikoku Tetsudō Chōsaka 鐵道大臣官房外國鐵道調査課 (Foreign railways research section of the Minister of Railways secretariat), **Shina no tetsudō 支那之鐵道** (The railways of China), revised edition, pub. by the ministry, 1923, 742 pp., maps, tables.

This is the best factual account of the subject in Japanese, giving a description of the history and management, routes, physical plant, traffic, rates, loans, or similar data for 38 lines in existence and further data on 61 (sic) lines not yet completed but contemplated (by someone). General statistics are added in 21 tables, with some further regulations and a chronology. This volume is a revised edition of a preceding version, and further revised editions seem to have been published since (not seen).

7.13 FISCAL ADMINISTRATION

Note: This is a somewhat miscellaneous collection of studies of Ch'ing fiscal organization and policy (a rather unexplored subject), with several articles on special aspects of taxation. The last five items (7.13.14–18) concern the customs system. In addition to general compendia in earlier sections, note the works of Hirase Minokichi (item 7.1.2) and Momose Hiromu (item 3.3.2) and the articles on tax-collectors (*ya-hang*) in sec. 7.16.

7.13.1 KATŌ Shigeshi 加藤繁, "Shinchō kōki no zaisei ni tsuite" 清朝後期の財政に就いて (On the fiscal administration of the late Ch'ing period), **Rekishi kyōiku** 歴史教育 (Historical education), pub. by the Society for Research in Historical Education, 14.2 (May 1939), 129–141.

Dividing Ch'ing history into two periods (153 years of expansion and consolidation to the end of the Ch'ien-lung period, 116 years of conservatism and decline thereafter), Dr. Katō summarizes briefly the financial vicissitudes of the imperial government in the second period, including the incidence of likin, contributions 捐納 and maritime customs duties. A brief lecture but by an outstanding expert in Chinese economic history, and one of the few studies of this rather unexplored topic. Also in the same author's **Shina keizaishi kōshō** (9.8.8), 2. 478–492.

7.13.2 MATSUI Yoshio 松井義夫, "Shinchō keihi no kenkyū" 清朝經費の研究 (A study of the expenditures of the Ch'ing dynasty), **Mantetsu chōsa geppō,** 14.11 (Nov. 1934), 1–39; 12 (Dec. 1934), 29–61; 15.1 (Jan. 1935), 41–82, many tables.

After briefly surveying the Ch'ing administrative institutions in general, with special attention to the fiscal agencies, the author discusses the Ch'ing public expenditures at length under the following heads: (1) a general account of the topic, (2) regulations governing expenditure and procedures followed, (3) divisions and types of expenditures, (4) the relationship between central and provincial expenditures, (5) relative percentages of the various items, (6) the increase of Ch'ing expenditures, (7) the imperial household, (8) salaries of civil and military personnel, (9) military expenditures, and (10) foreign loan payments. Though somewhat mechanically organized, this is a useful contribution.

7.13.3 SEKIGUCHI Takeo 關口猛夫, "Shinchō zaisei kikō no kenkyū josetsu" 清朝財政機構の研究序説 (An introductory study of Ch'ing dynasty fiscal organs), **Mantetsu chōsa geppō**, 20.1 (Jan. 1940), 1–36.

A rather elementary although well-informed study of the emperor's use of both Manchus and Chinese, his methods of controlling and using officials, and the offices and agencies of central and provincial administrative control, based largely on the **Ch'ing-shih kao**. The author tries to describe the institutional distribution of authority on fiscal affairs over the whole range of the governmental mechanism, central and provincial, but the result seems hardly worth it.

7. 13. 4 KASHIWAI Kisao 柏井象雄, "Nisshin Sensō ni okeru Shinchō no zaisei seisaku" 日清戰爭に於ける清朝の財政政策 (The Ch'ing fiscal policy in the Sino-Japanese War), **Tōa keizai ronsō**, 1.2 (May 1941), 20 pp.

This describes the Ch'ing fiscal system and position as of 1894 (pp. 1–8) and then traces the measures taken to finance the war and the indemnity, noting also the after effects. Use is made of the Tōa Dōbun Shoin volumes, **Shina keizai zensho**, 7. 2. 1, as well as the work of Chia Shih-i 賈士毅 and others.

7. 13. 5 FUJII Hiroshi 藤井宏, "Ichijōbempō no ichisokumen" 一條鞭法 の一側面 ("On the I-t'iao-pien-fa" ["the single-whip method"]), pp. 571–590 in **Wada hakushi kanreki kinen Tōyōshi ronsō**, 9. 8. 12.

A study which criticizes the previous interpretations of others and clarifies the history of this important tax reform, which forms an essential background element in Ch'ing fiscal institutions.

7. 13. 6 KITAMURA Hirotada 北村敬直, "Shindai ni okeru sozei kaikaku (chitei heichō)" 清代における租税改革 (地丁併徵) ("Reform of Taxation under the Ch'ing dynasty"), **Shakai keizai shigaku**, 15.3–4 (Oct. 1949) 1–38.

A close look at the oft-noted but little understood tax reform of 1712 by which the poll-tax and land-tax were combined, with a seemingly foolhardy promise that the land tax would never be increased. The author notes the background of this reform and its application in the provinces.

7. 13. 7 SUZUKI Chūsei 鈴木中正, "Shinchō no kenyakurei to sono seisakuteki igi" 清朝の儉約令とその政策的意義 ("Des lois somptuaires au temps des Ts'ing et ses significations politiques"), **Orientalica**, 1 (1948), 114–125.

A note on Ch'ing fiscal policy in the 18th century and early 19th century, with special reference to sumptuary laws, as compared with the similar regulations in the Tokugawa period in Japan. Though dealing with an interesting topic, the analysis in this note seems to be somewhat obscure.

7. 13. 8 KIMURA Masutarō 木村增太郎, **Shina zaisei ron** 支那財政論 (On the Chinese fiscal administration), Ōsakayagō Shoten, 1927, 643+102 pp.

A conventional survey of Chinese finance under the main academic categories: administration, receipts and expenditures, tax structure, debts, international control, and an appendix of documents, dating mainly from the 5th to 10th year of the Republic. The rather formalistic nature of this study seems evident from its treatment of the land tax in a mere 50 pages and the sources, Western and Chinese, are the usual sort. The author stresses the decentralized and chaotic condition of Chinese finance, and his approach to China's problems seems to have been influenced by the views of Naitō Torajirō (see ch. 7, "conclusion").

7.13.9 KASHIWAI Kisao 柏井象雄, **Kindai Shina zaisei shi** 近代支那財政史 (History of modern Chinese finance), Kyōto: Kyōiku Tosho Kabushiki Kaisha 教育圖書株式會社, 1942, 196 pp.

A brief and orthodox survey (written against a background of social history since the Opium War), of late Ch'ing financial collapse, Republican chaos and KMT reform efforts, with chapters on the tax structure and on the income tax.

7.13.10 HAYASHIDA Kazuo 林田和夫, "Shina no zaisei kikō to sono un'ei no tokushusei" 支那の財政機構と其運營の特殊性 (China's organs of fiscal administration and the special features of their management), **Mantetsu chōsa geppō**, 17.4 (April 1937), 31–75; 17.6 (June 1937), 105–155.

Subtitled "The conflict between state economy and social economy in China", this is a socio-historical study of the balance between central and local finance, taken as a key to the understanding of China's "semi-feudal" or "semi-colonial" social structure. While rather heavily sociological, in the German manner, it analyzes the details of the Ch'ing and Republican fiscal systems, including the balance between their revenue and expenditure, their collecting agencies, central supervision of local finance, the treasury and budget systems, etc. Though somewhat discursive, this appears to be one of the few studies which works out historical relationships between Ch'ing and Nanking government fiscal practices.

7.13.11 SHIOMI Saburō 汐見三郎, "Shina no sozei seido" 支那の租税制度 (The taxation system of China), **Tōa jimbun gakuhō**, 2.2 (July 1942), 149–173.

A brief survey by a specialist in finance. Subsections discuss rather theoretically, though with tables of data, (1) the historical background, (2) the national tax structure and (3) the local tax system of the Kuomintang Government.

7.13.12 SHIOYA Yasuo 鹽谷安夫, "Hoku-Shi zaisei shūnyū ni tsuite" 北支財政收入について (On the governmental income in North China), **Mantetsu chōsa geppō**, 18.7 (July 1938), 1–31.

Beginning with a five-page survey of statistical estimates and comments on government income in China since about 1850, this article analyzes the revenue sources of the Chinese Government in North China for 1935.

7.13.13 YAGI Yoshinosuke 八木芳之助, "Shina nōson no hōzei seido ni tsuite" 支那農村の包税制度に就いて (On the tax-farming system in the Chinese village), **Tōa keizai ronsō**, 1.1 (Feb. 1941), 25 pp.

A study of collection methods in Hopei, the history and structure of the system (with documentary forms), its evils and their reform.

7.13.14 TAKAYANAGI Matsuichirō 高柳松一郎, **Shina kanzei seido ron** 支那關稅制度論 (A study of the Chinese maritime customs administration), Kyōto: Naigai Shuppan Kabushiki Kaisha 内外出版株式會社, rev. and enlarged ed., 1926 (4th printing 1929, 612 pp.; first pub. 1920).

This systematic treatise, one of the best of its day, was compiled by a former member of the Chinese Maritime Customs, who quotes the Inspector General's circulars at length and is said to have made some use of other unpublished Customs materials. He deals with the origin of the Service, its special features, administrative organization and activities in China and in foreign areas, including such matters as special goods, harbor dues, and other ramifications. The last section discusses the financial, economic and political influence of the Customs in Chinese affairs and its future (pp. 440–490). The book was produced under the influence of the legal approach of Dr. Imai Yoshiyuki and the author was assisted by his constant advice.

7.13.15 MATSUZAWA Shigeichi 松澤繁一, and YASUDA Kaoru 安田薫, "Shina kaikan seido no rekishiteki hatten" 支那海關制度の歷史的發展 (Historical development of China's maritime customs system), **Mantetsu chōsa geppō,** 19.6 (June 1939), 23–56; 20.2 (Feb. 1940), 1–28 (unfinished, but no continuation published).

The first section, by Matsuzawa, summarizes the origin and growth of the Maritime Customs Service down to the Nationalist Government period without adding to common knowledge available in more detailed Western surveys, e.g. those of Stanley F. Wright. The second part, by Yasuda, summarizes tariff history, again thanks to Mr. Wright but with the addition of a few quotations from Marx and Engels' complete works.

7.13.16 IDE Kiwata 井出季和太, for the Taiwan Sōtokukambō Chōsaka 臺灣總督官房調査課, **Shina naikoku kanzei seido** 支那内國關稅制度 (The domestic customs system of China), printed for restricted use by the above agency, vol. 1, 1929 (not seen but **Shina kanzei seido no yurai,** 7.13.17, is a revised edition of this volume); vol. 2, 1929, 169 pp.; vol. 3, 1932, 422 pp.; vol. 4, 1934, 181 pp. (vols. 2–4 are nos. 176, 208 and 215 of **Minami Shina oyobi Nan'yō chōsa** 南支那及南洋調査).

Like the same author's studies of the South China ports, 7.17.2, these volumes are detailed descriptions of contemporary Chinese institutions, in this case the native customs (ch'ang-kuan 常關), including both their regular customs collections and the multifarions forms of likin 釐金 charges in the various provinces, on railways, on opium, etc., etc. This account of the likin system was compiled just before its abolition.

7.13.17 IDE Kiwata 井出季和太, for Taiwan Sōtokukambō Chōsaka 臺灣總督官房調査課, **Shina kanzei seido no yurai** 支那關稅制度の由來 (Origins of the Chinese customs system), printed for restricted circulation by the above agency, preface 1934, 80 pp. (no. 214 of **Minami Shina oyobi Nan'yō chōsa** 南支那及南洋調査 [Research on South China and Southeast Asia].

As a revised and re-christened edition of vol. 1 of the author's 4-volume account of the domestic customs and likin administration, 7.13.16, this booklet summarizes the history of the Chinese native customs arrangements through the various dynasties from ancient times.

7.13.18 MAITANI Eiichi 米谷榮一, **Kinsei Shina gaikoku bōekishi** 近世支那外國貿易史 (A history of the modern trade between China and foreign countries), Seikatsusha, 1939, 318 pp.

Begun in Peiping in 1938, this might more accurately be called a history of Sino-foreign relations over customs and tariff administration, since it is mainly concerned with a recital of the British establishment of the Maritime Customs during and after the two "opium" wars (to p. 70), further privileges secured by the Sino-Japanese War, China's movement for recovery of tariff autonomy (pp. 132-195), and changes in customs administration effected by the Manchurian, Shanghai, and China "incidents" down to 1938. The result is a secondary work which reprints many sets of trade regulations but adds little or nothing to economic history, although its data on the 1930s may be of use.

7.14 CURRENCY

Note: On the foreign silver and foreign dollars used in the China trade, see sec. 2.2 above.

7.14.1 MIYASHITA Tadao 宮下忠雄, **Chūgoku heisei no tokushu kenkyū, kindai Chūgoku ginryō seido no kenkyu** 中國幣制の特殊研究――近代中國銀兩制度の研究 (A special study of the Chinese currency system, a study of the silver tael system of modern China), Nippon Gakujutsu Shinkōkai 日本學術振興會, 1952, 618+indices 61 pp.

The author, now a professor at Kōbe University, was connected with the Shanghai Tōa Dōbun Shoin from 1932; returning to Japan in 1946, he developed this study of the Chinese silver tael currency in the period after the opening of the treaty ports, carrying forward the studies made by the late Katō Shigeshi 加藤繁 for the pre-treaty period. The result is a highly detailed, analytic and authoritative treatise. After first describing the origin and general features of the tael system, and posing the problem of its modernization in the late-Ch'ing and Republican periods, the author then examines modern currency developments region by region all over China, ending with the final process of the collapse and abolition of the tael system. In the process he naturally touches upon nearly all aspects of Chinese public finance in the modern century. While Customs reports are a valuable source in English, most of the story is derived from Chinese and especially Japanese monographic research. In this connection note Katō Shigeshi's monograph, "Kampōchō no kahei ni tsuite" 咸豐朝の貨幣に就いて ("On the Currency of the Hsien-feng Period"), pp. 2.421-449 in his **Shina keizaishi kōshō**, 9.8.8, which analyzes the reasons for the failure of the Ch'ing regime's efforts to introduce new copper coinage and other currencies, during and after the Taiping Rebellion.

7.14.2 MIYASHITA Tadao 宮下忠雄, **Shina kahei seido ron** 支那貨幣制

度論 (A treatise on the currency system of China), Ōsaka and Tōkyō: Hōbunkan 寶文館, 1938, 502+39 pp.

This volume, much of which appeared as articles in the Shanghai Tōa Dōbun Shoin organ, **Shina kenkyū,** seeks to describe both the facts of the Chinese currency system and the government policies and theories underlying it in the 1930s. Major sections concern the central banking system, the abolition of the tael in 1933, the question of the silver standard in China, the system of export duties and equalization charges on silver, the functioning of the new managed paper currency system instituted in November 1935, and the Maritime Customs gold unit system.

7. 14. 3 IIJIMA Banji 飯島幡司, **Shina heisei ron, sono kōhai to saiken** 支那幣制論その興廢と再建 (A treatise on the currency system of China, its vicissitudes and reconstruction), Yūhikaku, 1940, 389+index 6 pp.

Developed from lectures delivered at the Kōbe University of Commerce in 1939, this book briefly recounts China's currency history and touches on banks, currency reform, the end of the silver standard and emergence of the wartime paper currency (*fa-pi* 法幣), Nationalist Government currency policies and those of the Japanese puppet regimes in China. See the same author's **Shina heisei no kenkyū 支那幣制の研究** (Studies in the Chinese currency system), Yūhikaku, 1937, 409 pp., which refers especially to the American silver policy.

7. 14. 4 HOZUMI Fumio 穗積文雄, "Shindai kahei kō" 清代貨幣考 (A study of the currencies in the Ch'ing period), **Tōa keizai ronsō,** 2.3 (Sept. 1942), 31 pp.

A useful study of Ch'ing economic theorizing concerning currency, especially as indicated in the *ch'ien-pi* 錢幣 section of the economic treatise **(Shih-huo chih)** of the **Ch'ing-shih kao.** After describing the general development of currency policy and ideas about it, the author, as an economist, notes particularly seven methods advocated for a policy of increasing the currency and three methods for decreasing it, together with a number of other approaches. Acknowledging the advice of Abe Takeo and Saeki Tomi, Professor Hozumi has at least made a fresh, clear and illuminating approach to a confused and confusing subject. The text of pages 20–28 of this article is reproduced with the addition of 25 notes on pp. 226–239 of the author's **Shina kahei kō** of 1944, 7.14.5.

7. 14. 5 HOZUMI Fumio 穗積文雄, **Shina kahei kō 支那貨幣考** (Researches on Chinese currencies), Kyōto: Kyōto Inshokan 京都印書館, 1944, 392+index 17 pp., illus.

This volume by the recent head of the Kyōto University faculty of economics is based on studies of the theories about currency expressed in the treatises on economics **(Shih-huo chih)** of the dynastic histories, which he conducted at the Kyōto Institute. They range widely over Chinese history, discussing ancient currencies in kind (skins, shells, grain, etc.), forgery, minted coinage, T'ang dynasty coinage policies, Chinese and especially Han theories, and those of two officials of the Sung and one of the Ming. The study of Ch'ing currency policies, based on the pertinent section of the **Ch'ing-shih kao,** is however only a third the length of Professor Hozumi's article on the subject of 1942, 7.14.4.

7.15   BANKING AND CAPITAL

Note: In this section we list materials on Chinese banking, both modern and old-style, on capital formation, pawnbroking, the famous Shansi banks, and escort firms (hardly noticed in Western literature). Note the article on mining capital (7.9.1).

7.15.1 TOKUNAGA Kiyoyuki 德永清行, **Shina chūō ginkō ron** 支那中央銀行論 (A treatise on central banking in China), Yūhikaku, 1942, 609 pp. +index pp. 1–15 and bibliography pp. 17–58.

This standard work, by a Kyōto University professor, appears to go further into its subject, using many Western sources, than any Western work available. It is actually a history of modern central banking in China, arranged by periods from the early days of the Hu-pu bank of 1906, which became the Ta-Ch'ing bank in 1909, and the increasing need of a central bank (to p. 109); through the various institutions of the early Republican period (to p. 234); and the Nationalist Government central bank structure. The last part (from p. 433) describes the wartime banking system of occupied China.

7.15.2 MIYASHITA Tadao 宮下忠雄, **Shina ginkō seido ron** 支那銀行制度論 (A treatise on the Chinese banking system), Ganshōdō, 1941 (2nd printing, 1943, 483+index 24 pp., many tables).

The author, now a professor at Kōbe University, published most of this volume as articles in **Shina kenkyū.** They are mainly historical studies of the growth of the modern Chinese fiscal and banking institutions: late Ch'ing beginnings, banking under the Peking government after 1911 (to p. 102), under the Nationalist government subsequently, its organs of fiscal control, joint developments among the Chinese banks in Shanghai, the Shanghai clearing system old and new, and the developments from the late Ch'ing period in the field of currency (from p. 367).

7.15.3 KAGAWA Shun'ichirō 香川峻一郎, **Sensō shihon ron** 錢莊資本論 (Chinese banking and capital), Jitsugyō no Nipponsha, 1948, 306 pp.

This account of the inner workings of modern Chinese private banking is based on more than 10 years contact with banking circles. The author says he became interested in this mysterious subject through his North China banker-friend, the late Hu Pi-chiang 胡筆江, and studied it further with the help of researchers of the Hongkong-Shanghai Banking Corp. He disclaims academic preparation for such study, but develops his own interpretations, viewing Chinese private finance as having developed under the stimulus of foreign imperialism in a semi-colonial China where the money-shops (*ch'ienchuang* 錢莊) were a necessary link with the backward economy of the hinterland. Chinese modern bankers (*zaibatsu* 財閥) have relied on the ancient Chinese principle of the mutual-aid association (*pang* 幫). The interlocking directorates and power relations of the Northern and Southern banking cliques (each grouped around four banks) are therefore analyzed in detail, as well as their ramified connections with the old money-shops, warlords, the Four Banks of the Nationalist Government, and their various kinds of investments. While this book does not present unusual documentation, it offers an insider's view and provocative analysis of an enigmatic subject. Rev. by Ikeda Makoto 池田誠 in **Tōyōshi kenkyū**, 10. 6 (Feb. 1950), 511–515. The author is now a businessman in Tōkyō.

7. 15. 4 KATŌ Shigeshi 加藤繁, "Shindai ni okeru sempo sensō no hattatsu ni tsuite" 清代に於ける錢鋪錢莊の發達に就いて ("An inquiry into the development of banking during the Ch'ing dynasty"), **Tōyō gakuhō**, 31.3 (Dec. 1947), 335–344.

A survey by a leading specialist in the economic history of China which, though brief, quotes a considerable number of Ch'ing documents on the native banking establishments. Also printed in the same author's **Shina keizaishi kōshō** (9. 8. 8), 2. 463–477.

7. 15. 5 IZEKI Takao 井關孝雄, **Shina shomin kin'yū ron** 支那庶民金融論 (A treatise on native financial organs among the common people in China), Gakugeisha 學藝社, 1941, 249 pp.

Based on secondary materials, and using common knowledge concerning Chinese social and political institutions, this study describes the various financial agencies which function in the daily life of the common people in city and village: (1) the individual money-lender, pawnshop, mutual finance association (合會 etc.) and native bank of the old order; (2) the modern bank, credit cooperative, "agricultural warehouse" 農業倉庫 (an officially sponsored financing agency for farmers) and other credit agencies of more recent times. Numerous examples are given and there is a long 7-page bibliography.

7. 15. 6 ARIMOTO Kunizō 有本邦造, "Nimpō ni okeru kin'yū seido" 寧波に於ける金融制度 ("Le système monétaire à Nimpo"), **Tōa keizai kenkyū**, 15.4 (Oct. 1931), 579–596.

This study of the native banks and their methods of operation at Ningpo is one of the few which come anywhere near the question of the financial role of the Ningpo banks in the 19th century. Unfortunately this article concerns the modern period. See the author's similar piece, "Nimpō kachō seido no kenkyū" 寧波過賬制度の研究 ("Etude sur les opérations de virements à Nimpo"), ibid., 15. 1–2 (April 1931), 111–137.

7. 15. 7 TOKUNAGA Kiyoyuki 德永清行, for the Tōa Kenkyūjo, **Shina kyūshiki kin'yū ni okeru kankō, toku ni gingō sensō no tokushitsu o kadai to shite** 支那舊式金融に於ける慣行——特に銀號錢莊の特質を課題として (Customs concerning the old style Chinese money exchange, dealing especially with the characteristics of the *yin-hao* and *ch'ien-chuang*), in the series **Keizai ni kansuru Shina kankō chōsa hōkokusho** 經濟に關する支那慣行調査報告書 (Reports of investigation of economic customs in China), pub. by the Tōa Kenkyūjo, 1944, 167 pp.

This well-informed research into the native money market and institutions in North and Central China surveys briefly China's banking history but is chiefly concerned with the modern regulation of the native banks and their legal status and operations under the Nationalist Government and succeeding puppet regimes, together with capital formation and the working of the exchange generally.

7. 15. 8 MIYAZAKI Ichisada 宮崎市定, "Chūgoku kinsei ni okeru seigyō shihon no taishaku ni tsuite" 中國近世における生業資本の貸借について ("The loan of the funds for the small trades in the recent ages of China"), **Tōyōshi kenkyū**, 11.1 (Sept. 1950), 1–16.

Although its analysis is rather ramified and sometimes seems obscure, this article very usefully discusses new trends from the Sung period—the immigration of landlords and other rural population into the cities, the extensive circulation of money-capital among this augmented urban population, the separation between capital and management, the parcelling of loan-money into small sums and accounts, class differentiation among city dwellers (especially the ruthless control of debtors by creditors), and the emergence of a kind of middle-class in the form of land-agents and managers. See also 1. 4. 7.

7. 15. 9 ABE Takeo 安部健夫, "Shindai ni okeru tentōgyō no sūsei" 清代に於ける典當業の趨勢 ("Pawnbroking in the Ch'ing period"), pp. 1–36 in **Haneda hakushi shōju kinen Tōyōshi ronsō**, 9. 8. 3.

Using the Ch'ing edicts, memorials and compendia, mainly for the 18th century, the author tabulates a considerable body of data on the numbers of pawnshops, by region and historical period; their relationship to salt merchants, many of whom, as the salt trade declined, invested in pawnshops; and government funds supplied to merchants, in order to get interest with which to relieve impoverished bannermen.

7. 15. 10 HAYASHI Kōhei 林耕平, "Chū-Shi kakuchi ni okeru tentōgyō 中支各地に於ける典當業 (The pawnshop business in various parts of Central China), **Mantetsu chōsa geppō**, 23.3 (Mar. 1943), 69–101; 23.4 (April 1943), 43–89.

A detailed and statistical field report on all the ascertainable aspects of pawnbroking in Wusih, Hangchow and Nanking.

7. 15. 11 SUZUKI Sōichirō 鈴木總一郎, "Sansei hyōsō" 山西票莊 (The Shansi banks), **Keizai ronsō**, 50.2 (Feb. 1940), 232–254.

This study draws heavily on Ch'en Ch'i-t'ien's contemporary volume (陳其田, 山西票莊考略) in describing the origin, organization, capitalization, etc., of these important institutions. Other bibliography is also cited.

7. 15. 12 NAMPEI Masaharu 南平正治, "Hōten sengyō kōkai ni tsuite" 奉天錢業公會に就いて (On the Mukden bankers' gild), **Mantetsu chōsa geppō**, 14.1 (Jan. 1934), 122–142.

This description of the organization, scope, membership, meetings, finances, etc. of this gild is prefaced by a few pages tracing the origin of the "native banks" (ch'ien-chuang 錢莊) in Mukden from branches of the Shansi banks (Shan-hsi p'iao-chuang 山西票莊) during the 19th century.

7. 15. 13 KATŌ Shigeshi 加藤繁, "Hyōkyoku ni tsuite" 標局に就いて

("On the *Hyōkyoku* in Peking"), **Shakai keizai shigaku,** 4.6 (Sept. 1934), 694–696, photo.

> Report of a field interview concerning an old-style *piao-chü* or forwarding firm which escorted or convoyed goods and specie.

7. 15. 14 IMAHORI Seiji 今堀誠二, Hyōkyoku shōshi" 鏢局小志 (Notes on the armed escort firms, *piao-chü* 鏢局), **Tōyō bunka kenkyū,** 2 (Sept. 1946), pp. 51–57.

> Based on the author's field investigation of an old firm engaged solely in convoy work at Kuei-hua-ch'eng 歸化城, this brief article provides interesting information on the old-style overland escort business, evidently comparable to former activities of the Shansi-type banks (*p'iao-hao* 票號) which have now been superseded by modern transport, communication and banking facilities. The previous study of Katō Shigeshi is quoted.

7. 15. 15 TANAKA Tadao 田中忠夫, "Hyōkyoku ni tsuite" 鏢局ニ就テ ("*Piao-chü*"), **Tōa keizai kenkyū,** 8.4 (Oct. 1924), 109–123, illus.

> Though rather ancient, this note assembles data on the history, distribution, capital, organization, management, etc., of these forwarding and insurance firms which have hardly been studied in the West. See Lien-sheng Yang, **Money and Credit in China** (Cambridge, Mass.: Harvard University Press, 1952, 143 pp.), 81–2, sec. 9. 2–3.

7. 15. 16 AOTANI Kazuo 靑谷和夫, "Kahoku yūsei ninju hoken seido kōgai" 華北郵政人壽保險制度梗概 (An outline of the postal life insurance system in North China), **Tōa keizai ronsō,** 3.2 (May 1943), 44 pp.

> A brief history of life insurance in China and the methods, terms and other procedural arrangements and results of the Post Office insurance scheme, based on a number of life insurance trade journals and with a bibliography of 63 items.

7. 16 COMMERCIAL ORGANIZATION (EXCEPTING GILDS)

Note: Chinese gilds, which are taken up under sec. 8.5 (because they have been studied somewhat more extensively from the point of view of social organization) are no doubt the most famous type of Chinese commercial organization. Nevertheless, the other forms noted in this section seem to have considerable importance also. Note that compradors are dealt with in the following section (7.17), while we include here studies of the brokers (*ya-hang*) who figured in the taxation process (7.16.4–5). We add at the end two studies of local markets (7.16.6–7).

7. 16. 1 KŌSAKA Torizō 上坂西三, **Chūgoku kōeki kikō no kenkyū** 中國交易機構の研究 (A study of commercial mechanisms in China), Waseda University, 1949, 356+index 5 pp.

This unusual and valuable study makes use of the extensive monographic literature available in Japanese, together with Chinese sources, to describe major agencies of commercial interchange in the traditional Chinese economy: (1) North China junks (*min-ch'uan* 民船), their history, types, construction (illus.), management and functions, and role in the economy; (2) the North China junk- or native ship-firms (*ch'uan-hang* 船行), in all their various aspects; (3) Manchurian grain warehouses (*liang-chan* 糧棧), their forms and economic uses, etc.; (4) the so-called "brokers" or commercial middlemen (*ya-hang* 牙行), whose functions were developed into those of tax farmers or collectors, their institutional position and fiscal activities (e.g. regarding foreign shipping, p. 233 ff.); (5) compradors (*mai-pan* 買辦), their history, types, functions, etc. Needless to say, almost any one of these sections will be illuminating to the Western student of China. That on the *ya-hang* should be particularly of interest. Neither the large Western literature already available on junks nor the extensive Japanese publications on the comprador system seem to provide such an interesting study of the interconnections among these economic institutions.

7. 16. 2 NEGISHI Tadashi 根岸佶, for the Tōa Kenkyūjo 東亞研究所 (East Asia Research Institute), **Shōji ni kansuru kankō chōsa hōkokusho, gōko no kenkyū** 商事に關する慣行調査報告書——合股の研究 (Report of the investigation of customs and practices concerning commercial affairs, research study on partnership), Tōa Kenkyūjo, 1943, 623 + bibliography 15 pp.

This study, which is usually referred to simply as **Gōko no kenkyū,** was prepared for the academic subcommittee of the Sixth Research Committee of the Institute (which functioned as a wartime scholarly agency under semi-official auspices) and deals comprehensively with the subject of Chinese commercial organization on which Mr. Negishi has specialized (see his study of the comprador system, **Baiben seido no kenkyū, 7. 17. 11**). A 30-page table of contents indicates the extremely detailed analysis here given the subject of commercial partnership in its social, legal, and financial aspects and with respect to the change from partnership to other, e.g. company, forms of organization. It is safe to say that no comparably exhaustive field study exists in Western languages.

7. 16. 3 UBUKATA Naokichi 幼方直吉, "Chū-Shi no gōko ni kansuru shomondai" 中支の合股に關する諸問題 (Some problems concerning partnerships in Central China), **Mantetsu chōsa geppō,** 23.4 (April 1943), 91–116; 23.5 (May 1943), 1–31.

Based on field study of the dyeing and weaving industry in Wusih, this noteworthy study analyzes with examples the organization and functioning of business partnerships in their transitional stage in the process of modernization, their control and use of capital, methods of distributing profits, relation to society and future prospects. The author's approach is apparently influenced by comparable studies of European economic history. He is now a leading member of the Chūgoku Kenkyūjo.

7. 16. 4 UCHIDA Naosaku 內田直作, "Chūgoku ni okeru shōgyō chitsujo no kiso, gakō seido no saikentō" 中國における商業秩序の基礎——牙行制

度の再檢討 (Basis of the commercial order in China, a reexamination of the *ya-hang* [broker] system), **Hitotubashi ronsō** 一橋論叢, 22.2 (Aug. 1949), 362–386.

This article uses Ch'ing sources to analyze the functions of *ya-hang*, mainly by a study of the official regulations concerning brokers' licences (*ya-t'ieh* 牙帖), the collection of fees for them, and the like.

7.16.5 KONUMA Tadashi 小沼正, "Kahoku nōson shishū no 'gakō' ni tsuite, tokuni chōzei kikō to shite" 華北農村市集の牙行について，とくに徴税機構として ("The Ya-hang of the Agrarian Markets in North China, especially on the Taxation System"), pp. 221–236 in **Wada hakushi kanreki kinen Tōyōshi ronsō**, 9.8.12.

An interesting analysis of how this class of brokers became officially responsible for various aspects of the regulation and taxation of market trade and, eventually, full-fledged tax farmers, i.e. collectors. Though no reference is given, this article is apparently based on the reports of the famous field investigation by the SMR and the Tōa Kenkyūjo (8.1.1), in which the author participated.

7.16.6 KATŌ Shigeshi 加藤繁, "Shindai ni okeru sonchin no teiki-ichi" 清代に於ける村鎮の定期市 (The village markets held at fixed periods during the Ch'ing dynasty), **Tōyō gakuhō**, 23.2 (Feb. 1936), 153–204.

This outstanding factual study, based on the extensive use of local gezetteers, deals with the periodicity (frequency) of the village markets (usually every 10 days), their relation to the surrounding region, their products and personnel, relations with brokers et al., taxes paid, relation to officialdom, and the like. Gazetteers were used from 4 provinces of North China and 3 of South China and general conclusions are sought as to the role of these markets in the distribution of daily necessities among the populace. Also in the same author's **Shina keizaishi kōshō** (9.8.8), 2.505–556.

7.16.7 MASUI Tsuneo 増井經夫, "Kanton no kyoshi" 廣東の墟市 (Local markets of Kwangtung), **Tōa ronsō**, 4 (May 1941), 263–283.

Subtitled "An inquiry into the modernization of markets", this brief article gives a detailed list, drawn from the local gazetteers, of the local markets in and outside the city of Canton, and discusses their relation to the port of Canton. The author emphasizes the change visible in the late Ch'ing period when merchants on the spot (*tso-shang* 坐商) became more numerous than travelling merchants (*k'o-shang* 客商); he suggests that this change may have transformed the gild system.

7.17 TREATY PORTS AND COMPRADORS

Note: This continues from sec. 3.7 above. Compilations on the treaty ports (which should be compared with gazetteers in sec. 7.2) are followed by studies of foreign firms (*yang-hang*) and finally of their compradors.

This triumvirate of the foreigners' ports, firms, and compradors we believe should be studied as aspects of a single Sino-foreign trading community. Note also item 2.3.1 (Matsuda).

7.17.1 Tōa Dōbunkai Chōsa Hensambu 東亞同文會調查編纂部 (Research and compilation bureau of the Tōa Dōbunkai), **Shina kaikōjō shi** 支那開港場誌 (Gazetteer of the open ports in China), pub. by the above bureau, vol. 1, 1922, 1120 pp.; vol. 2, 1924, 1007 pp.

These detailed compendia record data on the history, public and private institutions, government and economic life, and miscellaneous aspects of each of the treaty ports. While much of this matter is no doubt duplicated in English language works, few if any of the latter provide a similar wealth of names and terms in Chinese, to say nothing of the Japanese point of view. Vol. 1 covers Shanghai, Soochow, Hangchow, Ningpo, and Wenchow; vol. 2 covers Chungking, Wanhsien, Ichang, Shasi, Changsha, Yochou, Hankow, Kiukiang, Wuhu, and Chinkiang. The further four volumes originally projected, covering other ports, were not published.

7.17.2 IDE Kiwata 井出季和太, for Taiwan Sōtokukambō Chōsaka 臺灣總督官房調查課 (Research section of the secretariat of the Taiwan Government General), **Minami Shina no kaikōjō** 南支那の開港場 (The South China treaty ports), printed by the above agency, prefaces 1930 and 1931, 3 vols., 228+286+212 pp. (nos. 184, 198 and 204 of **Minami Shina oyobi Nan'yō chōsa** 南支那及南洋調查 [Research on South China and Southeast Asia]).

Based on field-research by the author who visited the ports in 1930, these volumes describe somewhat painstakingly the physical site and installations, communication facilities, trade, and financial or other special features, problems or products of Canton, Hongkong (vol. 1), Amoy, Swatow (vol. 2), and Foochow (vol. 3).

7.17.3 OGURA Hirokatsu 小椋廣勝, **Honkon** 香港 (Hongkong), Iwanami Shoten, 1942, 207 pp., maps (Iwanami shinsho, 88).

A history principally of Hongkong's economic development, by a competent journalist-scholar who stayed there for some 3 years, this volume presents a good deal of data on trade and investments in capsule form. Reviewed by Uchida Naosaku 內田直作 in **Tōagaku,** 6 (Aug. 1942), 213–214.

7.17.4 UCHIDA Naosaku 內田直作, "Jiyū bōekikō to shite no Ei-ryō shokuminchi Honkon" 自由貿易港としての英領植民地香港 (The British colony of Hongkong as a free port), **Shina kenkyū,** 39 (Jan. 1936), 119–144; 40 (March 1936), 151–186.

This factual study takes up in succession the (1) origin and development, (2) special features (including harbor facilities and customs system), (3) trade (its historical variability, relations with China, Japan, etc.) and (4) future of Hongkong, using mainly British sources on the subject.

7.17.5 TONOKI Keiichi 殿木圭一, **Shanhai 上海** (Shanghai), Iwanami Shoten, 1942, 170+index 17 pp. (Iwanami shinsho, 89).

A well-condensed summary of Shanghai's modern history and institutions as of 1941, by a journalist who stayed there for three years; well indexed.

7.17.6 UCHIDA Naosaku 内田直作, "Yōkō seido no kenkyū" 洋行制度 の研究 (A study of the foreign firm system), **Shina kenkyū,** 50 (Mar. 1939), 187–211.

This brief article takes a quick look at the major foreign firms (*yang-hang*) and their position in China and also touches upon their internal organization.

7.17.7 UCHIDA Naosaku 内田直作, "Zai-Shi Eikoku shōsha I-wa [I-ho, Ewo] yōkō no hatten shiteki bunseki" 在支英國商社怡和洋行の發展史 的分析 (An historical analysis of the development of the British firm of Jardine, Matheson and Co. in China), **Shina kenkyū,** 51 (June 1939), 213–240: 52 (Nov. 1939), 151–192.

This survey analyzes the structure of subsidiary firms and interests which ramified from J. M. and Co. over the years, including some attention to the Chinese compradors associated with the firm. Evidently based on easily available, although scattered sources, in an effort to show British commercial imperialism at work, this article benefits from the author's earlier studies of the comprador system and foreign firms in general. See also his later article in a more interpretative framework: "Zai-Shi Eikoku keizai no kōsei" 在支英國 經濟の構成 (The composition of the British economy in China), **Hitotsubashi ronsō 一橋論叢,** 7.3 (March 1941). Prof. Uchida is now in the Seijō University 成城大學 in Tōkyō.

7.17.8 ŌI Senzō 大井專三, "Shina ni okeru Ei-Bei Tabako Torasuto no keiei keitai, zai-Shi gaikoku kigyō no hatten to baiben soshiki no ichikōsatsu" 支那に於ける英米煙草トラストの經營形態——在支外國企業 の發展と買辦組織の一考察 (The form of administration of the British American Tobacco Company in China, a study of the development of foreign enterprises in China and of the comprador system), **Tōa kenkyū shohō,** 26 (Feb. 1944), 1–47.

First tracing the growth of China's foreign trade and of the comprador system and its transformation, this study then looks at the B.A.T. as a prime example of this development. It notes the history of the growth of tobacco in China, the organization of the B.A.T., its branches and subsidiaries, form of capitalization, profits, and taxation, stressing the typically monopolistic activity of the B.A.T., operating through the transformed comprador system organized into its distributing network.

7.17.9 SUZUKI Sōichirō 鈴木總一郎, "Baiben seido" 買辦制度 (The comprador system), **Tōa keizai ronsō,** 1.1. (Feb. 1941), 20 pp.; "Baiben hassei no shakaiteki konkyo" 買辦發生の社會的根據 (Social causes of the emergence of compradors), ibid., 1.3 (Sept. 1941), 17 pp.; "Kakyō to baiben" 華僑と買辦, ibid., 2.4 (Dec. 1942), 20 pp.

These studies trace the origin of the comprador system, divide it into types and stages, and analyze its social causes and limitations, and finally compare the rise of the comprador with that of overseas Chinese merchants. Building on the previous Japanese studies of this important subject, the author presents an interesting, though somewhat rough and sketchy, interpretation and analysis.

7.17.10 UCHIDA Naosaku 內田直作, "Baiben seido no kenkyū" 買辦制度の研究 (A study of the comprador system), **Shina kenkyū,** 47 (July 1938), 19–36; 48 (Dec. 1938), 1–28; 49 (Jan. 1939), 1–24.

This essay surveys the origin of the comprador and his heyday in the later 19th century, the changes which overtook him in the 20th century, the comparable institutions in India and Japan, the functions of the comprador as a middleman, his security, profits and managerial system, and the various ways in which the system has changed, with the specialization of functions.

7.17.11 NEGISHI Tadashi 根岸佶, **Baiben seido no kenkyū 買辦制度の研究** (A study of the comprador system), Nippon Tosho Kabushiki Kaisha 日本圖書株式會社, 1948, 392 pp.

This comprehensive account of the origin, development and working of the comprador system summarizes many years of study. The author joined the Tōa Dōbun Shoin in Shanghai in 1901 and a volume on the comprador system was included in that institution's 12 volume work **Shina keizai zensho,** 7.2.1, which was completed in 1907 at a time when Japanese firms like Mitsui Bussan had already begun to dispense with compradors. Negishi analyzes not only the institutional origin of the treaty-port comprador system, as a necessary successor to pre-treaty arrangements, at Canton, but also the comprador's contractual and practical relations with his foreign employers and his important functions in the Chinese trading and financial communities. Compradors in India and Japan (*shōkan bantō* 商館番頭), comprador types, their guilds, their economic and social significance and the decline of the system are discussed in detail, although rather discursively and without footnotes. While British records mainly document the origin of the system, Japanese scholars have done the most extensive research on it, as part of their study of the British economic position in China. Bibliographical list, 61 items. See review by Ishikawa Shigeru 石川滋 in **Hitotsubashi ronsō,** 22.6 (Dec. 1949), 782–789.

7.18 FOREIGN INVESTMENTS AND FOREIGN TRADE

Note: These studies are mainly a product of a big wartime survey that made a general accounting of the Western investments, which the new order seemed to be inheriting. Studies of trade are at the end and a final item on boycotts; see also sec. 2.2 above on the Canton trade. For domestic trade, see the big compilations in sec. 7.2 (item 7.2.2 is on foreign trade as of 1892).

7.18.1 Tōa Kenkyūjo 東亞研究所, **Rekkoku tai-Shi tōshi to Shina kokusai shūshi 列國對支投資と支那國際收支** (Investments by the foreign powers in China and China's international receipts and payments), Tōa

Kenkyūjo, 1941, 258 pp.; republished with different preface by Jitsugyō no Nipponsha, 1944, 256 pp.

As part of the big Tōa Kenkyūjo wartime research program, this study was initiated in 1938, carried on both in Tōkyō and among commercial and financial records in the chief mainland cities (excepting Manchoukuo) and participated in by a number of scholars including Hirase Minokichi 平瀬巳之吉, Usami Seijirō 宇佐美誠次郎 and Matsuda Tomoo 松田智雄. It presents a summary of (1) Japanese investments in China, (2) other foreign investments, by types, (3) China's balance of foreign trade and (4) her balance of "invisible" payments other than trade (overseas Chinese remittances, etc.). In intensity and scope of sources, this study, and the larger volumes which followed it, 7.18.2–5, would appear to outweigh the standard Western work, of an earlier date, by Professor C. F. Remer, although it is doubtful that the latter's study is superseded insofar as analysis is concerned. The edition republished in 1944 by Jitsugyō no Nipponsha is a revised version and so can be considered as the final result of this extensive research project.

7.18.2 Tōa Kenkyūjo 東亞研究所, **Shogaikoku no tai-Shi tōshi 諸外國 の對支投資** (Investments of the foreign countries in China), printed by the Tōa Kenkyūjo, vol. 1 (classified confidential 秘), 1942, 679 pp.; vol. 2, 1943, 1217 pp.; vol. 3, 1943, 670 pp.; many tables and maps.

These massive volumes formed a report of the First Investigation Committee of the Tōa Kenkyūjo and provide in full form the data on which the summary volume of 1941, **Rekkoku tai-Shi tōshi to Shina kokusai shūshi**, 7.18.1, had already been based. Working systematically over a wide range of topics and materials, the research team put together a mass of evidence on each aspect of foreign investment in China: insurance, transportation, industry and mining, philanthropy, shipping, aircraft, posts, railroads, government funds, cultural works, etc. This seems to have amounted to an audit of the whole Western stake in China, as the Japanese were able to appraise it after their occupation.

7.18.3 SUGIMURA Kōzō 杉村廣藏, comp., for Tōa Kenkyūjo, **Shōwa jūichinen matsu genzai rekkoku tai-Shi tōshi gaiyō 昭和十一年末現在列 國對支投資概要** (A survey of investments by the foreign powers in China from the end of 1936 to the present), Tōa Kenkyūjo, 1943, 430 pp. (preface by the compiler dated Nov. 1940).

This special volume 別冊 among those published on this subject by the First Investigation Committee of the Tōa Kenkyūjo, deals with foreign investments in connection with Chinese finance and trade, foreign industrial concerns, government funds and cultural works. It is similarly detailed and includes 234 tables.

7.18.4 Tōa Kenkyūjo 東亞研究所, **Rekkoku no tai-Shi tōshi (bessatsu), Honkon ni okeru Eikoku no tōshi 列國の對支投資（別冊）──香港に於ける 英國の投資** (Investments of the foreign powers in China, special volume, British investment in Hongkong), Tōa Kenkyūjo, 1941, 83 pp., classified confidential 秘.

This factual wartime study by Matsuda Tomoo and others was later included in his book, **Igirisu shihon to Tōyō**, 2.3.1.

7.18.5 HIGUCHI Hiroshi 樋口弘, **Nippon no tai-Shi tōshi kenkyū** 日本の對支投資研究 (A study of Japanese investment in China), Seikatsusha, 1939, 704+index 16 pp.

The author was with the economic journal **Daimond ダイヤモンド** and prints in his preface two letters of encouragement from Prof. C.F. Remer, as a leading specialist in this field. His volume first surveys the history of foreign investment in China and the phases of Chinese government policy concerning it (to p. 224) and then proceeds to work out the various forms and aspects of Japanese investment in considerable detail (part 2, pp. 225-634): direct and indirect investments, loans of all kinds, local agencies and activities region by region, banks and *zaibatsu* enterprises, the historical position of Japan's China investment in her whole foreign investment, her profits and international competition. The last section is on Japanese investment in the new Asia as then foreseen. This appears to be the most comprehensive study of the subject.

7.18.6 TAMURA Kōsaku 田村幸策, **Shina gaisai shiron** 支那外債史論 (An historical study of China's foreign loans), Gaikō Jihōsha 外交時報社, rev. ed., 1936 (1st ed., 1935), 576+bibliog. 8 pp.

This systematic and comprehensive account of foreign loans runs from before the Sino-Japanese war, through the indemnities, railroad loans, and consortium activities, down to the 1930s. It is based largely on works in English, and stresses the legal aspect. Dr. Tamura is a former Foreign Service official who also spent time in China. He is now at Chūō University.

7.18.7 KASHIWAI Kisao 柏井象雄, "Shinchō makki ni okeru gaisai" 清朝末期に於ける外債 (Foreign loans in the late-Ch'ing period), **Tōa jimbun gakuhō**, 1.1 (Mar. 1941), 279–302.

A brief survey of Ch'ing fiscal policy as highly dependent (in the context of China's semi-colonial status) upon foreign loans from as early as 1865; based on commonplace materials.

7.18.8 USAMI Seijirō 宇佐美誠次郎, "Bōseki shihon no tai-Ka yushutsu ni tsuite" 紡績資本の對華輸出について (On the foreign investments in the cotton industry in China), **Tōhō gakuhō, Tōkyō,** 11.1 (Mar. 1940), 99–110, tables.

Brief interpretative notes, which pay special attention to the Japanese investment in China and its relationship to the whole economic structure of Japan. Formerly in the Tōhō Bunka Gakuin, the author is now at Hōsei University.

7.18.9 UCHIDA Naosaku 内田直作, "Shina taigai bōeki no judōsei no tokushitsu, kokunai kōgyōka katei no teitai to kanren shite" 支那對外貿易の受動性の特質──國内工業化過程の停滞と關聯して (The characteristic passivity of the foreign trade of China, in connection with the stationariness of the process of domestic industrialization), **Shina kenkyū,** 46 (March 1938), 67–88.

This analytic and statistical article explores (1) the arrested state of China's domestic industrialization as visible in the pattern of import and export of commercial goods and (2) the passivity of China's foreign trade and its significance for the domestic economy, concluding that Japanese trade must remedy the situation.

7. 18. 10 Tōa Kenkyūjo 東亞研究所, **Shina no bōeki shūshi** 支那の貿易收支 (China's commercial receipts and payments), pub. by Tōa Kenkyūjo, 1942, 458 pp., as a research report of the First Research Committee; **Shina no bōekigai shūshi** 支那の貿易外收支 (China's non-commercial receipts and payments), pub. by the same, 1942, 463 pp., as a research report of the same committee.

These two volumes formed a paralled product to the extensive and ramified study of foreign investments in China (see 7. 18. 1). Compiled by the economic historian Hirase Minokichi 平瀬巳之吉 and others, the first volume amasses data on receipts and payments of commercial goods and of currency for all of China and then presents tables arranged by regions. The second volume, compiled by Nambu Nobuharu 南部農夫治, Matsuda Tomoo 松田智雄 and others, seeks the balance of invisible payments through study of foreign loans, investments, Boxer indemnity payments, official, military and naval expenditures, overseas Chinese remittances, sea transport and insurance costs, and cultural and other expenditures.

7. 18. 11 IIDA Tōji 飯田藤次, "Chūgoku keizai to bōeki" 中國經濟と貿易 ("The Chinese Economy and Foreign Trade"), **Tōyōbunka kenkyū,** 7 (Feb. 1948), 18–33.

A competent technical economic study of the various factors affecting the contemporary Chinese balance of trade, relying on Maritime Customs figures and completed in August 1947. The author conducted research in Shanghai under the wartime Tōa Kenkyūjo, and is now a Foreign Office researcher.

7. 18. 12 Gaimushō Chōsakyoku 外務省調査局, **Chūkyō no bōeki** 中共の貿易 (The trade of the Chinese People's Republic), pub. by the Foreign Office for office use, 1951, vol. 1, **Enkaku seisaku hen** 沿革政策篇 (History and policy), 372 pp.; vol. 2, **Shiryō hen** 資料篇 (Source materials), 201 pp.

This study written by Kaneko Kazuo 金子一夫, a Foreign Office researcher, makes a careful record of the steps and pronouncements, agencies and procedures, of the new Peking regime concerning foreign trade and its various policies regarding commercial goods, barter and link arrangements, and the like. Vol. 2 gives statistics, and laws and regulations.

7.18. 13 UCHIDA Naosaku 内田直作, "Shina taigai boikotto no kenkyū, Shina taigai bōeki no higōrisei no tokushitsu" 支那對外ボイコットの研究 ——支那對外貿易の非合理性の特質 (A study of Chinese boycotts against

foreigners, a characteristic irrationality in the foreign trade of China),
Shina kenkyū, 47 (July 1938), 37–62; 48 (Dec. 1938), 29–46.

An account of the forms of boycott as they developed in China—concerted
stoppage of sales in the Canton period, refusal to buy (1905–1923), and breaking
off of economic relations as at Hongkong (1925–); the effect of the anti-Japanese boycott on Sino-Japanese trade; and the main causes of boycotts—economic, political, psychological, including their connection with gild psychology.

7.19   POST-WAR ECONOMIC DEVELOPMENTS

Note: While beyond our scope, strictly speaking, these articles offer
some guidance on matters generally obfuscated by propaganda.

7.19.1   Gaimushō Chōsakyoku, comp., **Sengo ni okeru Chūgoku keizai**
戰後における中國經濟 (The Chinese economy in the post-war period),
printed by the Foreign Office for office use, Dec. 1948, 249 pp.

This report by half a dozen staff members of the Fifth Section, while
based on Chinese publications rather than any exclusive information, is a
conscientious effort to appraise the state of Nationalist China's unhappy
financial system, industry and mining, agricultural and fishery productions,
foreign trade, smuggling trade (pp. 167–178), and the inflation problem up to
the fatal reform of 1948, with many tables and 30 pages of Chinese statistics.

7.19.2   Ishikawa Shigeru 石川滋, "Shin Chūgoku ni okeru kahei keizai
no seikaku" 新中國における貨幣經濟の性格 ("The Character of Monetary
Economy in New China"), **Hitotsubashi ronsō,** 23.5 (May 1950), 385–411.

Written in March 1950 by a highly competent journalist of the Jiji Press
時事通信社 foreign news staff, who was trained in economics, this research
article appraises the financial system inherited by the CCP from the KMT,
the elements from it which then persisted under the new regime (problems of
inflation, etc.), and CCP methods for dealing with them.

7.19.3   Ishikawa Shigeru 石川滋, "Chūgoku keizai hatten no ryōteki
haaku" 中國經濟發展の量的把握 ("The quantitative analysis of the
development of the Chinese economy"), **Tōyō bunka,** 8 (Feb. 1952),
44–75, charts.

This is a bold attempt at a comprehensive developmental analysis of the
Chinese economy. The article is divided into three sections entitled: 1) the
method of ascertaining the tempo of "development in backward nations";
2) an estimate of "the stationary structure" of China as of 1933; 3) an
appraisal of "the developmental structure" of China as of 1950.

7.19.4   Ōta Eiichi 太田英一, "Shina keizai to Daitōa senso" 支那經濟
と大東亞戰爭 (The Chinese economy and the Greater East Asia War),
Tōagaku, 6 (Aug. 1942), 135–180.

An essay which briefly but systematically surveys the economic impact of the war on aspects of the Chinese economy—the foreign concessions, occupied territory, Chungking—and tries to appraise future developments. Many such articles are of course available in contemporary periodicals; this one is not outstanding. See the author's volume, **Shina keizai no kōzō** 支那經濟の構造 (Structure of the Chinese economy), Nippon Hyōronsha, 1943, 332+5 pp., a survey work.

7.19.5 HATADA Takashi 旗田巍, "Shin Chūgoku ni okeru tochi kaikaku" 新中國における土地改革 ("The Land Reform Law under the Communist Party of China"), **Jimbun gakuhō** 人文學報 ("The journal of Social Sciences and Humanity"), ed. by "The Institute for Social Sciences and Humanity of Tōkyō Metropolitan University" 東京都立大學人文學會, 8 (Mar. 1953), 79–104.

Subtitled "with special attention to the preservation of the rich farmers' economy," this article by an experienced specialist on the Chinese rural society discusses the controversial problem of the rich farmer ("kulak") involved in the Land Reform Law promulgated in 1950. The author appraises the new policy of 1950 in its historical context and considers its possible consequences. See the author's background article on land reform (7.6.1).

7.19.6 MURMATSU Yūji 村松祐次, "Chūkyō tochi kaikaku no futatsu no jiki" 中共土地改革の二つの時期 ("Two Stages of the Land Reform in China"), **Hitotsubashi ronsō**, 23.5 (May 1950), 438–456.

A brief and sophisticated appraisal of the historical stages in the CCP land reform program (from the Kiangsi to the post-war periods), by a specialist in Chinese economic and sociological research.

7.19.7 HISAE Fukusaburō 久重福三郎, "Chūgoku no tochi kaikaku to kōchi menseki" 中國の土地改革と耕地面積 ("Land Reform and Cultivated Area in China"), **Kōbe gaidai ronsō** 神戸外大論叢 ("The Kobe City University Journal"), 3.1 (April 1952), 55–74, 8 tables.

This article offers a study of the land reform process (which is far from an "equal" division of land), an estimate of cultivated area (which is not enough to solve the agrarian problem), and a guess as to potentialities for increasing the cultivated area (which may amount to perphaps 30% in all of China proper).

8. CHINESE SOCIETY

Note: In one sense, which some seem to favor, all studies of China are studies of Chinese society in one or another aspect. Under sec. 1.5 we have called attention to the large body of writing that deals with the nature of Chinese society, particularly the Marxist interpretation of it. In the present section we note mainly professional academic studies by leading sociologists.

8.1 SOCIAL STRUCTURE: GENERAL STUDIES

Note: The first item below speaks for itself. Following it, are several items of professional sociological analysis in the strict sense of the term, among which attention should be called to the work of Fukutake Tadashi (item 8.1.8). Several volumes of somewhat broader scope are then noted, among which that of Muramatsu Yūji (item 8.1.12) seems particularly noteworthy. At the end are older studies which may have historical if not practical interest. Works of Shimizu Morimitsu are listed chiefly in sec. 8.2 and 8.3.

8.1.1 NIIDA Noboru 仁井田陞, representing Chūgoku Nōson Kankō Chōsa Kankōkai 中國農村慣行調查刊行會 (Society for the publication of the investigations concerning Chinese rural customs and practices), ed., **Chūgoku nōson kankō chōsa** 中國農村慣行調查 ("Rural Customs and Practices of China"), vol. 1, Iwanami Shoten, 1952, 320 pp., many photos, maps and tables, (2nd printing, 1953, with some comments added for photos.); vol. 2, 1954, 557 pp., many tables.

These initial volumes of a projected series (in 7 vols.) are the first formally published product of the justly famous rural research program carried out in Hopei and Shantung in the period 1939-43 by a team of field specialists drawn from the staff of the Research Department of the South Manchurian Railway Company. This field team under Suginohara Shun'ichi 杉之原舜一 collaborated closely with a group of legal scientists, mainly from Tōkyō University, who worked under the auspices of the wartime Tōa Kenkyūjo and were headed by Dr. Yamada Saburō 山田三良. These two academic and field-research groups met in a series of conferences and each formulated an analytic topic outline or plan of research categories (see 1. 33-44 for the Tōkyō group's outline, 1. 45-56 for the SMR outline). Both were inspired by the late Dr. Suehiro Itsutarō 末弘嚴太郎 (d. Sept. 1951), an outstanding specialist in modern legal science in Japan, some of whose original research suggestions are printed in

1. 17–32. The highly-organized result of these preparations was the compilation in Hopei and Shantung during 1940–41–42 of 123 volumes of data concerning "general features, family, village, land ownership, tenancy, irrigation, taxes and rates, purchase and sale of land, and finance and trade, etc". Copies of these volumes, circulated among the staff in duplicated form, have been used as the basis for some 50 or 60 articles and books published by individuals in recent years (listed on pp. 63–65 of vol. 1). Beginning with the present two volumes it is hoped to make the bulk of these materials available through publication.

Ably edited by Messrs. Niida Noboru 仁井田陞, Fukushima Masao 福島正夫, Ubukata Naokichi 幼方直吉, Andō Shizumasa 安藤鎮正, Uchida Chiyū 内田智雄, Konuma Tadashi 小沼正, Hatada Takashi 旗田巍, Honda Etsurō 本田悦郎 and Maeda Katsutarō 前田勝太郎, the first volume prints the report on Shun-i-hsien 順義縣 in Hopei and its sections on "general features", "village", "land purchase and sale" and "family" consist of full field records based on interviews conducted through interpreters on the spot in the villages. Thanks to careful methodological preparation, the data and statements derived from the Chinese informants are organized clearly and succinctly while retaining much of their original colloquial flavor. The detailed table of contents helps to locate subject matter. An "explanatory comment" by Hirano Yoshitarō 平野義太郎 (pp. 5–12) gives a somewhat critical appraisal of the project and its results, against the scholarly background of the research that went into it, and also compares it with the field investigation report by Mao Tse-tung of 1927. The second volume, by the same editors except Prof. Uchida, continues the report on Shun-i-hsien with sections on "tenancy," "land purchase and sale," "money-lending and trade," "water" (rivers, wells and ponds) and "taxes and dues (including landed property and land registration)".

8.1.2 SHIMIZU Morimitsu 清水盛光, **Shina shakai no kenkyū, shakai-gakuteki kōsatsu** 支那社會の研究──社會學的考察 (A study of Chinese society, a sociological investigation), Iwanami Shoten, 1939, 422 pp.

This volume is a revised compilation of the author's articles published in **Mantetsu chōsa geppō.** As a study of the old Chinese social structure, it analyzes the power of the old gild system, the traditional bases of the authoritarian power of the state, and self government in the village community. The last chapters on the Chinese family system were superseded by the author's later volume **Shina kazoku no kōzō,** 8.4.8. Now at the Kyōto Institute, the author was attached for many years to the SMR research department and concentrated on library research, abstaining on principle, it is said, from personal participation in any field survey. Though apparently impressed by the once fashionable Marixist theory of the Asiatic mode of production—especially its emphasis on the closed and self-sufficient village community as the basis of Asiatic despotism—Prof. Shimizu has followed the sociological approach of Durkheim rather than stressing class differentiation. His later writings have more and more inclined to point out, rather "a-historically", the importance of the traditional cooperative relationships in every aspect of Chinese social life. As a critique of his approach, see Hatada Takashi, "Chūgoku ni okeru senseishugi to 'Sonraku kyōdōtai riron'," 8.2.9.

8.1.3 USUI Jishō 臼井二尙, "Shina shakai no chiikiteki kitei" 支那社會の地域的規定 (Regional determinants of Chinese society), **Tōa jimbun gakuhō,** 1.2 (Sept. 1941), 325–369.

Based on well known materials such as thcse of A. H. Smith, D. H. Kulp, G. B. Cressey, H. B. Morse and Negishi Tadashi, this article by a sociologist discusses the many aspects of social life in the Chinese rural community which are derived from its closed and microcosmic situation. There is some mention also of cities and gilds.

8.1.4 USUI Jishō 臼井二尚, "Shina shakai no kaisōteki kitei" 支那社會の階層的規定 (Hierarchical determinants of Chinese society), **Tōa jimbun gakuhō,** 1.3 (Dec. 1941), 627–675.

Based on Western and Japanese secondary materials, this is an unappetizing sociological discussion of various aspects of the hierarchical social order inside the dispersed and closed rural communities of China. It also mentions the bureaucracy and the traditional examination system.

8.1.5 USUI Jishō 臼井二尚, "Shina no shakai to minzoku" 支那の社會と民族 (Society and nation in China), **Tōa jimbun gakuhō,** 4.1 (Mar. 1944), 34–92.

Supplementary to the author's two articles on Chinese society ("Shina shakai no chiikiteki kitei", 8.1.3, and "Shina shakai no kaisōteki kitei", 8.1.4), this article discusses how and why the Chinese nation has not been able to reach a modern stage based on an inter-local and inter-strata sense of national unity. He sees the regional and hierarchical fragmentation of Chinese society as the obstructive factors, and is sceptical of Sun Yat-sen's thesis that China can attain nationhood through an expansion and federation of clan institutions. Extensive bibliographical references are made to German and other Western studies of nationalism and the factors producing it.

8.1.6 SHIGEMATSU Toshiaki 重松俊明, "Shina no shakai to kokka" 支那の社會と國家 (Society and the state in China), **Tōa jimbun gakuhō,** 3.1 (Mar. 1943), 144–163; 2 (Oct. 1943), 267–323.

A German-trained sociologist here discusses the relation to each other of society and state in China, against their historical background. He applies to China the following sociological scheme of periodization: 1) folk society (genossenschaftliche Gesellschaft), 2) primitive class society (primitive Klassengesellschaft), 3) status society (Ständegesellschaft), 4) bourgeois society (bürgerliche Gesellschaft). According to the author, China entered in the Sung period into the fourth stage, which is also the stage of nationalism, but has remained almost in its beginning phase (hence China's so-called "stationariness"). This is probably because the enormous extent of her territory and population coincided with a low level of technology. The author seems to have been theoretically influenced by Tönnies and Durkheim, among others; as for Chinese history, he follows the interpretation of Dr. Naitō Torajirō.

8.1.7 SHIGEMATSU Toshiaki 重松俊明, "Shakai hattenshi yori mitaru Shina shakai no ichi" 社會發展史より見たる支那社會の位置 (The position of Chinese society in the history of social development), **Tōa jimbun gakuhō,** 2.1 (Mar. 1942), 27–70.

A sociologist here analyzes the Chinese ideas of face (*mien-tzu* 面子) and of the rites (*li* 禮), as an approach to understanding the nature of Chinese society. He distinguishes between *mien-tzu* of the "Gemeinschaft" type in a closed rural community, which is derived from the sense of shame and forms the prototype of *mien-tzu*, and *mien-tzu* of the "Gesellschaft" type in commercial relations, society and court life, which is based on the sense of vanity and is the formalistic and degenerate form of *mien-tzu*. Two kinds of *li*, a "Gemeinschaft" and a "Gesellschaft" type, are similarly distinguished. The interrelationship of *mien-tzu* and *li* and the position of *li* among the categories of social rules are then discussed. Finally, the present stage of Chinese society is defined as a "half-civilization of early Gesellschaft" (Halb-Zivilisation der Frühgesellschaft) stage, with an "egoistic" rather than "individualistic" basis for society. This Germanic sociological study, like many others, quotes A. H. Smith, *Chinese Characteristics*.

8.1.8 FUKUTAKE Tadashi 福武直, **Chūgoku nōson shakai no kōzō** 中國農村社會の構造 (The structure of Chinese village society), Yūhikaku, revised and enlarged ed., 1951, 507+51 pp., English summary 3 pp., first preface 1945 (1st ed., Taigadō, 1946, 507 pp.)

By a sociologist at Tōkyō University who is now specializing on Japan, this volume describes the rural social structure of China on the basis of the author's half dozen wartime field trips. He surveys the major aspects of the village society of Central China (as seen in Kiangsu and Chekiang) and the family and village structure of North China (Hopei and Shantung), with their various forms of organization. He denies the basic unity of the traditional rural family, where equal inheritance has been a disintegrating factor, and doubts also the group-consciousness and unitary cohesiveness of the old village, which has been neither as isolated, self-sufficient and microcosmic as is often supposed nor as tightly organized and exclusive. In this revised edition the critical comments of half a dozen other specialists are reprinted (app., pp. 19–51) as well as an essay comparing the Chinese and Japanese village (pp. 1–18). While criticized for an arbitrary selection of data and for relying on unrepresentative studies, this volume nevertheless makes a very cogent presentation.

8.1.9 FUKUTAKE Tadashi 福武直, **Chūgoku sonraku no shakai seikatsu** 中國村落の社會生活 (Social life of the Chinese village), Kōbundō, 1947, 161 pp. (Kyōyō bunko, 147).

A brief, clearly-organized but greatly-generalized sociological description of the Chinese village from all aspects, which condenses on a higher level the general picture already worked out in the same author's **Chūgoku nōson shakai no kōzō**, 8.1.8.

8.1.10 Nippon Jimbunkagakukai 日本人文科學會, ed., **Hōken isei** 封建遺制 (Institutional remnants of feudalism), Yūhikaku, 1951, 334 pp.

This volume prints the papers presented at the November 1949 meeting of the above society under a Ministry of Education subvention, feudalism proper having been discussed in the preceding year and "social tension" subsequently. All but the following three papers relate to feudalism in Japan or elsewhere than China:

(1) Amano Motonosuke 天野元之助, "Hōken isei to Chūgoku nōson" 封建遺

制と中國農村 (Institutional remnants of feudalism and the Chinese village), pp. 197-218. A factual and somewhat fragmentary appraisal.

(2) Adachi Ikutsune 安達生恒, "Chūgoku no shihon no seikaku, hikindaisei tono kanren ni oite" 中國の資本の性格—非近代性との關連において (The nature of Chinese capital, in connection with its non-modern character), pp. 219-234. Stresses the characteristic functioning of Chinese capital on the basis of concrete and personal connections.

(3) Hatada Takashi 旗田巍, "Chūgoku shakai no hōkensei, hōkenshakai no mura to Chūgoku no mura" 中國社會の封建性—封建社會の村と中國の村 (The feudal character of Chinese society, Chinese villages as compared with the villages in a [typically] feudal society), pp. 235-260. A brief but comprehensive reappraisal of the problems involved, using the reports of the field investigation in North China by the SMR and the Tōa Kenkyūjo.

8.1.11 TACHIBANA Shiraki 橘樸, **Shina shakai kenkyū** 支那社會研究 (A study of Chinese society), Nippon Hyōronsha, 1936 (4th printing, 1941, 588 pp., commentary postscript of 16 pp. by Ōkami Suehiro 大上末廣).

This influential socio-historical study treats the old class structure and the status of landlord and peasant, and then analyzes the special characteristics, in the Chinese scene, of the capitalist, laborer, and mandarin-bureaucrat (the last is based on the book, **Kuan-ch'ang hsien-hsing-chi 官場現形記**). It concludes with a chapter on the family system. This volume, like the two other volumes of the same author, was put together by his friends from articles scattered in many different periodicals. Tachibana Shiraki (1881–1945) was a member of the South Manchurian Railway Co. research staff for many years. Stimulated by the contemporary revolutionary movement, he began to study Marxist-Leninist literature in the 1920s and was deeply influenced by it but remained critical of the interpretions put forward by Radek, Trotsky, Bukharin and others, as well as the actual activities of the Chinese Communist Party. He established his own idea of periodization in Chinese history (see the commentary postscript by Ōkami Suehiro): 1) tribal community—Hsia 夏 and Yin 殷; 2) decentralized feudal society—Chou 周; 3) centralized feudal society—from Ch'in 秦 to T'ang 唐; this period saw a conflict between the centralized despotic state and the landed aristocracy; 4) the stage of semi-feudal merchant-capitalism—from Sung 宋 to Ch'ing 淸; this period saw the growth of a mammoth despotic bureaucracy based on the examination system (*k'o-chü* 科擧), as well as on the accumulation of huge merchant capital holdings; 5) the present stage of China is defined as "mandarin-precapitalistic". Tachibana's writings are useful also in providing much factual information.

8.1.12. MURAMATSU Yūji 村松祐次, **Chūgoku keizai no shakai taisei** 中國經濟の社會態制 (The social structure of China's economy), Tōyō Keizai Shimpōsha 東洋經濟新報社, 1949, 400 pp., 61 tables mainly of statistics. (Gendai keizaigaku sōsho, 24).

Completed in 1949 on the eve of the communist accession to power in China, and based on the extensive use of recent Japanese research on modern China including the reports of field investigation in North China by the South Manchurian Railway Co., this theoretical study by a scholar versed in modern economics and Max Weber's writings and also under the influence of W. Sombart, seeks to define systematically not only the objective conditions and general characteristics of the Chinese economy, but also its relation to political and social institutions. The result is one of the most comprehensive and penetrating analyses of the structure of Chinese society which has yet been attempted. The author's major sections deal with (1) problems of analytic method, (2) general conditions of agricultural and industrial production and trade, (3) the "external structure" (*gaibu taisei* 外部態制) or processes of the economy in its relation to all sorts of governmental administrative activities, village, family clan and gild associations of all kinds, and the Chinese type of market operations, (4) the "internal structure" of the economy (*naibu taisei* 內部態制) visible in the institutions of agriculture, trade and industry (e.g., landlord-tenant relations, commercial partnerships, labor contracting), and (5) prospects of capitalism and socialism. In so ambitious an effort the author is of course obliged to stick pretty much to general statements. His view of the promise of Mao's New Democracy may have been modified by later events. In any case this book has the merit of being a stimulating analytic work-out on a non-marxist and undogmatic basis. The author is now a professor of the Hitotsubashi University 一橋大學 (The former Tōkyō Commercial College 東京商科大學). See the reviews by Hashimoto Hideichi 橋本秀一, in **Tōyō bunka,** 2 (May 1950), 126–127; by Masubuchi Tatsuo 増淵龍夫 in **Hitotsubashi ronsō,** 22.5 (Nov. 1949), 692–697.

8.1.13 IIZUKA Kōji 飯塚浩二, **Sekaishi ni okeru Tōyō shakai** 世界史における東洋社會 (Oriental society in world history), Mainichi Shimbunsha 毎日新聞社, 1948, 253 pp.

This rather stratospheric sociological discussion deals in successive chapters with 1) the nomadic peoples (especially Chinghis Khan), from a geographico-anthropological point of view stressing the interdependent relationship between nomads and caravan traders, 2) the agrarian society of monsoon Asia, 3) its special features and essential character, 4) the social structure of Southeast Asia, and 5) the old question of oriental change and arrested development. Not many sources are available, or cited, for so broad an essay. See the reviews by Hanamura Yoshiki 花村芳樹 in **Tōyō bunka kenkyū,** 10 (Feb. 1949), 56–60: and by Momose Hiromu 百瀬弘 in **Shakai keizai shigaku,** 15. 3-4 (Oct. 1949), 108–109. The author is now in the Tōyō Bunka Kenkyūjo, Tōkyō University.

8.1.14 INABA Kunzan 稲葉君山, **Shina shakaishi kenkyū** 支那社會史研究 (An historical study of Chinese society), Daitōkaku 大鐙閣, 1922, 432 pp.

This early historical appraisal touches on many sociological questions since pursued by more recent research: special characteristics of Chiness society, familism, class struggle, cultural change, the antireligious movement, economic bases of Buddhism (80 pp.), taxes on merchants or brokers (56 pp.), international control, Chinese Mohammedans, etc. In these scholarly essays the author follows roughly the same approach as Dr. Naitō.

8.1.15 ITŌ Takeo 伊藤武雄, **Gendai Shina shakai kenkyū** 現代支那社會研究 (A study of contemporary Chinese society), Dōjinsha Shoten 同人社書店, 1927, 347 pp.

An early economic and sociological study of China's population, capital, labor, the KMT-CCP conflict, national liberation movement, and strikes in the textile industry, with a final essay contrasting north and south—a pioneer study important in its day. The author was a leading member of the SMR research bureau for many years, and influential as an organiser of research.

8. 1. 16 KATŌ Shigeshi 加藤繁, **Shina no shakai** 支那の社會 (The society of China), in **Iwanami kōza Tōyō shichō** (9. 8. 4), series 14, 1935, 54 pp.

A brief characterization of successive periods, ancient to modern, with six pages on Ming and Ch'ing which touch on the examination system and bureaucracy. This compact essay has been reprinted under the title of "Chūgoku no shakai" in the author's posthumous book: **Chūgoku keizaishi no kaitaku** 中國經濟史の開拓 (Pioneering in the study of Chinese economic history), ed. by Enoki Kazuo 榎一雄, Ōgiku Shoin 櫻菊書院, 1948, 268 pp. (including 102 page biography of Dr. Katō by Enoki and 18 page list of Katō's writings).

8. 1. 17 OZAKI Hotsumi 尾崎秀實, **Chūgoku shakai no kihon mondai** 中國社會の基本問題 (Basic problems of Chinese society), Sekai Hyōronsha, 1949, 306 pp.

The author of this volume, a Tōkyō University graduate formerly with the **Asahi shimbun** and the SMR and at one time an adviser of the Konoe cabinet, was executed in 1944 for involvement in the Sorge spy ring. His essays take up many commonly discussed aspects of modern Chinese society—questions of arrested development, periodization, bureaucracy, warlords, secret societies, foreign capitalism, the nationalist movement, KMT-CCP relations, and the like (to p. 156; this section was originally published as **Gendai Shina ron** 現代支那論, Iwanami Shoten, 1939, 215 pp., Iwanami shinsho, 10). The latter part (pp. 157–294), mainly on peasant society and the impact of capitalism, is the same as pp. 1–182 of the author's **Shina shakai keizai ron** 支那社會經濟論 (Seikatsusha, 1940, 240 pp., preface 1940). This single volume combining these earlier writings is slightly revised and abridged by a frient to adapt them to post-war circumstances. The author was a leading journalist on China problems; his views were fundamentally based on Marxism and his works widely read.

8. 2 VILLAGE STUDIES

Note: Compared with the preceding section, these materials are more in the nature of case-studies. They are closely related to those on landholding listed in sec. 7. 6 above, and both groups owe much to the voluminous reports of the SMR field investigation in North China (see under 8. 1. 1). The work of Hatada Takashi is outstanding among these field reports. The reader will note that one of the chief issues of interpretation revolves around the degree to which the traditional village communities should be considered "closed" or isolated socio-economic entities, as Marxist theory postulated. A good deal of evidence is presented against this view. Preceding items touching this question include 1.1.11, 1.4.1, 2.1.14, items in sec. 3.4, and 3.6.8. Note also 8.3.1–3 and items in sec. 8.4.

8.2.1 AMANO Motonosuke 天野元之助, **Shina nōson zakki** 支那農村襍記 (Notes on the Chinese village), Seikatsusha, 1942, 285 pp.

A long-time specialist on the Chinese farm economy records here a variety of impressions gained during travel and field study: on rail journeys through Shantung, in the silk-producing area around Wusih, in the Kiangning experimental hsien outside Nanking, etc. Many of his notes are worked up into brief essays on aspects of rural life, both economic and social (pp. 147–219) and another section is a field report on a Kirin village. The last section provides a critically annotated bibliography of Chinese periodicals on economics (some 50 items, in 27 pp.).

8.2.2 SHIMIZU Morimitsu 清水盛光, "Ōka (yanko) no rekishi" 秧歌の歴史 (History of the *yang-ko* ["planting song"]), **Chūgoku kenkyū,** 3 (Mar. 1948), 32–48.

An historical sketch giving examples and noting the early connection of the *yang-ko* with customs of rural labor-cooperation.

8.2.3 SHIMIZU Morimitsu 清水盛光, "Chūgoku kyōson no ṇōkō sagyō ni arawaretaru tsūryoku gassaku no keishiki" 中國鄉村の農耕作業に現はれたる通力合作の形式 ("Forms of the Cooperation for Agricultural Works in the Chinese Rural Community"), **Jimbun kagaku,** 2.3 (June 1948), 133–176.

This study draws extensively on gazetteers to provide an historically based survey of labor-exchange and other forms of rural cooperation.

8.2.4 SHIMIZU Morimitsu 清水盛光, "Chūgoku ni okeru kansei to ochibohiroi no zoku ni tsuite" 中國に於ける看青と落穗拾ひの俗について (On the customs of crop-watching [*k'an-ch'ing*] and picking up fallen grain in China), **Tōkō,** 7 (Jan. 1949), 57–68.

On the rise of modern crop-watching associations among the North China peasantry as a result of the *pao-chia* system's becoming ineffective. On the other hand, the customary right of picking up fallen grain goes back to remote antiquity, as indicated in many quoted passages. This is a product of careful library research.

8.2.5 SHIMIZU Morimitsu 清水盛光, "Chūgoku kyōson no chisui kangai ni arawaretaru tsūryoku gassaku no keishiki" 中國鄉村の治水灌漑に現はれたる通力合作の形式 ("Forms of the cooperation for irrigation in the Chinese rural community"), **Tōyō gakuhō, Kyōto,** 18 (Feb. 1950), 1–32.

Based on local gazetteers, field survey reports and classical writings, this sociological article treats many aspects of communal cooperation for purposes of irrigation, river conservancy, "praying for rain" (*ch'i-yü* 祈雨) and "rain-stopping" (*chih-yü* 止雨), stressing the role of these activities in strengthening the cohesiveness and collectivity of the rural community.

8.2.6 SHIMIZU Morimitsu 清水盛光, **Chūgoku no kyōson tōchi to son-raku** 中國の郷村統治と村落 (Governmental rural controls and the natural villages in China), **Shakai kōseishi taikei** (9.8.7), series 2, 1949, 123 pp.

This study by a leading sociologist in the China field singles out certain features among the traditional controls (police, taxation, admonition of the peasantry, education, etc.) exercised by Chinese governments over the natural villages in the countryside. Building upon the pioneer study edited by Dr. Wada Sei, 3.4.1, the author analyzes the principle of establishing the number of households 戶數編成, with extensive examples drawn from local gazetteers, and the principle of maintaining general oversight 統轄の原則, through various types of headmen 長, as in the *pao-chia* system. In both lines of analysis this sophisticated study ranges widely over the historical record. Reviewed by Muramatsu Yūji 村松祐次 in **Shakai keizai shigaku**, 16.4 (April 1950), 123–129.

8.2.7 SHIMIZU Morimitsu 清水盛光, **Chūgoku kyōson shakai ron** 中國鄉村社會論 (On the Chinese rural society), Iwanami Shoten, 1951, 659 pp.

Completed in 1948, this latest in Professor Shimizu's series of works on the traditional Chinese social structure considers (1) the question of natural and administrative villages, specifically, such matters as the functions of police, taxation, guidance of cultivators (*ch'üan-nung* 勸農, e.g. concerning sericulture) and moral instruction exercised by officialdom, vis-à-vis the collective life of the villagers; (2) the moral concepts based on community life 鄉黨道德思想 as compared with family ethics, their cultivation and forms; and (3) various institutional forms of cooperative effort in village life.

8.2.8 MATSUMOTO Yoshimi 松本善海, "Kyū-Chūgoku shakai no toku-shitsu-ron e no hansei" 舊中國社會の特質論への反省 ("Reflections on the theories of the characteristics of the old Chinese society"), **Tōyō bunka kenkyū,** 9 (Sept. 1948), 20–35; "Kyū-Chūgoku kokka no tokushitsu-ron e no hansei" 舊中國國家の特質論への反省 ("Reflections on the Theories of the Characteristics of the Old Chinese Country"), **Tōyō bunka kenkyū,** 10 (Feb. 1949), 37–51.

These two articles set forth the penetrating and somewhat cynical reflections of an historian who is deeply sceptical about the validity of the so-called stationary theory of Chinese history and also about the two popular theses that (a) dynasties reigned despotically over the dispersed and self-sufficient, microcosmic village-communities of China and that (b) the state and society in China have been separated from each other. The author takes as his starting point six theses from Sano Manabu's **Shinchō shakaishi,** 2.1.14, and so these articles may be considered a critique attacking the fundamental views asserted in that work.

8.2.9 HATADA Takashi 旗田巍, "Chūgoku ni okeru senseishugi to 'Sonraku kyōdōtai riron'" 中國における專制主義と「村落共同體理論」 ("Absolutism in China and The Theory of Village Community"), **Chūgoku kenkyū,** 13 (Sept. 1950), 2–12.

A useful critical and bibliographical article, mainly devoted to disagreement with Professor Shimizu Morimitsu's **Shina shakai no kenkyū,** 8.1.2. Professor

Hatada, as a specialist in field research, doubts the self-sufficiency and cohesiveness of village social structure. He is also sceptical about the allegedly necessary relationship between despotism and the village community which is implicit in the above-mentioned theory.

8.2.10 HATADA Takashi 旗田巍, "Hoku-Shi ni okeru sonraku jichi no ichikeitai, tokuni sonkōkai no kōsei ni tsuite 北支における村落自治の一形態──とくに村公會の構成について (A form of village self-government in North China, with special reference to the structure of village public meetings), pp. 615–635 in **Katō hakushi kanreki kinen Tōyōshi shūsetsu** (9.8.5).

By a specialist who worked in the SMR field study in North China, this is a case-study description of the oligarchical functioning of village meetings, and their selection of village headman (*ts'un-chang* 村長), based on the reports of the above-mentioned investigation. The author attempts to discriminate between governmental control from above through the *pao-chia* 保甲 system and the traditional and autonomous self-government of villages through the village meeting, so as to discuss their interrelated functioning. Reviewed by Kawakami Kōichi 河上光一 in **Rekishigaku kenkyū**, 98 (April 1942), 366–371.

8.2.11 HATADA Takashi 旗田巍, "Kahoku sonraku ni okeru kyōdō kankei no rekishiteki seikatsu, 'kansei' no hatten katei" 華北村落における協同關係の歴史的性格──「看吉」の發展過程 (The historical nature of cooperative relations in North China villages, the process of development of crop-watching [*k'an-ch'ing*]), **Rekishigaku kenkyū**, 139 (May 1949), 1–23.

An interesting study, from the field reports of several different villages, of the way in which peasant cooperation developed and was organized for the purpose of watching over and protecting growing crops. The author's point is that this development of cooperative relations did not mark the persistence of a traditional sense of community but resulted from the progressive impoverishment of village life; crop-watching is a conscious device to prevent thievery by hiring the poorest inhabitants who would otherwise steal the crops and create disturbances. He distinguishes various types of crop-watching as representing different developmental stages of this kind of cooperation.

8.2.12 HATADA Takashi 旗田巍, "Kahoku no nōson ni okeru 'Kaiyōshi' no kankō, sonraku kyōdōtai teki kankei eno saikentō" 華北の農村における開葉子の慣行──村落共同體的關係への再檢討 ("The Custom of *K'ai-yeh-tzu* 開葉子 in Rural Communities of Northern China—Re-examination into circumstances of Village Community"), **Shigaku zasshi**, 58.4 (Oct. 1949), 43–54.

Using the famous reports of the SMR and Tōa Kenkyūjo field investigation, 8.1.1, this article deals with the rather unnoticed custom, in North China rural communities, which permits anyone to pick kaoliang leaves, but only during a specified period, and prohibits picking by anyone, even the owners, at any other time. Considering the strikingly low level of communal cooperation in agricultural production in North China, Professor Hatada concludes that this peculiar custom is not a persisting feature of the so-called "communal" village life, but a calculated means to prevent theft by poor peasants. It is thus a phenomenon in the process of decline of traditional relations in rural communities.

8.2.13 NAOE Hiroji 直江廣治, "Kahoku sonraku no denshō umpansha" 華北村落の傳承運搬者 (The folklore conveyers in North China villages), **Chūgoku hyōron**, 2.1 (Jan. 1947), 36–44.

A first-hand account of the itinerant medicine-sellers (*mai-yao ti* 賣藥的), brush-sellers (*mai-pi ti* 賣筆的), ironsmiths (*t'ieh-chiang* 鐵匠), story-tellers (*shuo-shu ti* 說書的), and geomancers (*feng-shui hsien-sheng* 風水先生) of various types and the function they perform in transmitting folk tales and legends (with 6 examples)—a function which, the author notes, formerly played a part in the activity of secret societies. This article was intended to deemphasize the so-called isolated and closed nature of the Chinese villges.

8.2.14 MURAMATSU Yūji 村松祐次, "Ranjō ken to Jihokusai son" 欒城縣と寺北柴村 (Luan-ch'eng hsien and Ssu-pei-ch'ai village), **Hitotsubashi ronsō**, 22.1 (July 1949), 180–207.

Subtitled "A Chinese district town and village", this case-study of the relationship of hsien to village is based on more than 10 volumes of field-investigation materials compiled under the big wartime project of the SMR in North China. The hsien in question, in Hopei, is studied by Professor Muramatsu in its various functions that affect the village, particularly as concerns police and taxes.

8.2.15 KAINŌ Michitaka 戒能通孝, **Hoku-Shi nōson ni okeru kankō gaisetsu** 北支農村に於ける慣行概説 (A general account of the customs in North China villages), Tōa Kenkyūjo, 1944, 117 pp.

Using the reports of the SMR field investigation, a flexible-minded legal scientist provides a suggestive case-study view of village life in North China, including (1) the organization of the village, (2) economic and family life of the peasants, (3) forms of land ownership, (4) management and labor in agriculture, (5) legal customs concerning tenancy, and (6) land-mortgaged loans and buying and selling of land. The author is now at the Tōkyō Metropolitan University as a professor of civil law.

8.2.16 KAMIMURA Shizui 上村鎮威, "Sanseishō Rinfun-ken Kōkaten seisan kōzō bunseki" 山西省臨汾縣高河店生産構造分析 (An analysis of the production structure of a small village named Kao-ho-tien in Lin-fen-hsien, Shansi province), **Tōa jimbun gakuhō**, 1.4 (Feb. 1942), 1113–1133.

A theoretical analysis based on the reports of the field survey by the North China Research Branch of the South Manchurian Railway Co., in which the author himself participated. He is particularly interested in the persistence of the so-called "natural economy" based on garden-farming (*Gartenbau*) and family labor and its transformation into small-scale merchant-goods production ("pre-capitalistic" stage). Finally the relationship between these and the type of landownership in this village is discussed.

8.2.17 MOMOSE Hiromu 百瀬弘, "Shimmatsu Chokureishō sonchin kokō shōkō" 清末直隷省村鎮戸口小考 ("A Brief Statment of the Household Number and Population of Villages in Chihli Province toward the End of the Ch'ing Dynasty"), **Tōhō gakuhō, Tōkyō,** 12.3 (Dec. 1941), 99–112, tables.

A brief but elaborate statistical study, based on the official maps of two villages and a local gazetteer.

8.2.18 HIRANO Yositarō 平野義太郎, **Hoku-Shi no sonraku shakai** 北支の村落社會 (The village society of North China), pub. by the author, 1944, 289 pp., mimeo.

A report made by the author for the Tōa Kenkyūjo and SMR's "research on customs" in North China, this is a rather factual study of local government controls, village associations and markets, popular Taoist ideology, forms of leadership and cooperation (including crop-watching), and village ethics and justice, all arranged apparently on the general assumption that the Chinese village has traditionally been a closed and microcosmic entity under an "Asiatic despotism". This view was challenged by other researchers in the same project. See Kainō Michitaka's article, "Shina tochihō kankō josetsu", 3.6.8, which, though it does not mention Mr. Hirano, is apparently a criticism of his approach.

8.2.19 HIRANO Yoshitarō 平野義太郎, "Hokuchū-Shi ni okeru nōson shūraku no chōkan" 北中支における農村聚落の鳥瞰 (A birdseye view of rural villages in North and Central China), **Tōa kenkyū shohō,** 10 (June 1941), 364–397, illus., charts.

On the basis of a field trip to Hopei and Kiangsu, the author first contrasts the terrain, soil, crops and general ecology of the regions north and south of the Huai River, then compares the two regions in regard to the forms of village structure and joint family organization, and finally notes certain cooperative institutions and social practices such as crop-watching (*k'an-ch'ing* 看青) and village drama.

8.2.20 ISHIDA Seiichi 石田精一, "Nan-Man no sonraku kōsei" 南満の村落構成 (Village structure of South Manchuria), **Mantetsu chōsa geppō,** 21.9 (Sept. 1941), 1–65, many tables.

Subtitled as centering on the old "official village" and based on village studies in Liao-yang 遼陽 hsien, this field report is cast in the context of the European (Marxist) developmental stages of social structure (slavery to feudalism to capitalism, etc.). It studies historically during the previous century the gradual shift in the factors of social cohesion (*ketsugō* 結合) in the official village, from a kinship basis to an economic and individualistic basis.

8.2.21 TANAKA Tadao 田中忠夫, **Kakumei Shina nōson no jisshōteki kenkyū** 革命支那農村の實證的研究 (A concrete factual study of the Chinese village in revolution), Shūjinsha 衆人社, 1930, 444 pp.

A series of studies, long since professionally outdated, based on 6 or 7 years work in North and Central China in the 1920s. Major topics include the forms of capitalist development, village organization, tenancy, farm labor, and the peasant movement. Within this framework there may be some value for the historian in sections on the Red Spear Society 紅槍會 (pp. 240–275) or other aspects of agrarian discontent as then observed.

8.3 THE GENTRY AND MILITARY

Note: These few writings, largely supplementary to larger works already noted, touch on generally neglected subjects.

8.3.1 MOTOMURA Shōichi 本村正一, "Shindai shakai ni okeru shinshi no sonzai" 清代社會に於ける紳士の存在 (The existence of the gentry in Ch'ing society), **Shien 史淵**, 24 (Nov. 1940), 61–78.

An attempt to define the rather elusive "gentry class" (including both mandarins in active service and local gentry) by noting (1) its responsibility for leading local corps (*t'uan-lien* 團練 and *hsiang-yung* 鄉勇) in the Taiping Rebellion period, (2) its inseparable relationship with the examination system, (3) the semi-official status of local gentry 鄉紳 as a class of degree holders and assistants to the official class, (4) their economic status and cultural position, etc. A percipient essay but unfortunately wholly without a statistical basis or much documentation.

8.3.2 NEGISHI Tadashi 根岸佶, **Chūgoku shakai ni okeru shidōsō, Chūgoku kirō shinshi no kenkyū** 中國社會に於ける指導層──中國耆老紳士の研究 (The leadership-stratum in Chinese society, a study of the Chinese elders and gentry), Heiwa Shobō 平和書房, 1947, 278 pp.

This socio-historical study tries to appraise the status and functions of the elders and gentry, their history in the Chinese state and relation to the central power. The author takes examples from various provinces, personal histories, writings, incidents, and customs to illustrate some interesting but often inconsistent interpretations. No references, footnotes or bibliography are given, but the book is apparently based, in part, on the report of the famous field survey in North China sponsored by the South Manchurian Railway Co. and the Tōa Kenkyūjo. See the critical reviews by Matsumoto Yoshimi 松本善海, in **Chūgoku kenkyū**, 3 (March 1948), pp. 85–90; and by Kitamura Hirotada 北村敬直, in **Tōyōshi kenkyū**, 10.3 (July 1948), 84–88. The author has been an outstanding specialist on Chinese commercial organization, see 7.16.2, 8.5.4–7.

8.3.3 SAKAI Tadao 酒井忠夫, "Kyōshin ni tsuite" 鄉紳について ("Hsiang-shen, the Chinese gentry in the Society of Ming and Ching period"), **Shichō**, 47 (Dec. 1952), 1–18.

This review of the status and nature of the gentry class discusses it first with reference mainly to the 17th century and then deals with its power situation, emphasizing its landlord-official-merchant-usurer character. Quoting many examples, the author discusses the use and scope of the term *hsiang-shen*

鄉紳 as compared with similar terms such as *shen-shih* 紳士 or *shih-jen* 士人.
He detects through this analysis a change in Chinese social structure during
the late-Ming and early-Ch'ing period.

8.3.4 ŌTANI ("Ohtani") Kōtarō 大谷孝太郎, "Jushō, Chūgoku buji
shisō no shōseki" 儒將——中國武事思想のしょう跡 (The Confucian
general, some traces of Chinese military thought), **Hikone ronsō 彦根論
叢**, ed. by Shiga University Economics Society 滋賀大學經濟學會, 3
(Feb. 1953), 24–64.

An interesting, though rough and sometimes obscure, review of classical
thought concerning the place of the military, and its gradual modification, as
background for an historical perspective on militarism (warlords, etc.) in
modern China. Following the general lead of Ojima Sukema, **Chūgoku no
kakumei shisō**, 6.1.1, the author traces the development in the Sung period
and after of the concept of the Confucian scholar-general, a tradition inherited
by Tseng Kuo-fan, et al.

8.3.5 KATŌ Shigeshi 加藤繁, "Shina to bushi kaikyū" 支那と武士階級
("China und sein Ritterstand"), **Shigaku zasshi,** 50.1 (Jan. 1939), 1–19.

Although this interesting comparison of the warrior class in China with
the Japanese samurai class concerns mainly the pre-modern period, it is of
some sociological interest for later times.

8.3.6 HATANO Yoshihiro 波多野善大, "Hokuyō gumbatsu no seiritsu
katei" 北洋軍閥の成立過程 ("The growth-process of the Peiyang militarist
party"), **Nagoya daigaku bungakubu kenkyū ronshū,** 6 (1953), 211-262.

A well-documented broad survey of the vicissitudes of the personal armies
in the late Ch'ing period, from the local corps of Tseng Kuo-fan, through Li
Hung-chang's Huai Army and Pei-yang Navy, and the establishment of the
New Army after the Sino-Japanese War, down to the rise to power of Yuan
Shih-k'ai and his manoeuvres in confronting the revolution in 1911. This article
forms a useful bibliographical introduction to a neglected aspect of modern
Chinese history.

8.4 FAMILY SYSTEM

Note: On this subject the work of Makino Tatsumi is outstanding. The
first item, by Niida Noboru, represents the influence both of the wartime
field studies in North China and of the author's legal studies (see sec.
3.6 above). Under that section note especially 3.6.1–2 (Niida), 3.6.7
(Shiga), 3.6.9 (Hayashi), 3.6.16 (Makino).

8.4.1 NIIDA Noboru 仁井田陞, **Chūgoku no nōson kazoku 中國の農村
家族** (The Chinese rural family), Tōkyō University, Tōyō Bunka Kenkyūjo,
1952, 406+index 14 pp., illus.

This highly sophisticated and critical research, benefitting by the author's legal and sociological studies and by the results of the wartime SMR field investigations in North China, analyzes the structure and functioning of the peasant family from many different angles: joint-family relationships, family organization and splitting of families (equal division of patrimony); discipline of intra-family labor (the pater familias and rather slavish intra-familial relationships); family-communism intertwined with patriarchal authority; Schlüsselgewalt (the status and function of the housewife); divorce; and bloody acts of revenge between joint-family villages. This book is a basic work in this field and requires careful perusal by all scholars interested in the social structure of China. The influence of Dr. Niida's views is visible in the work of other Japanese scholars; see, e.g., Hatada Takashi ("Chūgoku tochi kaikaku no rekishiteki seikaku" 7. 6. 1) and Imahori Seiji ("Chūgoku ni okeru hōkenteki shōkōgyō no kikō", 1. 4. 10). See reviews by Otake Fumio in **Shichō**, 47 (Dec. 1952), 52–54: by Ōtake Hideo 大竹秀男 in **Kōbe hōgaku zasshi** 神戸 法學雜誌 ("Kobe Law Journal"), ed. by the Faculty of Law, Kōbe University, 2. 4 (Mar. 1953), 800–811: and by Yokoyama Suguru 横山英 in **Shigaku zasshi,** 62. 8 (Aug. 1953), 785–790. See also Niida's "Chūgoku no kafuchō kenryoku no kōzō" 中國の家父長權力の構造 ("The Structure of the Patriarchal Power in China"), **Hōshakaigaku** 法社會學 ("Sociology of Law"), pub. by Japanese Association of Sociology of Law, with offices in the Faculty of Law, Tōkyō University, 4 (July 1953), 1–36, an analysis by comparison with *manus* and *patria potestas* in Roman law.

8. 4. 2 MAKINO Tatsumi 牧野巽, **Shina kazoku kenkyū** 支那家族研究 (Studies of the Chinese family), Seikatsusha, 1944, 724 pp., photos.

This collection of articles published over many years is concerned mainly with the Han period but includes a chapter comparing statistics on family structure in China and Japan, and another on the modern Chinese family (pp. 565–618). The first 37 pages also give a general account of the Chinese family system. Dr. Makino is not only a trained sociologist but also unusually competent at classical Chinese, having been the son of a leading classicist and sinologist. His writings are basic for students of the Chinese family system. He is now a professor of the Faculty of Education, Tōkyō University.

8. 4. 3 MAKINO Tatsumi 牧野巽, **Kinsei Chūgoku sōzoku kenkyū** 近世 中國宗族研究 (A study of the clan in modern China), Nikkō Shoin, 1949, 329 pp.

Six of these nine essays are articles previously printed elsewhere; all deal with traditional China, mainly in the Ming and Ch'ing periods. The author has pioneered in the exploration and use of family genealogies, as a hitherto unexploited historical source, and several of his chapters demonstrate their value in specific instances. He also examines the persistent influence of familocentric doctrines preserved in certain classic works like **Chu-tzu chia-li** 朱子家禮. In general, the author stresses the recent development in the clan system since the Sung period, as distinguished from the earlier period.

8. 4. 4 MAKINO Tatsumi 牧野巽, "Sōshi to sono hattatsu" 宗祠と其の 發達 ("On *Tsung tz'u* 宗祠 [the Ancestral Hall] and its Historical Development in China"), **Tōhō gakuhō, Tōkyō,** 9 (Jan. 1939), 173–250.

(Listed as to be continued, but no continuation has been published.)

A sociological study making extensive use of Ch'ing sources, with copious quotations, to present the evidence on the actual social function of ancestral shrines, where worship is conducted by all the members of a clan.

8. 4. 5 MAKINO Tatsumi 牧野巽, "Kanton no gōzokushi to gōzokufu" 廣東の合族祠と合族譜 (Joint-family shrines and genealogies in Kwang-tung), in **Kindai Chūgoku kenkyū**, 1948, (9. 8. 6), 89–129.

This case study by a leading sociologist is based particularly on genealogies of the Su 蘇 clan preserved in the Wu-kung shu-yüan 武功書院 in Canton. The introductory 17 pages are enlightening and suggestive, stressing the significance of joint-family shrines as centers of the inter-village activities of a clan, and closely connected with the examination system (*k'o-chü* 科擧).

8. 4. 6 MAKINO Tatsumi 牧野巽, "Kanton no gōzokushi to gōzokufu" 廣東の合族祠と合族譜 ("Ho-tsou-ts'you et ho-tsou-pou de Canton, les temples ancestrals et les livres généalogiques des associations des clans qui portent le même nom"), **Orientalica**, 2 (1949), 163–186.

A futher exploration, by a leading sociologist who visited South China during the war, of the research potentialities of this little-used type of material. This is a companion article to the one published under the same title in pp. 89–129 of Niida Noboru, ed., **Kindai Chūgoku kenkyū**, 9. 8. 6.

8. 4. 7 MAKINO Tatsumi 牧野巽, **Shina ni okeru kazoku seido** 支那に於ける家族制度 (The family system in China), in **Iwanami kōza Tōyō shichō** (9. 8. 4), series 10, 1935, 44 pp.

A neat roundup of a dozen different aspects of the family system, ancient and modern, with reference notes. The author stresses the actual conditions and functioning of the system rather than the traditional family ethics of the philosophers.

8. 4. 8 SHIMIZU Morimitsu 清水盛光, **Shina kazoku no kōzō** 支那家族の構造 (The structure of the Chinese family), Iwanami Shoten, 1942, 582 pp.

This sociological study, once a standard work, treats the various aspects of the Chinese kinship system both historically and structurally, including the collective consciousness or psychology 全體意識 of family life and emphasis on hierarchic relations. Reviewed by Obata Tatsuo 小畑龍雄 in **Tōyōshi kenkyū**, 7. 6 (Nov.-Dec. 1942), 55–57.

8. 4. 9 SHIMIZU Morimitsu 清水盛光, **Chūgoku zokusan seido kō** 中國族産制度攷 (A study of the system of clan or joint-family property in China), Iwanami Shoten, 1949, 219 pp.

This socio-legal study analyzes the kinds, forms, origins, uses and various other aspects of clan-held property, as distinguished from the patrimony of a single family, considering it historically as well as in its social context. Reviewed by Moriya Mitsuo 守屋美都雄 in **Shigaku zasshi**, 58.4 (Oct. 1949), 406–411.

8.4.10 Kɪᴛᴀᴍᴜʀᴀ Hirotada 北村敬直, "Shindai kaitō no ichikōsatsu" 清代械闘の一考察 ("The conflict among the Clanish Groups in Ch'ing [清] Dynasty"), **Shirin**, 33.1 (Jan. 1950), 64–77.

Starting from the legal-sociological article by Dr. Niida on the same topic (now included in the latter's **Chūgoku no nōson kazoku**, 8.4.1), the author approaches this topic from a somewhat different angle and attempts to analyze the phenomena of blood feuds between joint-family villages (especially in Fukien and Kwangtung) in the context of the local socio-economic circumstances. He notes especially the impact of foreign trade (and acceleration of money economy), as well as the class-differentiation within joint-family groups.

8.9.11 Hɪʀᴀɴᴏ Yoshitarō 平野義太郎, "Hoku-Shi sonraku no kisoyōso to shite no sōzoku oyobi sonbyō" 北支村落の基礎要素としての宗族及び村廟 (The family clans and the village shrines as the basic elements in the North China villages), pp. 1–145 in Tōa Kenkyūjo, ed., **Shina nōson kankō chōsa hōkokusho** 支那農村慣行調査報告書 (Reports of investigations of rural customs in China), pub. by the same, series 1 第一輯, 1943, 329 pp.

Based on the reports of the SMR field investigation, this is a study of joint-family social structure and the ancestral shrines and village temples in North China. The latter are compared functionally with shrines and temples in Japan, and the author then concentrates on aspects of village temples (*ts'un-miao* 村廟) and their tutelary deities—popular Taoism, uses of the various gods, the Confucian temple, local deities, temple associations (*miao-hui* 廟會) and the like.

8.4.12 Uᴄʜɪᴅᴀ Chiyū 内田智雄, **Chūgoku nōson no kazoku to shinkō** 中國農村の家族と信仰 (The family system and religion in the Chinese village), Kōbundō, 1948, 303 pp.

The author had seven year's experience living in China, mainly in Peking and Dairen, and participated in 1947 in the S.M.R. project on **Rural Customs and Practices of China**, 8.1.1. This volume reports 1) observations on the family, particularly the marriage system, derived from that project, and 2) results of village research in Liao-yang 遼陽 hsien in South Manchuria, particularly on religious practices such as prayers for rain (*ch'iu-yü* 求雨).

8.4.13 Uᴄʜɪᴅᴀ Chiyū 内田智雄, "Kahoku nōson ni okeru dōzoku no saishi gyōji ni tsuite" 華北農村における同族の祭祖行事について ("Ancestral Worship in Agricultural Communities in North China"), **Tōhō gakuhō, Kyōto,** 22 (Feb. 1953), 59–94.

Based on the reports of the SMR field survey, in which the author participated, this factual article studies the actualities of the joint-family relationship in North China. After a careful analysis of several examples of customary ceremonies of ancestor worship, the author concludes that the looseness of the joint-family tie is much greater than assumed in once fashionable and still widespread theories about it.

8.5 GILDS

Note: These materials are closely related to those in sec. 7.16 above. The studies of Niida, Negishi, and Imahori are the major recent work on gilds. Note item 7.15.12 on the Mukden banker's gild.

8.5.1 NIIDA Noboru 仁井田陞, **Chūgoku no shakai to girudo** 中國の社會とギルド (The society and gilds of China), Iwanami Shoten, 1951, 289 +index 29 pp., illus.

This research report prepared at the Institute for Oriental Culture of Tōkyō University is the result of long-continued study, including first-hand investigation among gild members in Peking during the war years. It is mainly concerned with the craft and merchant gilds of Peking, as compared with European medieval gilds, and analyzes their social context, structure and socio-economic functions in authoritative detail (pp. 1–244). There follows a briefer study of moslem gilds in Peking. This volume sums up much previous work by the author, Katō Shigeshi 加藤繁, Negishi Tadashi 根岸佶, J. S. Burgess and others. Along with the author's **Chūgoku no nōson kazoku,** 8.4.1, this book sets forth the results of his attempt to analyze Chinese society from a combined legal-sociological and historical point of view. The aim is to provide an understanding of the entire social structure and its transformation, including the problems of mentality, law, ethics and religion. See the reviews by Saeki Yūichi 佐伯有一 in **Rekishigaku kenkyū,** 159 (Sept. 1952), 43–49; by Imahori Seiji 今堀誠二 in **Shigaku zasshi,** 61.2 (Feb. 1952), 174–180; by Muramatsu Yūji 村松祐次 in **Shakai keizai shigaku,** 18.6 (May 1953), 541–548.

8.5.2 NIIDA Noboru 仁井田陞, "Shindai no Kankō Sansensei kaikan to Sansen-hō (girudo)" 清代の漢口山陜西會館と山陜帮 (ギルド) ("Shan-shan-hsi-hoi-kuan and Shan-shan Guild in Hankow under the Ch'ing Dynasty"), **Shakai keizai shigaku,** 13.6 (Sept. 1943), 497–518.

A study of the history and regulations of the Shansi-Shensi provincial gild at Hankow in the late 19th century, and its connections with Shansi merchant organizations, based on the gild gazetteer and similar sources.

8.5.3 NIIDA Noboru 仁井田陞, "Pekin no kōshō girudo to sono enkaku" 北京の工商ギルドと其の沿革 (Merchant and craft-gilds in Peking and their historical development), part 1.1, **Tōyōbunka kenkyūjo kiyō,** 1 (Dec. 1948), 239–358, illus. (to be continued, but no continuation has been published).

Mainly based on first-hand knowledge, this article provides a detailed case-study, from a socio-legal viewpoint, of two gilds in Peking (out of six projected in the table of contents). Though no continuation in this detailed form has appeared, other results of Dr. Niida's research on gilds in Peking have been published in several articles and then revised and put together in a single volume entitled **Chūgoku no shakai to girudo,** 8.5.1. The present article gives details of jade merchants' and book merchants' gilds, with some documents.

8.5.4 NEGISHI Tadashi 根岸佶, **Chūgoku no girudo** 中國のギルド (The gilds of China), Nippon Hyōron Shinsha 新社, 1953, 488 pp.

This latest study in the field of Chinese commercial organization, by one of its most experienced researchers, is an over-all survey which puts many earlier surveys out of business. After dealing with general questions of the place of gilds in the Chinese state and society, including the question of *ya-hang* 牙行 (pp. 60–69), Dr. Negishi takes up native-place gilds or "associations of fellow-provincials" (*t'ung-hsiang* 同鄉), including their history, the Shansi banks organization (*pang* 幫), the Ningpo, Fukien, and Kwangtung *pang* (pp. 101–127), gild halls, and provincial gilds connected with overseas Chinese communities. The last 200 pages deal with economic (mainly commercial) gilds, their history, management, functions, etc., concluding with the rather prudent early policy of the CCP regarding them. In the bibliography, pp. 479–488, articles are cited without the names of the relevant periodicals. This is an authoritative summary of a large subject, but since no footnote references are provided and it covers many topics briefly, it leaves something to be desired as a guide for further research.

8.5.5 NEGISHI Tadashi 根岸佶, **Shina girudo no kenkyū** 支那ギルド の研究 (A study of Chinese gilds), Shibun Shoin 斯文書院, 1932, 442 +index 22 pp.

Developed from his early work at the Tōa Dōbun Shoin in Shanghai before 1921, this now has been partly superseded by writings of Niida Noboru and others, but like Negishi's other studies, it provides a detailed analysis, based on field investigation as well as extensive bibliography (pp. 421–442). Main sections of the book deal with (1) fellow-provincial (*t'ung-hsiang* 同鄉) assocations, with case-examples from Shanghai, (2) economic associations, including merchant and artisan gilds, with examples from Canton (the Cohong, pp. 377–404) and Shanghai. Reviewed by Shikimori Tomishi 式守富司 in **Rekishigaku kenkyū**, 1.1 (Nov. 1933), 77–79; and by Ochi Motoji 越智元治 in **Shakai keizai shigaku,** 2.12 (March 1933), 1342–1346.

8.5.6 NEGISHI Tadashi 根岸佶, **Shanhai no girudo** 上海のギルド (The gilds of Shanghai), Nippon Hyōronsha, 1951, 412 pp.

As a veteran specialist on Chinese economic organization and long-time resident of Shanghai, the author brings the maximum insight and background to these case studies of (1) the Ningpo gild at Shanghai, (2) the Haining 海寧 gild at Shanghai, (3) the Shanghai native bankers' (*ch'ien-chuang* 錢莊) gild, (4) the Shanghai bankers' association 銀行公會, and the gilds for (5) the the rice trade, (6) the silk and satin trade, (7) handicrafts and construction, concluding with (8) the Shanghai Chamber of Commerce as an overall organization, and some further essays. Bibliography pp. 402–412. See reviews by Otake Fumio 小竹文夫, in **Shigaku zasshi,** 60.9 (Sept. 1951), 846–850; and by Imahori Seiji 今堀誠二, in **Chūgoku kenkyū,** 15 (Jan. 1952), 73–77.

8.5.7 NEGISHI Tadashi 根岸佶, "Shanhai sengyō girudo" 上海錢業ギ ルド (The Shanghai native-bankers' gild), **Hitotsubashi ronsō,** 22.1 (July 1949), 150–179.

The history, structure, and functions of this important gild, by a pioneer specialist in Chinese commercial organization.

8. 5. 8 IMAHORI Seiji 今堀誠二, "Chūgoku ni okeru girudo māchanto no kōzō" 中國におけるギルドマーチャントの構造 (The structure of the gild merchant in China), pp. 205–232 in Niida Noboru, ed., **Kindai Chūgoku no shakai to keizai,** 5. 7. 1.

Enlarging somewhat on the scope of the term as used by H.B. Morse and following the lead of Negishi Tadashi's study of gilds, 8. 5. 5, the author here puts forward an analysis of gilds merchant (i.e., federative organizations of gilds with semi-official governing powers including justice, defence and police). He describes how they have been formed, their social connections, internal organization, and practical functions. This analysis is necessarily in rather general terms and is mainly based on the author's field investigation in Inner Mongolia, like his other article entitled "Chūgoku ni okeru hōkenteki shōkōgyō no kikō," 1. 4. 10.

8. 5. 9 IMAHORI Seiji 今堀誠二, "Katō engyō dōgyō kumiai no kenkyū" 河東鹽業同業組合の研究 (A study of the gilds of salt producers and merchants in the Ho-tung region), **Shigaku zasshi,** 55. 9 (Sept. 1944), 945–978; 10 (Oct. 1944), 1060–1111; 56. 1 (Jan. 1945), 20–48.

Based on wartime field investigation by the author as well as on library research, both conducted under the constant advice of Dr. Niida Noboru, this article provides full details on the gilds of salt producers and merchants in the so-called Ho-tung producing region (i.e. An-i 安邑 hsien, in Shansi province). The first section deals rather factually with the separate gilds, their historical background and current activities. The second section gives a general view of these gilds as a whole, discussing their organization and activities such as regulation-making, maintenance of gild-halls, religious services, activities connected with justice and the police, liaison between merchants and government authorities, and their relation to various enterprises and social philanthropies.

8. 5. 10 IMAHORI Seiji 今堀誠二, "Chūgoku ni okeru shōkō girudo no sobyō" 中國に於ける商工ギルドの素描 (A rough sketch of merchant and artisan gilds in China), pp. 345–379 in Hiroshima Bunrika Daigaku Shigakka Kyōshitsu, ed., **Shigaku kenkyū kinen ronsō** 史學研究記念論叢, Kyōto: Yanagiwara Shoten 柳原書店, 1950, 602+16 pp.

Still seeking to generalize concerning China on the basis of research in Inner Mongolia, the author attempts to sum up the social context, organization, major activities (with government, adjudicative, managerial, mutual aid, relief, etc.), and administration of gilds, with results that seem rather theoretical. He sees gilds as tools for controlling and exploiting production on behalf of commercial capital, cloaked in the disguise of communal activities.

8. 5. 11 IMAHORI Seiji 今堀誠二, "Chūgoku ni okeru yōman girudo no kōzō" 中國に於けるヨーマンギルドの構造 ("Organization of Yeoman Gilds in China"), **Shakai keizai shigaku,** 18. 1 (April 1952), 23–49; 18. 2 (June 1952), 124–159.

This latest study by a leading Marxist analyst of Chinese gild structure finds yeoman gilds developing in China from the 14th century (they disappeared at the end of the Ming period and reappeared from the K'ang-hsi era). Mainly on the basis of evidence collected in Inner Mongolia (at Kweihua 歸化, Paotow 包頭, and elsewhere in Suiyuan), he analyses, in an intensively interpretative but somewhat painstaking manner, the history, membership, organization, and activities of these yeoman gilds.

8.5.12 Hiroshima Bunrika Daigaku, **Tōyō no shakai** 東洋の社會 (Oriental society), Meguro Shoten, 目黒書店, 1948, 397 pp., Hiroshima Bunrika Daigaku Tōyōshigaku kenkyūshitsu kiyō, dai-issatsu, 廣島文理科大學東洋史學研究室紀要第一冊 (Memoirs of the research department for the study of Oriental history of Hiroshima Bunrika University [now Hiroshima Kyōiku 教育 Daigaku], vol. 1).

Among the five articles in this collection, the fourth by Oshibuchi Hajime 鴛淵一, is entitled "Shinchō zenki shakai zakkō" 清朝前紀社會雜考 (Miscellaneous notes on the early society of the Ch'ing period), pp. 257–350. The last, by Imahori Seiji 今堀誠二, is entitled "Kindai ni okeru Kaihō no shōgyō girudo ni tsuite" 近代に於ける開封の商業ギルドに就て (On the commercial gilds in Kaifeng in the modern period), pp. 351–397. It is a report of the author's field investigation (in 1944) of four merchant firms dealing in money, cotton textiles, silk and imported goods (including their imitations) chosen from among some 50 commercial gilds in Kaifeng. The author adds a separate section of general observations which are made in the context of the "feudal" society of China in its process of decline.

8.5.13 KATŌ Shigeshi 加藤繁, "Shindai ni okeru Pekin no shōnin kaikan ni tsuite" 清代に於ける北京の商人會館に就いて (On merchant-gilds in Peking in the Ch'ing period), **Shigaku zasshi**, 53.2 (Feb. 1942), 151–181.

Based on first-hand investigation by the author, this article gives data on nine merchant-gilds in Peking (of merchants dealing in silver, dyes, tobacco, drugs, oil, silk, imported rarities, and jade and books) with special reference to their activities in the Ch'ing period. The regional background of the merchants concerned is also noted, e.g. the decline of Shansi influence in Peking around the Tao-kuang period. Also printed in the same author's **Shina keizaishi kōshō**, 9.8.8, 2.557–584.

8.5.14 NAKAMURA Jihei 中村治兵衞, "Shindai Kokōmai ryūtsū no ichimen" 清代湖廣米流通の一面 ("Circulation of Rice Produced in the District of Hu-kwang under the Ching Dynasty"), **Shakai keizai shigaku**, 18.3 (Aug. 1952), 269–281.

Subtitled as "viewed from the Hunan provincial gild-hall in Nanking" 南京の湖南會館よりみた, this study starts by describing the gild-hall founded by the contributions of Tseng Kuo-fan and other Hunanese after their victory over the Taipings. It then gives an informed account of the Hunan-Hupei rice trade and merchants centering around Nanking from the late 18th century, and their bitter competition with the influential Anhwei rice merchants,—an economic aspect or background element in the political rivalry between the Hunan and Anhwei cliques.

8.6 SECRET SOCIETIES, ETC.

Note: By "etc." we refer to the fact that the so-called secret societies of China were often indistinguishable from fraternal associations and/or religious cults. It is difficult to differentiate between these three types of associations, which tend to interpenetrate. At least it seems well established that few "secret societies" have been secret. Most of these items are raw material, compiled by persons in the know, rather than by research scholars. One of the latter, however, is Suzuki Chūsei (8.6.1, 8.6.6). The final item by Kubo Noritada is closely related to studies of popular Taoism by him and others noted in sec. 6.4 above.

8.6.1 SUZUKI Chūsei 鈴木中正, "Shindai ni okeru shūkyōteki hanran no seikaku" 清代に於ける宗教的叛亂の性格 ("The characters of sects rebellious under the Dynasty Ching"), **Tōyō bunka kenkyū,** 9 (Sept. 1948), 36–52.

> This article was virtually a *résumé* of the author's then unpublished larger study on the White Lotus Rebellion (**Shinchō chūkishi kenkyū,** 1952, 2.4.1) with some added general appraisal of the religious rebellions of the Ch'ing period.

8.6.2 SAKAI Tadao 酒井忠夫, **Kindai Shina ni okeru shūkyō kessha no kenkyū** 近代支那に於ける宗教結社の研究 (A study of religious societies in modern China), Tōa Kenkyūjo, 1944, 309 pp.

> Distinguishing these modern religious organizations from the better-known religious (or religion-tinged) secret societies, the writer first surveys their role in Chinese history and then notes the increase of such bodies in the early Republican period, devoting special attention to the Taoist Red Swastika Society (Hung wan-tzu hui 紅卍字會). Another chapter is devoted to a study of oracles (*fu–chi* 扶乩), pp. 31–53. A good deal of bibliography is also noted. The author specialized on this topic in Shanghai. See also 6.4.4.

8.6.3 SAKAI Tadao 酒井忠夫, "Gendai Chūgoku ni okeru himitsu kessha (hō-kai)" 現代中國に於ける秘密結社〔幫會〕(Secret societies [*pang-hui*] in contemporary China), in **Kindai Chūgoku kenkyū,** 1948, 9.8.6, pp. 133–164.

> A summary, by a scholar who has much first-hand knowledge, of the origins and activities of the Red Spear Society (Hung-ch'iang hui 紅槍會), Elder Brother Society (Ko-lao hui 哥老會 together with the Hung-pang 洪帮), and Ch'ing-pang 青帮 (together with the Ch'ing-men lo-chiao 清門羅教), sometimes called the Green Gang. Footnotes give bibliography.

8.6.4 SAWASAKI Kenzō 澤崎堅造, "Sekai Kōmanjikai ni tsuite" 世界紅卍字會について (On the World Red Swastika Society), **Tōa jimbun gakuhō,** 2.3 (Dec. 1942), 450–466.

A brief survey of this religious and philanthropic association founded in

China in the early 1920s, the doctrine of which is a kind of syncretism linked with an oracle (*fu-chi* 扶乩) system. The writer cites various publications of the society.

8.6.5 TSUKAMOTO Zenryū 塚本善隆, "Rakyō no seiritsu to ryūden ni tsuite" 羅敎の成立と流傳について ("On the birth and growth of the Lo-tsu-chiao 羅祖敎 or the Wu-wei Sect 無爲敎 in Ming era"), **Tōhō gakuhō, Kyōto,** 17 (Nov. 1949), 11-34.

A study of the origin of one of the secret societies which flourished during the Ch'ing period, by an eminent specialist in Buddhism.

8.6.6 SUZUKI Chūsei 鈴木中正, "Rakyō ni tsuite, Shindai Shina shūkyō kessha no ichirei" 羅敎について──清代支那宗敎結社の一例 (On the Lo-chiao, as an example of the Chinese religious societies in the Ch'ing period), **Tōyōbunka kenkyūjo kiyō,** 1 (Dec. 1948), 441-501.

Mainly based on the memorials published in **Shih-liao hsün-k'an** 史料旬刊 and on G. G. M. de Groot's **Sectarianism and Religious Persecution in China,** this article discusses many aspects of the Lo-chiao in the Ch'ing period—names and locations of societies, personal background of members, their organization, dogma, relationship with the governmental authorities, and the like.

8.6.7 HIRAYAMA Shū 平山周, "Shina kakumeitō oyobi himitsu kessha" 支那革命黨及秘密結社 (The Chinese revolutionary party and the secret societies), **Nippon oyobi Nipponjin,** 569 (Nov. 1, 1911), special supplement, 98 pp. (+10 pp. on revolutionary martyrs).

With a preface by Chang Ping-lin 章炳麟, this journalistic compilation assembles a considerable account of the Triad Society (San-ho hui 三合會) and Elder Brother Society (Ko-lao hui 哥老會), with many illustrations of esoteric seals and symbolism, including the secret language of the arrangement of teacups, reputedly so dear to the heart of Dr. Sun, and other long vocabularies and regulations. An account of Sun's activities then follows (from p. 75). This study was plainly the product of a man in revolutionary circles, which interpenetrated the secret societies, and the wealth of contemporary, inside detail here presented seems to have been generally overlooked by later researchers.

8.6.8 INOUE Kōbai (Susumu) 井上紅梅 (進), **Hito** 匪徒 (Bandit gangs), Shanghai: Nippondō Shoin 日本堂書院, 1923 (3rd printing, 1924, 347 pp.).

This popularly written account of modern Chinese secret societies describes the history, organization, symbolism, terminology, etc. of the Ch'ing-pang 青幇 (known to Westerners as the "Green Gang") and Hung-pang 紅 ("Red Gang") and then describes four other groups of phenomena: thieves and burglars, scoundrels, beggars and heterodox religious cults. Sources are not cited; this volume by an old China hand appears to be based rather indiscriminately on contact with the Chinese scene rather than scholarly research.

8.6.9 Mantetsu Pekin Kōsho Kenkyūshitsu 滿鐵北京公所研究室, comp., "Kōsō kai" 紅槍會 (The Red Spear Society), **Pekin Mantetsu geppō** 北京

滿鐵月報 4.5 (Dec. [?] 1927), 15-94, (whole number 25).

This study of the Red Spears is approached in the context of their use in the Kuomintang-Communist conflict in Hunan and therefore as a factor in the civil strife of the late 1920s and the "failure" of the people's revolution. A good deal of information and contemporary comment are assembled, with speculation as to the social potentialities of such groups.

8.6.10 Mantetsu Pekin Kōsho Kenkyūshitsu 滿鐵北京公所研究室, comp., "Chūgoku ni okeru dohi" 中國に於ける土匪 (Bandits in China), **Pekin Mantetsu geppō,** 4.6 (Dec. [?] 1927), 1-82, (whole number 26).

This continues the study of the Red Spear Society, 8.6.9, discussing the causes and forms of banditry and assembling rather miscellaneous data on various groups, their names and special language, etc.

8.6.11 SUEMITSU Takayoshi 末光高義, **Shina no himitsu kessha to jizen kessha** 支那の秘密結社と慈善結社 (China's secret societies and charitable associations), Manshū Hyōronsha 滿洲評論社, 1932 (3rd printing, 1939, 395 pp., preface by Tachibana Shiraki 橘樸 11 pp., illus.)

Written by an experienced police officer, this is a useful survey of certain aspects of the major Chinese "secret societies", including in that term the modern city gangs, religious and benevolent societies and traditional rural anti-dynastic orders. While not documented nor based on historical research so much as on compilation and some special information, the volume assembles a great deal of "common knowledge", quoted statements, liturgy, by-laws, and documents. It gives connected accounts of the Ch'ing-pang 青幫, Hung-pang 紅, Tsai-chia-li 在家裡, Red Spears 紅槍會, and Big Swords 大刀會, and mentions briefly some two score other secret groups, as well as ten religious or charitable societies, winding up with the Tao-yuan 道院 and the World Red Swastika Society 世界紅卍字會.

8.6.12 SHIBUTANI Gō 澁谷剛, "Minami Shina no kaizoku oyobi dohi ni kansuru chōsa" 南支那の海賊及び土匪に關する調査 ("Enquête relative aux pirates et aux rebelles de la Chine du Sud"), **Tōa keizai kenkyū,** 12.1 (Jan. 1928), 124-145.

The author, who was living in Canton and running a shipping firm, summarizes the various causes of banditry, the various kinds of bandits on land and water, their organization, activities, secret language, source of armament, and distribution in Kwangtung and Kwangsi. An unusual feature of this article is his presentation of tables (pp. 135-145) of leaders, gangs, etc. in some detail.

8.6.13 KUBO Noritada 窪德忠, "Ikkandō ni tsuite" 一貫道について ("I-kuan-tao—One of the Religious Secret Societies in Modern China"), **Tōyōbunka kenkyūjo kiyō,** 4 (Mar. 1953), 173-249.

A thorough and very interesting study of a modern cult (put together in the late 1920s from old traditional elements), which explores its synthetic ideology, real and legendary history, practices, literature, and organization as far as the written record will permit. The author expresses indebtedness to Li Shih-yü 李世瑜 and Fr. W.A. Grootaers who have pioneered in this field; this article is in fact mainly based on the testimony of Mr. Li, who was once a member of this sect.

8. 7   OVERSEAS CHINESE COMMUNITIES

Note: These selections deal with the history of the overseas Chinese, their position in Southeast Asia, and their communities in Japan, in that order. It seems evident that Japanese research on this strategic group has been more intensive than studies by Western scholars.

8. 7. 1   NARITA Setsuo 成田節男, **Kakyō shi** 華僑史 (History of the overseas Chinese), Keisetsu Shoin 螢雪書院, 1941, 445 pp.

This study, by a historian then attached to the South Seas Bureau of the Foreign Office, takes a long historical look at the early Chinese trade relations with Southeast Asia, from pre-T'ang to 1842. Following chapters are on the coolie trade, the overseas Chinese in Sun Yat-sen's revolution (pp. 156-181), their growth in the 20th century, and Ch'ing and Republican policy toward them (pp. 193-210). Sections are then devoted to each of 6 overseas areas, which are again treated historically. While not heavily documented nor very detailed, this intelligent survey provides perhaps more perspective on the place of the Chinese of Southeast Asia in Chinese history than any other work available. It is one of the few comprehensive studies yet attempted.

8. 7. 2   FUKUDA Shōzō 福田省三, **Kakyō keizai ron** 華僑經濟論 (On the overseas Chinese economy), Ganshōdō Shoten, 1939, 500 pp.

This volume was completed in 1937 after several years study at the Tōa Dōbun Shoin (the T'ung-wen College) in Shanghai, where it was planned as the first of a series on China's international economic position. The author discusses the various causes of Chinese overseas migration (demographic, political, commercial, etc.), the forms of migration (coolie trade, etc.), the economic bases established by migrants and their position especially in the various countries of Southeast Asia (*nan-yang* 南洋, to p. 392). The final chapters then take up the relations of China and Japan to the overseas Chinese. Western and Asian sources are used in roughly equal proportions.

8. 7. 3   FUKUDA Shōzō 福田省三, "Marai ni okeru Kakyō rōdōsha" マライに於ける華僑勞働者 (The overseas Chinese laborers in Malaya), **Tōagaku,** 7 (Mar. 1943), 21-56, tables.

By a specialist in this field, this is a competent wartime study of the status, provenance, and organization of Chinese laborers in Malaya, including how they got there and a comparison with local laborers.

8. 7. 4   NEGISHI Tadashi 根岸佶, **Kakyō zakki** 華僑襍記 (Notes on the overseas Chinese), Asahi Shimbunsha 朝日新聞社, 1942, 267 pp.

This small volume written with the help of Professor Uchida Naosaku attempts to summarize quickly all aspects of the Chinese position in Southeast

Asia—history, economic status, conditions in each country (including the Philippines), organizations, relations with China (including useful detail on channels of overseas remittances), and wartime policies concerning them. A compilation of available information, including a number of miscellaneous bits and pieces.

8.7.5 Kakyō 華僑 (The overseas Chinese), vol. 13, number 12 of **Keizai shiryō** 經濟資料, pub. by Tōa Keizai Chōsakyoku 東亞經濟調査局 of the SMR, Dec. 1927, pp. 1-218.

This issue, aside from some general bibliography, is devoted to one of the earliest comprehensive surveys of the overseas Chinese all over the world, their history in each place and contemporary numbers and activities (pp. 1-142, of of which pp. 51 ff. concern Southeast Asia). A second section then discusses the local problems of the overseas Chinese in each area. While brief, this study provided a useful picture at the time.

8.7.6 (Tōa Kenkyūjo) Daisan Chōsa Iinkai (東亞研究所) 第三調査委員會 (The third research committee [of the Tōa Kenkyūjo]), "Nanyō Kakyō chōsa no kekka gaiyō" 南洋華僑調査の結果概要 (An outline of the results of the investigation of the overseas Chinese in Southeast Asia), **Tōa kenkyū shohō**, 13 (Dec. 1941), 857-902, charts.

Over a two-year period this research group studied (1) the historical development of the overseas Chinese communities in Southeast Asia, (2) their economic position and (3) the anti-Japanese movement among them. This report summarizes the findings, necessarily in rather general terms and without source reference.

8.7.7 TSUKIMURA Ichirō 月村市郎, "Nanyō Kakyō shimbun ni kansuru ichichōsa" 南洋華僑新聞に關する一調査 (An investigation of the newspapers published by the overseas Chinese in Southeast Asia), **Tōa kenkyū shohō**, 11 (Aug. 1941), 642-675.

This informative survey gives a history of the overseas Chinese press beginning from 1815 at Malacca, listing major papers by date of founding with further details; KMT policy toward the overseas press is documented and current conditions appraised. Data on some 74 publications is given in an appendix.

8.7.8 IWAO Seiichi 岩生成一, "Gekō (Bantam) no Shina machi ni tsuite" 下港 (Bantam) の支那町について ("A Historical Study of Chinese Town in Bantam"), **Tōyō gakuhō**, 31.4 (June 1948), 440-471.

Like the author's study of Amboyna (8.7.9), this article uses mainly Dutch sources to describe the early Chinese immigration, setting up of the Chinese quarter and its subsequent development and regulation down into the Ch'ing period. Formerly at Taihoku University, Taiwan, where he specialized on Japan's early relations with Southeast Asia, the author is now at Tōkyō University in the field of Japanese history.

8.7.9 IWAO Seiichi 岩生成一, "Amboina no shoki Shina machi ni tsuite"

アンボイナ (Amboina) の初期支那町について ("On the Chinese Quarter at Amboyna in the Early Days"), **Tōyō gakuhō**, 33. 3–4 (Oct. 1950), 269–311.

While it barely reaches the 19th century, this study based on old Dutch accounts gives a most illuminating picture of the process by which Chinese went to Amboyna, their numbers, the growth of the Chinese quarter and its administration, and Dutch policies developed for regulating it. A companion article to Iwao's study of Bantam, 8. 7. 8.

8. 7. 10 UCHIDA Naosaku 内田直作, "Kakyō shihon no zenkiteki seikaku, Marē no Rikuyū zaibatsu o chūshin to shite" 華僑資本の前期的性格——マレーの陸佑財閥を中心として ("Pre-capitalistic traits of enterprises among the Chinese in the Straits Settlements", with special reference to the Loke Yew Konzern in Malaya), **Tōyō bunka**, 7 (Nov. 1951), 27–51.

This article by a well known economic historian of China is apparently based on the author's field research in Malaya during the war. The first six sections provide an interesting factual account of the coming of a certain Chinese, Loke Yew 陸佑, to Singapore in 1845, the associations or gild (*pang* 幫) connections of Loke Yew, the struggle between two rival firms, Ghee Hin Kongsi (*I-hsing kung-ssu* 義興公司) and Hai San 海山 Kongsi in the Larut tin-mining district, the process of capital accumulation on the part of Loke Yew, the later decline of his industrial-capitalist activities, and the genealogy of his family. The remaining two interpretative sections discuss the elementary or "pre-capitalistic" nature (in Max Weber's sense) of Loke Yew's capital-forming activities and the persistence of pre-capitalistic traits among the Chinese abroad. This appears to be one of the few such case studies of overseas Chinese enterprise.

8. 7. 11 UCHIDA Naosaku 内田直作, "Nippon ni okeru Kakyō shakai no keisei" 日本に於ける華僑社會の形成 ("The Formation of the Overseas Chinese Communities in Japan"), **Tōyōbunka kenkyū**, 10 (Feb. 1949), 18–36.

Subtitled "On the establishment of the gild organizations" 公所團體の成立經過について, this interesting article describes how Chinese coming to early-Meiji Japan, following the example of the four Chinese merchant gilds of the Nagasaki period, developed similar organizations in the various treaty ports. This article was later included in the author's book **Nippon Kakyō shakai no kenkyū**, 8. 7. 12.

8. 7. 12 UCHIDA Naosaku 内田直作, **Nippon Kakyō shakai no kenkyū** 日本華僑社會の研究 (A study of the communities of overseas Chinese in Japan), Dōbunkan 同文館, 1949, 392 pp.

Drawing his source materials from Nagasaki, Kōbe, Ōsaka, Yokohama, and Hakodate (see bibliography, pp. 383–92), the author presents a systematic and factual account of 1) the population figures and economic status of the Chinese in Japan, 2) their organization in the Edo period and 3) since the beginning of Meiji, including the establishment of (a) merchant gilds (*kung-so* or *kōsho* 公所), (b) Chinese Community associations (*Chūka kaikan* 中華會館), (c) the officially inspired Chinese General Chamber of Commerce (*Chūka sōshōkai* 中華總商會), after 1903, and (d) professional associations of students and the like. See review by Negishi Tadashi 根岸佶, in **Hitotsubashi ronsō,** 22.6 (Dec. 1949), 775–782.

8. 7. 13 UCHIDA Naosaku 內田直作, and SHIOWAKI Kōshirō 鹽脇幸四郎, ed., **Ryū Nichi Kakyō keizai bunseki** 留日華僑經濟分析 (An analysis of the economy of the overseas Chinese in Japan), Kawade Shobō, 1950, 204 pp.

A study, based on field research, of the position of the Chinese communities in Japan after World War II viewed against their historical background. Covering mainly Tōkyō, Kōbe, Ōsaka and Yokohama, the authors analyze the rather abnormal florescence and then decline in Chinese capital resources and their rather stationary situation in numbers and occupations. The book stresses also the social and economic differences between the Chinese from the mainland and those from Taiwan.

8. 8 SOCIAL PSYCHOLOGY

Note: This heading has been inserted mainly to offer evidence of the fact that this important social science is relatively undeveloped and neglected. See also the legal and sociological studies of Kainō Michitaka and Niida Noboru and item 2. 3. 4.

8. 8. 1 TSUKISHIMA Kenzō 築島謙三, "Kazokushugi shakai no dōtoku, Chūgoku nōson no chōsa kekka o megutte" 家族主義社會の道德——中國農村の調査結果をめぐつて ("The morals of a society based on the family", with special reference to the results of surveys of Chinese villages), **Tōyō bunka,** 5 (April 1951), 40–71.

This article, by a social psychologist who has published several articles on primitive psychology, is mainly based at second hand on the field survey reports published years ago by Fei Hsiao-t'ung 費孝通 (**Peasant Life in China,** 1939, about a village near Soochow) and D. H. Kulp (**Country Life in South China,** 1925, about a village near Swatow), as well as recent work of Niida Noboru and others. The author stresses the permeation of the so-called Confucian morality into the lower social strata of China and the persistent influence of the traditional morality even on persons seemingly rather detached from the old social order. Some comparisons are suggested between the Chinese idea of "face" (*mien-tzu* 面子) and the modern Western ideas of "freedom" and "honor."

8. 8. 2 ŌTANI Kōtarō 大谷孝太郎, **Gendai Shinajin seishin kōzō no kenkyū** 現代支那人精神構造の研究 (A study of the psychological structure of the contemporary Chinese), Shanghai: Tōa Dōbun Shoin Shina Kenkyūbu 支那研究部, 1935, 878 pp.

This is perhaps the largest and certainly one of the most comprehensive foreign analyses of "the Chinese" and why they are so. While lacking the more recent vocabulary of "character structure" studies, the author examines the Chinese national character, recites at length what other foreigners have said about it (pp. 131-627, not excepting the Rev. Arthur H. Smith), and puts forth his own descriptive analysis. See also his article "Shina shakai no hirenzokusei", 8.8.3.

8.8.3 ŌTANI Kōtarō 大谷孝太郎, "Shina shakai no hirenzokusei" 支那社會の非連續性 (The discontinuous nature of Chinese society), **Shina kenkyū,** 10 (May 1926), 1-25; 11 (Sept. 1926), 1-24; 12 (Dec. 1926), 1-35; 13 (Mar. 1927), 1-32; 14 (July 1927), 1-31; 16 (April 1928), 1-68; 17 (July 1928), 1-56.

This lengthy treatise, which evidently was not published in book form, is a systematic but rather abstract exploration of Chinese institutions and attitudes, including social and national consciousness, ideals and virtues, habits of mind, psychological orientations and other intangibles which should gladden the heart of any sociologist. The author seems to have been under the scholarly influence of Dr. Takata Yasuma 高田保馬, a leading Japanese sociologist. Unfortunately the author, as a pioneer, made this effort at scientific analysis of Chinese social psychology without benefit of the most recent frames of reference.

8.8.4 KUMANO Shōhei 熊野正平, "Shinateki kangaekata" シナ的考へ方 (The Chinese way of thinking), **Hitotsubashi ronsō,** 22.2 (Aug. 1949), 340-361.

Based on the personal observations of the author during his long stay in China (mainly in Shanghai and Peking), this rescript of a lecture discusses the empirical characteristics of the Chinese mentality, stressing its contradictory extremes: practical and realistic but also formalistic (e.g. "face" *mien-tzu* 面子); conservative but also enthusiastic over novelties; persevering but also passively resigned (the so-called *mei-yu fa-tzu* 沒有法子 psychology). According to this authority, the Chinese are inclined to solve such a dualistic dilemma by an eclectic compromise, rather than through a dialectic and thorough-going mental struggle.

8.8.5 OTAKE Fumio 小竹文夫, "Shina kakuchi minjō ron (miteikō)" 支那各地民情論 (未定稿) (A treatise on the temperament of the populace in various places in China [undefinitive draft]), **Shina kenkyū,** 53 (Feb. 1940), 97-157.

An unusual, if not unique, selection from the main Ch'ing geographical work, **Ta-Ch'ing i-t'ung chih** 大清一統志, of the characterizations of the temperament, human nature, abilities or tendencies of the people of each prefecture (*fu* 府) or other subdivision (*chou* 州, *t'ing* 廳) of the empire. These characterizations were in turn drawn by the Ch'ing compilers from earlier local or general gazetteers (though without page references). While Baron von Richthofen recorded similar observations in the provinces he traversed, Professor Otake notes that this is a rather neglected subject of study—even though generally accepted in common parlance, like the difference in temperament between Kantō 關東 and Kansai 關西 people in Japan.

8.8.6 NAKAMURA Hajime 中村元, **Tōyōjin no shii hōhō** 東洋人の思惟方法 (The ways of thought of the Eastern peoples), Misuzu Shobō みすず書房, vol. 1 (India and China), 1948, 554 pp.; vol. 2 (Japan and Tibet), 1949, 526+index 35 pp.

The China section of this complicated and difficult pioneer effort (1. 291–554) takes up various features of the Chinese language from the point of view of judgment 判斷, inference 推理, and other logical and grammatical problems, and then discusses certain cultural influences of Buddhism and other aspects of Chinese thought. This is a work of logic and philosophy not easy for the mere historian to follow, but there are interesting sections on the Chinese traditions of conservatism, individualism, naturalism, social status, and the like, which should be of interest in connection with any study of Chinese social psychology. Prof. Nakamura is in the Faculty of Letters, Tōkyō University.

8.8.7 NAGAO Ryūzō 永尾龍造, **Shina minzoku shi** 支那民俗誌 (Gazetteer of Chinese popular customs), Maruzen, vol. 1, 1940, 672+31 pp; vol. 2, 1941, 883+52 pp; vol. 6, 1942, 847+52 pp., color prints, photos. (The other volumes of the originally projected 13 vols. have not been published, the manuscripts having been destroyed by fire).

These elaborate volumes, though incomplete, describe in detail many aspects of the social customs and formal activities, ceremonies and rituals of traditional Chinese life. The first 1455 pages, for example, describe the Chinese New Year. Interesting color reproductions of door gods and other objects and many photographs, together with extensive indices, should make these volumes of value for research.

9. REFERENCE WORKS

Note: Sections 1. 2 and 7. 2 above list a number of gazetteers and compilations which in a sense have more reference value for the historian than materials in this section. Here we list mainly the aids to scholarship of which beginning students should be aware.

9. 1 BIBLIOGRAPHICAL GUIDES

Note: Items 9. 1. 1 by Wada Sei and 9. 1. 2 by Abe Takeo are the best starting points in this field as of 1953. Note that the journal **Shigaku zasshi, 9. 9. 17,** has useful annual surveys of new publications, by periods.

9. 1. 1 WADA Sei 和田清, **Chūgokushi gaisetsu** 中國史概說 (An outline of Chinese history), Iwanami Shoten, 1951, 2 vols., 486+index of 44 pp. (Iwanami zensho series, 120, 121.)

> Some 130 pp. of volume 2 are on the modern period, since the rise of the Manchus. Extensive lists of references (totalling more than 200 items) are inserted to accompany the brief sections of text. Since these have been judiciously selected by a master of the subject, they make an invaluable bibliographical guide. The selections in Chinese and Japanese are somewhat more comprehensive than those in Western languages, some of the latter materials having been still unavailable in Japan. The index and a chronology of 46 pp. are also useful, but the main help for Western students lies in Professor Wada's brief but outspoken characterizations of a tremendous number of Japanese and Chinese works. The concluding general bibliographical essay alone touches on another 100 items in 25 pages. The second printing (1952) includes a short additional essay of 6 pp., touching on newly published works totalling 50 items. As of 1953 this is the best single guide to historical work on China. See review by Uemura Seiji 植村清二, in **Shigaku zasshi,** 60. 4 (April 1951), 357–363.

9. 1. 2 ABE Takeo 安部健夫, "Shin" 清 (Ch'ing), pp. 415–503, in **Chūgoku shigaku nyūmon** 中國史學入門 (Guide to Chinese historical studies), ed. by Tōhō Gakujutsu Kyōkai 東方學術協會 (Society for Oriental studies), Kyōto: Heian Bunko 平安文庫, 1951, 693 pp.

> The author is in charge of Ch'ing researches at the Jimbunkagaku Kenkyūjo, Kyōto. He first surveys the major phases of China's history from 1644 to about 1830 and thence to 1912; and then notes the main bodies of source

258

materials and major works of scholarship in Chinese, Japanese and Western languages. Nine subsequent sections then discuss principal sources and writings by periods, including nineteenth century foreign relations in general, the Canton trade and Opium War, the Taiping Rebellion and Anglo-French expedition, the 1898 Reform Movement, and the Boxer Rebellion, with references to some 50 Chinese and Japanese works under these headings. While major Western works are included, this bibliographical essay is chiefly useful for its judicious selection of Chinese and Japanese items. The author acknowledges the assistance of Hatano Yoshihiro 波多野善大. See review by Enoki Kazuo 榎一雄 in **Shigaku zasshi**, 61.9 (Sept. 1953), 851–853.

9.1.3 MOMOSE Hiromu 百瀬弘, "Min-Shin" 明清 (Ming and Ch'ing), in **Shakai keizai shigaku** 社會經濟史學, 10.11–12 (March 1941), special issue 特輯, "The Development of Social and Economic History", part 2 下卷, pp. 63–76; and reprinted on pp. 345–358 in the later book of the same title (Iwanami, 1944, 690 pp.).

Somewhat more extensive than the companion article of Usami Seijirō on Republican China, 9.1.4, this is a scholarly survey of recently published research on the Ming and Ch'ing periods.

9.1.4 USAMI Seijirō 宇佐美誠次郎, "Chūkaminkoku" 中華民國 (The Chinese Republic), in **Shakai keizai shigaku,** 10.11–12 (March 1941), special issue, part 2 下卷, pp. 77–84; and reprinted on pp. 359–366 of **Shakai keizai shigaku no hattatsu** 社會經濟史學の發達 (The development of social and economic history studies), ed. by the society for the same, Iwanami Shoten, 1944, 690 pp.

A brief bibliographic list and critique of then recent publications, East and West, on Republican China.

9.1.5 Rekishigaku Kenkyūkai 歷史學研究會 ("The Historical Science Society"), **Rekishigaku nempō, Shōwa jūnen ban** 歷史學年報——昭和十年版 (Historical studies annual, edition for the year 1935), special number, **Rekishigaku kenkyū,** 5.1, whole no. 25 (Nov. 1935), 416 pp.

This special issue is more comprehensive than the annual bibliographical survey articles published by this journal. It includes a section on studies of Chinese history by major topics (pp. 114–146), a list of books (pp. 181–185), and a series of essays on academic work in major fields (pp. 293–377), including politics and foreign relations, the "Asiatic mode of production" アジア的生產様式 (pp. 306–313), economic history, and cultural history (pp. 344–354 by Itano Chōhachi 板野長八).

9.1.6 Rekishigaku Kenkyūkai, ed., **Rekishigaku no seika to kadai** 歷史學の成果と課題 (Topics and results in historical studies), Iwanami Shoten, 3 annual vols.: (1) **1949 nen rekishigaku nempō** 一九四九年歷史學年報 (Historical studies annual for 1949), 1950, 229 pp. (later printings marked as vol. no. 1); (2) ibid. vol. 2 for 1950, pub. 1951, 112+bibliog. 45 pp. of materials pub. between Jan. 1, 1949 and July 1950; (3) ibid. vol. 3 for

1951, pub. 1952, 202 + 64 pp. bibliog. of materials pub. between Aug. 1950 and Oct. 1951.

These bibliographical annuals provide in each case a series of survey articles by rather young historians on various periods of history (generally by writers interested in a Marxist interpretation). Vol. 1 is divided into Japanese, Oriental and Western sections, with no bibliographic listings; vol. 2 into ancient, medieval, modern and contemporary periods; vol. 3 into ancient, feudal and modern periods, with sections on "Feudal China, later period, Ming and Ch'ing (pre-history of the Taiping Rebellion)", pp. 72–83, by Saeki Yūichi 佐伯有一; and on "The modern Orient, modern China", pp. 131–146, by Satoi Hikoshichirō 里井彦七郎. Some 47 books and 220 articles are listed in the 1951 volume under Oriental history, which may indicate the current volume of production in Japan. See also 1.5.5.

9.1.7 ISHIHARA Michihiro 石原道博, "Shina jihen irai Nisshi kankei no shokenkyū" 支那事變以來日支關係の諸研究 ("Studies on Sino-Japanese Relations since the outbreak of China Conflict"), **Rekishigaku kenkyū,** 8.12, whole number 61 (Dec. 1938), 1357–1373.

This bibliographical survey of contemporary Japanese publications on China, while it devotes only three pages to those on the modern period, presents them rather interestingly in the context of Sino-Japanese friction.

9.1.8 BABA Akio 馬場明男, **Shina mondai bunken jiten** 支那問題文献辭典 (Dictionary of documentation on the problems of China), Keiō Shobō 慶應書房, 1940, 350 pp.

This list of selected books and passages in books, quickly compiled for beginners, is arranged by topics under social, economic and political sections and might seem at first glance to be a useful guide to Japanese writings on China. A brief perusal reveals, however, that academic scholarship is barely represented and several works are cited repeatedly under different topics, including a number of translated works. Further examination divulges that the works from which most references are drawn, most of them secondary surveys, have a rather marked similarity of terminology and assumptions and offer little in the way of scholarly research. The Great Soviet Encyclopaedia is also cited from time to time.

9.1.9 MURAMATSU Yūji 村松祐次, "Chūgoku kindai keizaishi" 中國近代經濟史 (Modern Chinese economic history), pp. 175–214 in Tōkyō Shōka Daigaku (now called Hitotsubashi Daigaku) Hitotsubashi Shimbunbu 東京商科大學一橋新聞部 (Hitotsubashi [students'] press bureau, Tōkyō Commercial College), ed., **Keizaigaku kenkyū no shiori** 經濟學研究の栞 (A guide to the study of economics), vol. 3, Shunjūsha 春秋社, 1950, 214 + index 18 pp.

A useful and critically descriptive bibliography of works on modern China, emphasizing economics but mainly devoted to English-language materials.

9. 2 BIBLIOGRAPHICAL LISTS

Note: The annual volumes of the Kyōto Institute (9.2.1) provide the best general coverage for academic purposes, although other items listed below offer additional information. Note that item 9.4.1 is a bibliography of biographical works. Bibliographies on special topics have been inserted in preceding sections (when published as separate books or articles) or have been noted in our descriptions of certain entries (when published as parts of various books or articles). For such bibliographies, see the following entries (the subjects of which in each case should be evident from the entry numbers): 1.1.1–2, 1.4.1, 1.5.3, 2.4.3, 2.4.11, 2.4.28, 2.6.1, 2.6.15, 3.5.10, 3.6.1, 3.7.4, 4.1.9, 4.2.1, 4.2.10, 4.2.11, 4.2.13, 4.5.1, 4.5.4, 4.7.6, 5.1.9, 5.2.14, 5.3.2, 5.5.2, 5.6.1, 5.6.21, 5.11.5, 6.1.4–6, 6.2.1, 6.5.3, 6.7.6–8, 6.7.12, 6.8.8, 6.8.17, 6.9.18, 6.11.7, 7.2.11, 7.4.1, 7.5.2, 7.8.1, 7.9.3, 7.11.1, 7.15.5, 7.15.16, 7.17.11, 8.1.16, 8.2.1, 8.2.9, 8.3.6, 8.5.4–6, 8.6.2, 8.7.5, 8.7.7, 9.5.7, 9.9.30.

9.2.1 Tōhō Bunka Kenkyūjo 東方文化研究所 ("The Institute of Oriental Culture") and its successor (see below), Kyōto, comp. and pub., **Tōyōshi kenkyū bunken ruimoku** 東洋史研究文献類目 ("Annual Bibliography of Oriental Studies"), 8 annual or biennial volumes for the period 1934–1945 and a combined volume for 1946–50, as follows:

(1) **Shōwa kyūnendo Tōyōshi kenkyū bunken ruimoku** 昭和九年度 ("Bibliography of Oriental Studies 1934"), pub. by Tōhō Bunka Gakuin Kyōto Kenkyūjo ("The Academy of Oriental Culture, Kyōto Institute"), 1935, 83 pp.+6 pp. Western languages, compiled in April 1935.

(2) The same for 1935, pub. 1937, 97 pp. Oriental+39 pp. Western languages, compiled in July 1936.

(3) The same for 1936 ("Annual Bibliography of Oriental Studies for the year 1936"), pub. by Tōhō Bunka Kenkyūjo ("The Institute of Oriental Culture, Kyōto"), 1938, 91+60 pp., comp. July 1938.

(4) The same for 1937, pub. 1939, 77+53 pp., comp. Oct. 1939.

(5) The same for 1938 and 1939, pub. 1941, 122+92 pp., comp. Sept. 1940.

(6) The same for 1940 and 1941, pub. 1945, 279 pp. all languages.

(7) The same for 1942 and 1943, pub. by Kyōto Daigaku Jimbun Kagaku Kenkyūjo, 1949, 222 pp.+index 43 pp. (pp. 3–52 are a survey of developments in various fields of Oriental history by a dozen specialists. Very few Western language entries.)

(8) The same for 1944 and 1945, 100 pp., n.d., preface dated March 1946.

(9) The same for 1946–1950, pub. by "The Research Institute of Humanistic Sciences (Formerly The Institute of Oriental Culture), Kyōto University, June 1952," pub. Sept. 1952, 150+index 28+Western languages 137+index 21 pp.

This series forms a notable contribution to the bibliographic art, having steadily developed its techniques of classification and indexing as well as its international coverage of materials. As indicated above, the Kyōto Institute was originally affiliated to the Tōhō Bunka Gakuin, like its sister institute in Tōkyō, but became independent and changed its name. Being a center of research on the classical and humanistic tradition, its bibliographies are stronger in such studies than in modern social science or current politics, but there is no comparable bibliography which specializes adequately and on an annual basis in the latter areas. The international coverage of these volumes is of course a notable feature. They were produced by Fujieda Akira 藤枝晃, Kurata Junnosuke 倉田淳之助, and others.

9.2.2 Ōtsuka Shigakkai Kōshi Bukai 大塚史學會高師部會, comp., **Tōyōshi rombun yōmoku** 東洋史論文要目 (Essential articles in Oriental history), pub. by the society, 1936, rev. and enlarged ed., 362 pp.

A simple list of Japanese articles appearing in some 139 periodicals and collectanea from 1868 to Dec. 1935 in the field of Oriental history. Items on China (pp. 1–239) are arranged under a variety of topics, those on foreign relations covering pp. 209–225. Though the method of compilation is rather simple (e.g. no index) and there are some inaccuracies, this is the only single bibliography which comprehensively covers this period. For practical purposes it is continued in the bibliographical series of the Kyōto Institute, 9.2.1. Production of this bibliography was supervised by Dr. Nakayama Kyūshirō 中山久四郎, whose surname had been Nakamura 中村 (see 6.8.14).

9.2.3 Shigakukai 史學會 (Society for historical studies, with offices in Tōkyō University), comp., **Shigaku bunken mokuroku, 1946–1950** 史學 文献目録 **1946–1950** (Bibliography of historical materials, 1946–1950), Yamakawa Shuppansha 山川出版社, 1952, 204 pp.

A useful list of 9022 books and articles, with an author index, based in part on combing some 350 pepiodicals. Among the 2000 items on Oriental history (numbers 5001 to 6978), the totals are roughly as follows: 22 general books, ca. 17 primarily historical books on the Ch'ing period and 48 on modern China, plus books in other categories; 97 primarily historical articles on the Ch'ing and 71 on contemporary China, plus articles in other categories.

9.2.4 Kokusai Bunka Shinkōkai 國際文化振興會 ("The Society for International Cultural Relations"), ed., **K. B. S. Bibliographical Register of Important Works written in Japanese on Japan and the Far East published during the year 1933,** pub. by KBS, 1938, 180 pp.; the same for 1934, pub. 1940, 211 pp.; the same for 1935, pub. 1942, 211 pp.; for 1936, pub. 1942, 203 pp.; for 1937, pub. 1943, 217 pp.; for 1938, pub. 1949, 111 pp.

This useful English-language series began with a volume for 1932 (not seen by us). Each volume presents a careful selection of books and articles under topics, the sections on China running usually to 25 or 30 pages, mainly on cultural subjects with rather little on modern history. Note the indices of authors' names in English spelling, at the back of each volume, which are generally reliable guides to this enigmatic aspect of Japanese scholarly life.

9.2.5 Yü Shih-yü 于式玉, comp., **Jih-pen ch'i-k'an san-ship-pa-chung chung Tung-fang-hsüeh lun-wen pien-mu fu yin-te** 日本期刊三十八種中東 方學論文篇目附引得 ("A Bibliography of Orientological Contributions in Thirty-Eight Japanese Periodicals"), Peiping: Harvard-Yenching Institute

Sinological Index Series, with offices in Yenching University Library, 1933, 30+343 pp. (Supplement no. 6).

The entries on history in this bibliography, covering pp. 16–40, are in rather miscellaneous order, but indices are provided of authors' names in *rōmaji* and by the 4-corner system, and of titles by the latter, to which in turn are provided Wade-Giles and radical indices. Only the major historical periodicals are covered, but they are usually covered from their inception.

9.2.6 Yü Shih-yü 于式玉 and Liu Hsüan-min 劉選民, comp., **I-pai-ch'i-shih-wu-chung Jih-pen ch'i-k'an-chung Tung-fang-hsüeh lun-wen pien-mu fu yin-te** 一百七十五種日本期刊中東方學論文篇目附引得 ("A Bibliography of Orientological Contributions in One Hundred and Seventy-five Japanese Periodicals"), Peiping: Harvard-Yenching Institute, 1940, 44+198+131 +124+36 pp. (Harvard-Yenching Institute Sinological Index Series, Supplement no. 13).

This useful volume supplements "Supplement no. 6" of 1933 by covering a great many more periodicals and up-dating the coverage of those used previously, up to about 1939. For History, see pp. 26–65. It should be noted that Yenching Library holdings were incomplete in many cases, and the user is advised to take account of the listed issue numbers of periodicals covered (see pp. V–XVIII). In addition to topical arrangement and 4-corner indices of titles and authors, the compilers courageously provided, as before, an extremely useful index of Japanese authors by *rōmaji* (alphabetic) readings of their names. This list compiled in Peiping will be found, however, to contain some incorrect readings of names, a circumstance with which we sympathize.

9.2.7 (Tōa Kenkyūjo Dairoku Chōsa Iinkai) Shina Keizai Kankō Chōsabu (東亞研究所第六調査委員會) 支那經濟慣行調査部 (Research section on Chinese economic customs [of the sixth research committee of the Tōa Kenkyūjo]) with offices in the Faculty of Economics of Kyōto Imperial University, **Shina kōkōgyō ni kansuru shuyō bunken mokuroku** 支那鑛工業に關する主要文獻目錄 ("Bibliography on the Industry and Mining in China"), mimeographed with printed covers by the above research section for limited distribution by the Tōa Kenkyūjo, dated on covers Dec. 1940, vol. 1 第一, 470 pp., Japanese language materials; vol. 2 第二, 428 pp., Chinese; vol. 3, part 1, 221 pp., and part 2, 333 pp., European languages.

These products of wartime assiduity list an enormous amount of printed matter, both books and articles, on the various aspects of Chinese industry, labor, and mining, not excluding Miss Gulielma F. Alsop's **My Chinese Days** (Boston, 1918). Topically arranged, these lists present almost a plethora of references and constitute undoubtedly the largest single compilation of this rather raw and undigested type. Other volumes which we have seen in this series are the following:

Shina shōgyō ni kansuru shuyō bunken mokuroku 商業 ("Bibliography on the Commerce in China"), Dec. 1940, vol. 1, 63+33 pp., in Japanese and European languages; vol. 2, 299 pp., in Chinese.

Shina kin'yū ni kansuru shuyō bunken mokuroku 金融 ("Bibliography on the Finance in China"), Dec. 1940, vol. 1, 98+17 pp., Japanese and European; vol. 2, 218 pp., Chinese.

Shina nōgyō ni kansuru shuyō bunken mokuroku 農業 (Bibliography of important materials concerning Chinese agriculture), Dec. 1940, vol. 1, 104 pp., Japanese materials.

All these materials are well enough arranged by topics and with adequate reference data, although listed indiscriminately as to quality.

9.2.8 Tōa Kenkyūjo 東亞研究所, comp., **Rekkoku no tai-Shi seiryoku shintōshi bunken mokuroku** 列國の對支勢力滲透史文獻目録 (A bibliography concerning the history of the infiltration into China of the influence of the foreign powers), Tōa Kenkyūjo, 1942, 146 pp.

Based on holdings of books in all languages in some 21 libraries and research offices, this wartime bibliography is arranged by a useful and detailed topical breakdown and gives details of content (chapter headings), although indices of authors or titles are lacking. Japanese, Chinese, and Western materials are included as well as their call numbers in Japanese libraries.

9.2.9 Gaimushō Bunshoka 外務省文書課 (Foreign Office, Section of Archives), **Mantetsu kankōbutsu mokuroku** 滿鐵刊行物目録 (Bibliography of SMR publications), mimeographed by the Foreign Office and classified confidential 秘, April 1933, alphabetic アルハベット list of titles, 57 pp.; classified list of titles, 68 pp.

Including items labelled confidential and secret 極秘, this useful list gives titles, channel of publication (e.g. as documentation 資料, pamphlet パンフレット, or report, or in a periodical or series), and year and month of publication for somewhere around 600 items dated between 1909 and 1931. Authors or producing sections are not indicated and the user will still confront the problem of finding files of SMR publications.

9.2.10 IWAMURA Shinobu 岩村忍 and FUJIEDA Akira 藤枝晃, **Mōko kenkyū bunken mokuroku, 1900–1950** 蒙古研究文獻目録一九〇〇——一九五〇年 ("Bibliography of Mongolia for 1900–1950"), Kyōto: Jimbun-kagaku Kenkyūjo (formerly The Institute of Oriental Culture), Kyōto University, 1953, 46 pp., not for sale. (**Annual Bibliography of Oriental Studies,** Special Number I).

A useful list of some 1800 items compiled by two leading specialists. It is arranged by authors and includes both books and articles appearing in some 200 Japanese periodicals. The compilers explain that little of note appeared before 1900. Some items were purposely omitted; this list presents the essential results of Japanese scholarship.

9.2.11 SAKURAI Yoshiyuki 櫻井義之, **Meiji nenkan Chōsen kenkyū bunkenshi** 明治年間朝鮮研究文獻誌 (Bibliography of materials for research on Korea published in the Meiji period), Keijō: Shomotsu Dōkōkai 書物同好會, 1941, 421 pp.

An outstanding bibliography, with pertinent notes, of 579 Japanese works on Korea arranged according to the appropriate topics and with title and author indices. The author worked at the library of Keijō University for many years and applied himself, under the stimulus of Dr. Yoshino Sakuzō, to bibliographical research. A succeeding volume for the Taishō-Shōwa period is now being prepared. Mr. Sakurai is now in the Library of the Tōkyō Metropolitan University.

9.3 STUDIES OF SOURCES

Note: These are scholarly writings which deal with major historical sources, sometimes in translations (which we normally have eschewed). They might also be called textual critiques or bibliographical studies. Note that the last item is a list of Chinese periodicals.

9.3.1 MOMOSE Hiromu 百瀬弘, "Dai-Shin eten no hensan ni kansuru ichikōsatsu" 大清會典の編纂に關する一考察 (A study of the compilations of the **Ta-Ch'ing hui-tien**), **Tōhō gakuhō, Tōkyō,** 11.1 (Mar. 1940), 360–369.

A careful bibliographical note which evaluates these important compendia on the basis of an analysis of how they were compiled. A useful table dates the stages of this process in each case.

9.3.2 MOMOSE Hiromu 百瀬弘, "Shimmatsu no Keiseibumpen ni tsuite" 清末の經世文編について (On the **Ching-shih-wen pien** of the late Ch'ing), pp. 877–892 in **Ikeuchi hakushi kanreki kinen Tōyōshi ronsō** 池內博士還曆記念東洋史論叢 (Essays on Oriental history collected to commemorate the 60th birthday of Dr. Ikeuchi), Zauhō Kankōkai 座右寶刊行會, 1940, 912 pp.

A useful bibliographical study of the important series of "Collected essays on administration" or "statecraft" which began with the **Huang-ch'ao ching-shih-wen pien 皇朝** compiled by Ho Ch'ang-ling 賀長齡 (the actual work of compilation being done by Wei Yüan 魏源) in 120 *chüan* in 1827, a work which brought together some 2100 essays arranged under 8 classes and 65 sub-heads. The writer then notes the several supplementary publications of this type and analyzes the scope of their content.

9.3.3 SAEKI Tomi 佐伯富, comp., "Tōkaroku mokuji" 東華錄目次 (A chronological table of contents for the **Tung-hua-lu**), **Tōa jimbun gakuhō,** 2.2 (July 1942), 288–297.

This is simply a chronologically-arranged table indicating what period (in the lunar calendar) is covered by each volume of the **Tung-hua-lu,** as a standard compilation of Ch'ing imperial annals, including the edition of the Kuang-hsü period.

9.3.4 ŌTANI Takeo 大谷建夫, "Shinshikō no seiritsu to sono seikaku" 清史稿の成立と其の性格 (The production of the **Ch'ing-shih kao** and its nature), **Mantetsu chōsa geppō,** 22.9 (Sept. 1942), 171–182.

The author, of the Dairen Library, describes the arrangements for compilation of the Ch'ing history, the editorial personnel and program, assignments of sections to specific compilers, the process of compilation, and the financial background of publication (grants from Yüan Shih-k'ai and then from Chang Tso-lin). He then discusses the book's characteristics as a traditional work and quotes some criticisms of it from a modern point of view.

9.3.5 OTAKE Fumio 小竹文夫, "Shinshikō seigohyō" 清史稿正誤表 (A table of errata for the **Ch'ing-shih kao**), **Shina kenkyū,** 48 (Dec. 1938), 123–152.

Professor Otake here notes down almost 1200 textual errors, with corrections, drawn mainly from the biographical section of the **Draft History of the Ch'ing,** but representing all parts of it. Many are misprints or errors in names.

9.3.6 AMAGAI Kenzaburō 天海謙三郎, "Shindai Shokkashi yakkō, ichi, kokō" 清代食貨志譯稿————一, 戶口 (A draft translation of the **Shih-huo chih** for the Ch'ing period, part one, population), **Mantetsu chōsa geppō,** 22.9 (Sept. 1942), 139–170; continued on other topics in 23.3 (Mar. 1943), 163–211; and 23.5 (May 1943), 151–174.

A carefully and fully annotated translation of the essay on population in the **Shih-huo chih** of the **Ch'ing-shih kao,** with a list of pertinent references to population in 8 major compendia on Ch'ing institutions, including **Shinkoku gyōseihō,** 1.2.1, as the most convenient guide.

9.3.7 TAKIGAWA Masajirō 瀧川政次郎, "Kobu sokurei kō" 戶部則例考 ("Considerations on ''Hu-pu-tse' (sic)" [Regulations of the Board of Revenue]), **Shakai keizai shigaku,** 13.6 (Sept. 1943), 519–578.

A detailed study of the compilation, contents, sections, etc. of this important but little used Ch'ing compendium, **Hu-pu tse-li,** last revised in 1874.

9.3.8 MIKUNIYA Hiroshi 三國谷宏, "Shinki Tōa gaikōshi ni kansuru kinkan no kambun shiryō ni tsuite" 清季東亞外交史に關する近刊の漢文史料について ("The materials for study of foreign relations towards the end of the Ch'ing Dynasty, recently published in China"), **Tōhō gakuhō, Kyōto,** 4 (Dec. 1933), 425–443.

A rather critical appraisal of the following newly published collections of documents:
1. **Ch'ing-chi wai-chiao shih-liao** 清季外交史料,
2. **Ch'ing Kuang-hsü-ch'ao Chung-Jih chiao-she shih-liao** 清光緒朝中日交涉史料,
3. Weng T'ung-ho's diary 翁文恭公日記,
4. The collected writings of Chang Chih-tung 張文襄公全集,
5. The collected writings of Hsü Ching-ch'eng 許文肅公遺稾.

The author calls attention to Western war origin documents and pertinent studies of diplomacy available at the time of writing. While now rather ancient, this is one of the few articles which tries to compare the Chinese and Western documents in question.

9.3.9 IWAI Hirosato 岩井大慧, supervisor 監修, **Shina sōhō kaisetsu** 支那叢報解說 (An explanatory guide to **The Chinese Repository**), Maruzen, 15 vols., 1942-1944, each vol. ca. 150-200 pp., illus.

These volumes provide a summary and running comment on the contents of **The Chinese Repository** (Macao and Canton, monthly, 1832-1851, ed. by E. C. Bridgman and S. Wells Williams). They regroup the contents of each annual volume under main headings—"Reviews", "Articles", "Journal of Occurrences", etc., with extensive quotation and translation from the original. This was done to provide a guide to the original English text of the **Repository,** the first 15 volumes of which were reproduced and published in Japan at the same time. Only vols. 1-15 out of the **Repository's** 20 volumes are accordingly covered in these guidebooks; even so, the modest history-making enterprise of Messrs. Bridgman and Williams would appear to be honored in Japan even more than in China or the West. Each volume was the work of different commentators. Among the commentators were Numata Tomoo 沼田鞆雄, Momose Hiromu 百瀬弘, Yazawa Toshihiko 矢澤利彦, Ichiko Chūzō 市古宙三, and Kawabata Genji 河鰭源治.

9.3.10 ENOKI Kazuo 榎一雄, and ICHIKO Chūzō 市古宙三, "Shina ni okeru bunken no genson jōtai" 支那に於ける文獻の現存狀態 (The state of preservation of documents in China), **Tōa ronsō,** 2 (Jan. 1940), 189-231.

Report of a survey, in which the two authors, accompanying Professor Wada Sei of Tōkyō Imperial University, looked at archives and libraries in Peking, Paoting, Taiyuan, Kaifeng, Nanking, Soochow, Hangchow, Shanghai, Kalgan and Kuei-hua-ch'eng, in August and September 1939. Library publications and statistics are extensively quoted. This is perhaps the only such expert appraisal on so broad a plane in recent times.

9.3.11 SANETŌ Keishū 實藤惠秀, comp., "Chūgoku zasshi nempyō" 中國雜誌年表 (Chronological table of Chinese periodicals), **Chūgoku bungaku,** 74 (July 1941), 176-190; 76 (Sept. 1941), 272-285.

This list is arranged by date of first issue from 1896 through 1919 and includes some 200 items through 1912 and almost as many more for 1913-1919. The compiler began his list in 1937 and had the benefit of the standard Chinese work, Ko Kung-chen 戈公振, **Chung-kuo pao-hsueh shih 中國報學史,** to which this list appears to form a valuable supplement.

9.4 BIOGRAPHICAL DICTIONARIES AND DIRECTORIES

Note: These are in roughly chronological order. The following entries in preceding sections have also been noted as having who's who material: 4.2.10-16, 4.6.7, 5.2.14, 5.3.6-7, 5.5.7, 5.6.4, 6.10.4, 6.11.5, 8.1.16, also 9.5.3-4, 9.8.1.

9.4.1 HIRAOKA Takeo 平岡武夫, comp., **Rekidai meijin nempu 歷代名人年譜** (Chronological biographies of famous men of successive Chinese dynasties), Kyōto: Jimbunkagaku Kenkyūjo, index no. 10, 人文科學研究所索引第十, 1951, 163 pp.

For our purposes, pages 72–105 list **nien-p'u** of about 400 persons born between 1735–1899, arranged by order of birth. Wade and other indices are added.

9.4.2 Shina Kenkyūkai 支那研究會 (China research society), ed., **Saishin Shina kanshinroku** 最新支那官紳錄 (Leading personnel of modern China), Fuzambō, 1918, 83+76+798+434 pp.

Produced by a research agency located in Peking, with the help of local Japanese officers and legation officials, this who's who lists over 5,000 Chinese in all walks of life, with both a Japanese *kana* index and a character index (in both of which Wade-Giles readings are also supplied). An appendix (434 pp.) gives extensive tables of Chinese official posts plus a useful who's who of Japanese who had been active in various lines in China. The table of contents of the appendix lists a long section on Chinese officials holding posts in the period 1911–18, but these pages (89–305) and other sections were not bound in the volume we have seen.

9.4.3 SAWAMURA Yukio 澤村幸夫 and UEDA Toshio 植田捷雄, comp., **Shina jinshiroku** 支那人士錄 (A who's who of China), Osaka: Mainichi Shinbunsha 每日新聞社, 1929, 148+79 pp.

A convenient compilation of data on nearly 1000 Chinese of note in the 1920s, with indices by characters and by Wade romanization. Connections with Japan are noted.

9.4.4 NEGISHI Tadashi 根岸佶, supervisor, and AMAGAI Kenzaburō 天海謙三郎, compiler, **Chūkaminkoku jitsugyō meikan** 中華民國實業名鑑 (Directory of business firms in the Chinese Republic), Tōa Dōbunkai, 1934, 1270+132+127 pp.

This enormous compilation of business data records the name, place, founding, capital, balance-sheets, equipment, productive capacity, market, trademark, manager, directors, etc., etc. of some 6,200 banking, insurance, trading, manufacturing and other enterprises. While similar data on Western firms in China has been regularly available in various China "hong lists" and directories, this directory includes a multitude of modern Chinese firms not similarly recorded. Indices are supplied (1) by *kana* readings and (2) by English name or Wade-Giles romanization.

9.4.5 Gaimushō Jōhōbu 外務省情報部 (Information Bureau of the Foreign Office), ed., **Gendai Chūkaminkoku Manshūteikoku jimmeikan** 現代中華民國滿洲帝國人名鑑 (Who's who of the contemporary Republic of China and Empire of Manchoukuo), Tōa Dōbunkai, 1937 (2nd printing, 1939, 78+144+699 pp., charts).

A who's who of personnel in both (Nationalist) China and Manchoukuo, with indices by Japanese *kana* readings and Wade-Giles romanization and over 7,000 entries listed by their Japanese readings; also six charts of government organization.

9.4.6 HASHIKAWA Tokio 橋川時雄, **Chūgoku bunkakai jimbutsu sōkan** 中國文化界人物總鑑 (A who's who of persons in Chinese cultural circles), Chūka Hōrei Inshokan 中華法令印書館, 1940, 82+815+28+12+8 pp., illus.

This valuable biographic dictionary gives data on more than 4000 Chinese alive at some time after 1912 in the fields of art and letters, including writers and artists of all sorts, researchers, teachers and educational personnel, scientists, religious personnel, members of the press, lawyers, officials, actors, musicians, and in short members of the educated and professional or "cultured" classes in a broad sense. Data was obtained from a wide combing of directories, publisher's lists and periodicals as well as the compiler's first hand information and includes the subject's publications as far as possible, as well as his educational history. A chronology of cultural events since 1839 (28 pp.), and tables of Chinese and foreign universities are added. Character index only.

9.4.7 Gaimushō Ajiakyoku 外務省アジア局, **Gendai Chūgoku Chōsen jimmeikan** 現代中國朝鮮人名鑑 (Biographical dictionary of contemporary Chinese and Koreans), pub. by the Foreign Office for office use, 1953, 28+50+313+6 pp., charts.

This latest who's who of Chinese and Koreans (of all camps) has useful Wade-Giles and Japanese syllabary indices for some 2000 living Chinese, including many recently risen to prominence. Vital statistics are less detailed than the official experience of the subjects.

9.5 ENCYCLOPAEDIAS AND CHRONOLOGIES

Note: Gazetteers, as distinct from encyclopaedias, have been listed above in sections 1.2 and 7.2. In this present section the last four items are chronologies. Note also the following entries in earlier sections which contain chronologies on topics which should be evident from these entry numbers: 4.1.4, 4.2.14, 4.5.4, 4.9.7, 5.2.11, 5.3.1, 5.6.1-2, 5.6.7-8, 5.6.10, 5.7.2, 6.8.17, 6.10.4, 6.11.1, 7.12.9, 9.1.1, 9.4.6, 9.8.1.

9.5.1 Tōyō rekishi daijiten 東洋歴史大辭典 (Encyclopaedia of Oriental history), pub. by Heibonsha, 9 vols., 1937-1939, ca. 550 pp. each vol., index (vol. 9) 322 pp., maps, illus.

Supervised by Ikeuchi Hiroshi 池內宏, Hashimoto Masukichi 橋本增吉, Hamada Kōsaku 濱田耕作, and Yano Jin'ichi 矢野仁一, and actually edited by Suzuki Shun 鈴木俊, Aoki Tomitarō 青木富太郎 and other younger historians, this invaluable reference compendium covers all of Asian history. Contents are arranged in phonetic order by the Japanese syllabary and include a great many names of persons and places and a good deal of bibliography. While many items are now out of date or lack continued interest, the longer monographic entries on countries (e.g. India, 1. 149-175, many photos; or China, "Shina", 3. 485-514 and 4. 1-111) or dynasties (e.g. Ch'ing, "Shin", 4. 455-502) constitute authoritative treatises organized for handy reference. Treatment

of Chinese subjects is more extensive though less interpretative in this encyclopaedia of 1939 than in the recent **Sekai rekishi jiten,** 9.5.2. It would be
foolhardy to suggest that the latter has put this earlier work off the shelf,
although it must be noted that a good deal of the authoritative Japanese work
on modern China has been published since 1939. Reviewed by Kishibe Shigeo
岸邊成雄 in **Shigaku zasshi,** 48.6 (June 1937), 810–812; by Toyama Gunji 外山
軍治 in **Shirin,** 22.2 (April 1937), 177–179.

9.5.2 Sekai rekishi jiten 世界歴史事典 (Encyclopaedia of world history),
pub. by Heibonsha, 21 vols. (including supplement and index) and 4 vols.
of selected documents and bibliography, of which vols. 1–21 have been
published in 1951–1954, ca. 350 pp. each vol., maps, illus.

Niida Noboru, Egami Namio 江上波夫, Mori Shikazō 森鹿三, Nohara Shirō,
Matsumoto Yoshimi and Nishijima Sadao served as the editors for the items
concerning Oriental history, in this ambitious, comprehensive and profusely
illustrated work, which is intended to be a reference work for the high school
level. Building on experience gained from the famous **Tōyō rekishi daijiten,**
9.5.1, it provides extensive bibliography, good maps and interesting illustrations
over an enormous range of subject matter (e.g. India and related subjects, 2.
104–148; China, "Chūgoku", and related subjects, 12.226–280; Ch'ing dynasty,
"Shin", 10.82 only). Comparision with the former work would indicate that
it is not by any means superseded in content, although it is outdated in those
respects where duplication of subject matter occurs. The new work includes
more interpretative articles. As is to be expected in large compilations by
many hands, the quality of articles varies considerably. Where the earlier
work treated the origin of the Opium War, for example, entirely with reference
to the growth of the Anglo-Indian-Chinese trade in opium (1.49–52), the later
work presents two topics, one on "the opium question" and the other on the
war. Yet the earlier account cites a bibliography of basic Chinese and British
sources while the more recent account of this war quotes Marx at length and
cites his works together with a 1932 article of Hani Gorō, 1.5.6.

9.5.3 HIRANO Yoshitarō 平野義太郎, KAJI Wataru 鹿地亘, ISHIHAMA
Tomoyuki 石濱知行, ITŌ Takeo 伊藤武雄, supervisors, **Zōho gendai
Chūgoku jiten** 増補現代中國辭典 (Encyclopaedia of contemporary China,
enlarged edition), pub. by the association for the purpose, 1952, 620+98
+43+39 pp.+index 21 pp., maps, illus.

A product of the Chūgoku Kenkyūjo in Tōkyō, this large compilation, first
issued in 1950, has been up-dated by the addition of materials relating to the
three years 1950–51–52. The 72 contributors include some party-line propa-
gandists, a number of competent research specialists, and some leading scholars.
The contents touch upon all aspects of the CCP regime—natural environment
(pp. 1–84), international relations, government and politics (163–300), economics,
society and culture (427–500), learning and arts, and a long section on all
periods of history (pp. 561–620). An appendix gives a 98 page who's who of
modern and contemporary Chinese, arranged by Japanese pronounciation with
a Wade-Giles index added; a collection of 14 laws and regulations; and a
chronology for 1839–1952 (39 pp.). The history section is by 9 authors, including Iwamura Shinobu of the Kyōto Institute (on early East-West contact),
Itano Chōhachi of Hokkaidō (on the history of Confucian thought) and Ima-

hori Seiji of Hiroshima (on the history of gilds). History of the Chinese revolution is dealt with under politics (pp. 165–178). On the whole, while containing many useful contributions, this work must be regarded as a mixture of wheat and tares and used with circumspection. See the review of the first edition by Kobori Iwao 小堀巖 in **Tōyō bunka,** 5 (April 1951), 103–107.

9. 5. 4 Shina mondai jiten 支那問題辭典 (Encyclopaedia of the China problem), pub. by Chūō Kōronsha, 1942, 776 pp.+who's who 101 pp.+ chronology of cultural events 88 pp.+index 56 pp.

This comprehensive work on the geography, foreign affairs, politics, economics, finance, society and culture of modern China has a long list of contributors and a seemingly comprehensive coverage of topics. It is, however in effect a series of some 62 survey articles, each with a brief bibliography, by the various specialized contributors, rather than a dictionary-type encyclopaedia, and the contributors vary rather widely both in angle of approach and in competence. Makino Tatsumi, Amano Motonosuke, Ueda Toshio, Tachibana Shiraki, Hirano Yoshitarō are included among them. The who's who is arranged by the *kana* syllabary, i.e. Japanese readings, only. The chronology covers 1497–1940 too rapidly to be of much use but includes both Japan and China. This volume has the appearance of a rather quick job.

9. 5. 5 Yamaguchi Keizai Semmon Gakkō, Tōa Keizai Kenkyūkai 山口經 濟專門學校東亞經濟研究會, ed., **Shina shakai ˌkeizai daijiten** 支那社會 經濟大辭典 (Social and economic encyclopaedia of China), Taigadō, vol. 1 of 8 vols. planned, A ア to KA カ, 1944, 507 pp. (No other volumes published).

This first volume of an uncompleted wartime project had Professor Nishiyama Yoshihisa 西山榮久 as its chief editor and Dr. Kimura Masutarō 木村 增太郎 as editorial advisor. Its entries are arranged by *kana* readings but with characters and Wade-Giles romanizations added. Dates are given in both Chinese and Western forms. Places, persons, book titles, and administrative terms are all included, with a good deal of historical data but no index. Among the contributors were Niida Noboru, Nishiyama Yoshihisa, Amano Motonosuke, Ichiko Chūzō and many other scholars well known in this field. While its fragmentary nature makes it difficult to use, the contributions in this volume appear to be of good quality.

9. 5. 6 Tōa Kenkyūjo 東亞研究所, **Shina kindai hyakunempyō sōkō** 支那 近代百年表草稿 (Draft chronological tables for a century of modern China), pub. for limited circulation by Tōa Kenkyūjo, 1941, 401 pp., maps.

Covering 1834–1937, this useful reference volume first gives a year-by-year date-list of political, economic, and social-cultural events, with Japanese and some Western events also noted (to p. 160). It then presents a summary account of developments during this century (with a running start from 1516 to 1839), in which periods of a few years each are taken as units and discussed in detail, with extensive statistical tables inserted in the text and a breakdown by topics within these periods. Source references are not given. This volume makes a great body of data available. See review by Ichiko Chūzō 市古宙三, in **Shigaku zasshi,** 52. 6 (June 1941), 718–721.

9.5.7 Kojima Shōtarō 小島昌太郎, **Shina saikin daiji nempyō** 支那最近大事年表 (Chronological tables of major events in modern China), Yūhikaku, 1942, 980 pp.

This substantial and useful reference volume first gives an 80-page chronological list of main events from 1840 to 1941, marking them as domestic 內, foreign 外, social 社, commercial 貿, etc., etc. The bulk of the book (to p. 854) is then devoted to concise factual summaries of these events, with the addition of source references, e.g. to works of Prof. Yano Jin'ichi or to the **Ch'ing-shih kao** 清史稿, by volume and page. Founding dates of all sorts of institutions are included, as well as the usual dates for wars, treaties, politics and calamities. The result is a formidable research tool of scientific value. A long character-and-*kana* index is followed by one of Western names. The bibliography of works referred to runs to 20 pages.

9.5.8 Baba Akio 馬場明男, **Shina seiji keizai nempyō** 支那政治經濟年表 (Chronological tables on Chinese politics and economics), Keiō Shobō 慶應書房, 1943, 674 pp.

Like the similar volume by Kojima Shōtarō, 9.5.7, this reference work summarizes on a chronological framework (though somewhat rhetorically) major events from 1838 to 1940, with indices of persons and subjects (pp. 657–674). However, it is neither so concise, so well arranged nor so comprehensive as the work by Kojima, and its source references are fewer and lack page numbers.

9.5.9 Andō Hikotarō 安藤彥太郎, Saitō Akio 齋藤秋男 and Koma Masaharu 駒正春, "Gendai Chūgoku kyōikubunkashi nempyō, 1949-nen jūgatsu igo no bu" 現代中國敎育文化史年表——一九四九年十月以後の部 ("Chronological Table of Educational-Cultural History of Contemporary China, October 1949 and after"), **Chūgoku kenkyū,** 16 (Sept. 1952), 94–128.

Compiled by the above members of the Chūgoku Kenkyūjo in Tōkyō, this long chronology lists a great variety of the formal activities in communist Peking—convening of congresses, arrivals of delegations, exhibitions, etc.—and also includes dates for the inauguration of institutions, journals and newspapers.

9.6 TECHNICAL DICTIONARIES AND TRANSLATION AIDS

Note: These reference works supplement the encyclopaedias in the preceding section.

9.6.1 Naitō Kenkichi 內藤乾吉, published by, **Rikubu seigo chūkai** 六部成語註解 (**Liu-pu ch'eng-yü** [The established terminology of the Six Boards], with explanatory notes), printed by Kōbundō (in Kyōto), 1940, 149+index 25 pp.

The **Liu-pu ch'eng-yü** originally produced in the Ch'ien-lung period was a handbook for official clerks which gave the equivalent Chinese and Manchu terms used in administration. From it was produced in the late Ch'ing period the present work which added "explanatory notes", i.e. definitions of terms, to the Chinese part of the original, thus creating one of the few manuals which explain the all-but-insoluble intricacies of the official terminology used in the dynastic records. Having become justly prized by modern scholars acquainted with it (e.g. those working on the **Shinkoku gyōseihō**, 1.2.1), this little volume, with a long and interesting history already behind it, was republished with the help of Professor Miyazaki Ichisada for the use of modern historians in 1940. See the 20-page introduction by Mr. Naitō, who is a son of the late Dr. Naitō Torajirō, and his article in **Tōyōshi kenkyū**, 5.5 (July-Aug. 1940), 336–350. The index supplied with this Japanese edition is in itself worth the price of admission.

9.6.2 OKANO Ichirō 岡野一朗, **Shina keizai jiten** 支那經濟辭典 (Economic dictionary for China), Tōyō Shoseki 東洋書籍 pub. assoc., 1931, 797 pp.

This valuable compilation of more than 7,000 phrases, names, terms, etc., was put together by a professor at the Tōa Dōbun Shoin in Shanghai with the help of his colleagues, including Negishi Tadashi 根岸佶, who wrote the preface. Both the book and its index or table of contents (59 pp.) is arranged by *kana* (i.e., Japanese) readings of Chinese characters; no separate character index is provided. Wade-Giles romanizations are given for each entry, as well as *kana* readings. Sources are not given. Nevertheless this collection of terms from the first three decades of the century undoubtedly caught many technical or esoteric meanings which would otherwise be hard to trace. Political and administrative items have of course crept into the mainly economic subject matter.

9.6.3 HIGASHIKAWA Tokuji 東川德治, **Tenkai** 典海 (A dictionary of institutional phrases), Hōsei Daigaku 法政大學, 1930, 1130 pp.

Originally planned as a dictionary of Chinese legal terms, this reference compilation not unnaturally expanded to include key terms and phrases from the whole range of Chinese institutional history. Omitting proper names of persons and places, the compiler concentrated upon the terminology relating to economic, administrative, ceremonial, examinational and other activities in Chinese society as recorded in the classics, histories and other relevant works. The character index (89 pp.) and contents are both arranged by *on* readings, which are printed in *kana* beside all entries. Sources are indicated without page references, e.g. "[論語]" or "[清國行政法卷六]", and are nowhere listed separately with full data concerning them. However, the vast scope of the entries and the rather full explanations of them should make this volume of help in the study of Ch'ing institutions, in which terminological exactitude is the first essential, even though it has been criticized as containing a considerable number of errors. This volume originated as a by-product of the author's participation in the compilation of **Shinkoku gyōseihō**, 1.2.1.

9.6.4 KATŌ Tetsuya 加藤鐵矢, supervisor, for the Manshūteikoku Kyōwakai Chiseki Seirikyoku Bunkai 滿洲帝國協和會地籍整理局分會, comp., **Tochi yōgo jiten** 土地用語辭典 (Dictionary of terminology relating to the land), Ganshōdō Shoten, 1939, 169+676+50+27 pp.

This dictionary of technical terms takes on added importance from the fact that it was widely used by Japanese researchers in wartime China, whose reports now constitute so large a part of the literature on the Chinese land system and rural society. Somewhere around 8000 terms are dealt with; *kana* and Wade-Giles readings, and Western language equivalents are added; the area of use, whether China, Japan, Europe or Manchuria, is indicated; and indices are supplied by *kana* readings, by radicals, and by subjects. The greater part of the terms appear to refer to China but translations of Western terms are included.

9.7 GEOGRAPHICAL AIDS

Note: As indicated in sec. 7.4, we have not attempted to penetrate the very considerable body of materials on geography and historical geography, which can be pursued through the files of several specialized journals. There appears to be no counterpart, for the late-Ch'ing period, of Aoyamo Sadao 青山定男, comp., **Tokushi hōyo kiyō sakuin Shina rekidai chimei yōran** 讀史方輿紀要索引支那歷代地名要覽 (Guide to Chinese place names of successive dynasties, an index to **Tu-shih fang-yü chi-yao** [Geographical essentials for the reader of history, comp. by Ku Tsu-yü 顧祖禹]), pub. by the Tōhō Bunka Gakuin Tōkyō Kenkyūjo, 1933, 721 pp.+indices, which covers Chinese place names historically down to the early Ch'ing period, identifying them with the present names. For a place-name dictionary, see also 5.1.7.

9.7.1 YANAI Watari ("Wataru") 箭内亙, original compiler, WADA Sei ("Kiyoshi") 和田清, supplementary editor, **Tōyō dokushi chizu** 東洋讀史地圖 (Atlas for the student of Oriental history), Fuzambō, 2nd revised edition, 1941 (4th printing, 1943, 33 maps and 64 pp. explanatory text). Prefaces by Yanai 1912 and Wada 1940.

A standard work for Japanese students. Text gives useful supplementary data. Maps cover all of Asia, with special attention to Korea and China, omitting Japan. No index.

9.7.2 Gaimushō Jōhōbu 外務省情報部, ed., **Shina chimei shūsei** 支那地名集成 (A compilation of Chinese place names), Nippon Gaiji Kyōkai 日本外事協會, 1936, 626 pp.

This list of about 10,000 place names has the merit of being arranged according to English spellings, which follow the Chinese Post Office romanization generally but include variants (pp. 1–445). These place names are also listed by their phonetic transcriptions into *kana* (pp. 451–618). Unfortunately the information on each place is limited to a brief statement of its location as on the "south bank" of a certain river or "northwest" of a certain hsien city, etc. Although 24 Chinese and Japanese atlases and compilations were used, including the **Shen-pao 申報** atlas, the data are not as precise as in an older work such as that of G.M.H. Playfair, **The Cities and Towns of China,** 1910 (1879).

9.7.3 HIBINO Takeo 日比野丈夫, "Shimmatsu yori genzai ni itaru Shina no sokuryō chizu" 清末より現在に至る支那の測量地圖 (Measured maps of China from late Ch'ing to the present), **Tōhō gakuhō, Kyōto,** 13.3 (June 1943), 439–468.

> This history of modern Chinese cartography describes in succession the revision of the **Ta-Ch'ing hui-tien** 大清會典 and publication of provincial maps, cartography conducted for the New Army 新軍, and by the General Staff 參謀本部, in each case with lists of maps. Maps of Hopei province and the new Chinese atlas of 1934 are also noted.

9.8 COLLECTIONS OF ESSAYS AND ARTICLES

Note: Arranged alphabetically *by title*, these materials include chiefly the *festschrift* produced for leading scholars, symposia, and other collections of writings from which we have extracted certain articles or fascicles in preceding sections. In other words, this is a list of sources from which earlier entries have been drawn, placed here for reference to complete the data on those entries. These sources, however, may also have interest as containing further items that we have not chosen to cite above.

9.8.1 Sōgensha 創元社, **Ajia mondai kōza アジア問題講座** (Lectures on Asian problems), Sōgensha, 1939 (vol. 12, 1940), 12 vols., ea. vol., 400–550 pp., illus.

> With Ishida Mikinosuke 石田幹之助, Wada Sei 和田清, Tachibana Shiraki 橘樸, Hatano Ken'ichi 波多野乾一, and others as editorial advisers, and Ozaki Hotsumi 尾崎秀實 and others as editors, this popular series provided the general public with a broad coverage of current knowledge and opinion concerning China and adjacent areas, just before the China Incident became enlarged into the Pacific War. Each volume has a few interesting contemporary photographs at the front (including Ch'en Tu-hsiu in vol. 11, wearing a hat) and a symposium of a dozen or even a score of articles by various authors on current developments on the mainland and their background. Vols. 1–3 are entitled "Politics and military"; 4–6, "Economy and production"; 7–8 "Peoples and history"; 9, "Society and customs"; 10–11, "Thought and culture"; and 12, "A who's who of Asia and a general Asian chronology". Nearly every China specialist of the day is represented by one or more articles. Written for the wartime public, they appear to have rather little research value.

9.8.2 Tōa Dōbun Shoin Shina Kenkyūbu 東亞同文書院支那研究部 **Gendai Shina kōza 現代支那講座** (Lectures on present day China), pub. by the same in Shanghai, 1939, 6 vols., each vol. ca. 200–350 pp.

This popularized and systematic symposium of China studies by the professors of the Tōa Dōbun Shoin deals with: vol. 1) geography (by Ueda Shinzō 上田信三), communications (by Baba Kuwatarō 馬場鍬太郎), contemporary history (by Otake Fumio 小竹文夫); vol. 2) politics and constitutions (by Harumiya Chikane 春宮千鐵), penal law (by Matsui Toshiaki 松井利明), civil law and judicial process (by Narumiya Yoshizō 成宮嘉造), diplomacy (by Shigemitsu Osamu 重光藏); vol. 3) fiscal policy (by Ōta Eiichi 太田英一), currency and banking (by Miyashita Tadao 宮下忠雄); vol. 4) general characteristics of economics (by Hozumi Fumio 穗積文雄), agriculture (by Hisae Fukusaburō 久重福三郎), live stock, fishing, and forestry (by Ueda Shinzō 上田信三), mining (by Hisae Fukusaburō 久重福三郎); vol. 5) industry (by Toda Yoshirō 戶田義郎), commerce (by Kubota Shōzō 久保田正三), foreign trade (by Uchida Naosaku 內田直作); vol. 6) education (by Fukuda Katsuzō 福田勝藏), social structure (by Nozaki Shumpei 野崎駿平), annual customs (by Kageyama Takashi 影山巍), literature (by Suzuki Takurō 鈴木擇郎), language (by Sakamoto Ichirō 坂本一郎), thought (by Kumano Shōhei 熊野正平), and religion (by Hayashi Tetsuo 林哲夫).

Although this compilation seems to be inevitably a mixture of uneven quality, it includes competent surveys such as that of Prof. Toda on modern industry (108 pp. in vol. 5).

9.8.3 Tōyōshi Kenkyūkai 東洋史研究會, Kyōto University, comp., **Haneda hakushi shōju kinen Tōyōshi ronsō** 羽田博士頌壽記念東洋史論叢 ("Asiatic Studies in honour of Tōru Haneda on the occasion of his sixtieth birthday, May 15, MCMXLII"), Kyōto: Tōyōshi Kenkyūkai 東洋史研究會, Kyōto University, 1950, 1003+xiii+23 pp.

A fine memorial volume, mainly on pre-Ch'ing China, items from which we note elsewhere under the authors' names.

9.8.4 Iwanami Shoten 岩波書店, ed., **Iwanami kōza Tōyō shichō** 岩波講座東洋思潮 (The Iwanami series on Oriental trends of thought), series 1 to 18 第一回 to 第十八回, 1934–1936 (each fascicule or box containing 3 to 6 booklets of 50 to 200 pp.).

These thin volumes, as related to China, are noted elsewhere separately under the authors' names. This ambitious series brought together several score of booklets on Japanese, Indian, Chinese, Buddhist and other forms of Oriental thought, but not many refer to modern China. Those we have noted under the authors' names have background usefulness in the field of intellectual history.

9.8.5 WADA "Kiyoshi" [Sei] 和田清, ed., **Katō hakushi kanreki kinen Tōyōshi shūsetsu** 加藤博士還曆記念東洋史集說 (Studies of Oriental history collected to commemorate the 60th birthday of Dr. Katō), Fuzambō, 1941, 952 pp.

A memorial volume devoted mainly to earlier periods, items from which we cite by authors' names. Note the more recent collection, 9.8.8.

9.8.6 NIIDA Noboru 仁井田陞, ed., for the Gakujutsu Kenkyū Kaigi, Gendai Chūgoku Kenkyū Tokubetsu Iinkai, 學術研究會議現代中國研究

特別委員會 (Special committee for research on modern China of the Scientific Research Conference), **Kindai Chūgoku kenkyū** 近代中國研究 (Researches on Modern China), Kōgakusha 好學社, 1948, 361 pp.

An interesting symposium of 11 articles, most of which we list separately under their various authors. Each article provides a brief critical survey of its topic, usually without footnotes. This research group was organized in order to make a joint study of the modernization process in China. It is one of those special committees for joint study of urgent problems which have been sponsored and financed by the Ministry of Education.

9.8.7 WATANABE Yoshimichi 渡部義通, HIRANO Yoshitarō 平野義太郎, ŌTSUKA Hisao 大塚久雄, and others, ed., **Shakai kōseishi taikei** 社會構成史大系 (A comprehensive and systematic corpus of histories of social structure), Nippon Hyōronsha, 1949–1951, 9 series, each containing 3 articles in separate booklets (the originally projected continuing series have not been published).

The Second Section 第二部 of this broad project was devoted to booklets on "the development of Oriental social structure" 東洋社會構成の發展 and we list six of its products under the author's names.

9.8.8 KATŌ Shigeshi 加藤繁, **Shina keizaishi kōshō** 支那經濟史考證 ("Studies in Chinese economic history"), vol. 2, Tōyō Bunko 東洋文庫, 1953, 924+index 32 and English summaries 24 pp., photos (Tōyō Bunko ronsō, 34).

Professor Katō, 1880–1946 (erroneously called Katō Shigeru in most bibliographies), taught Far Eastern history at Tōkyō University from 1925 to 1941 and was a pioneer in the field of economic history. Since his articles are arranged by periods, vol. 1 (1952) of this work does not concern us. Vol. 2 presents articles ("revised and augmented by the author") on coinage and other aspects of the Sung (to p. 420), followed by ten items on aspects of the late Ch'ing economy and several more pieces of a more general nature. Most of these are listed by us elsewhere above as seen *in situ* in their original published form, with references to this volume added. See also 9.8.5.

9.8.9 Tōa Kenkyūkai (Society for East Asian studies), **Tōa kenkyū kōza** 東亞研究講座 (Lectures in East Asian studies), published by the same, vol. 1 輯, 1925—seen through vol. 105 (1942), ea. vol. ca. 40–100 pp. (with exceptions), illus.

This long series written by old China hands, journalists and scholars, and bound in various colors of paper, sought to make the results of scholarship available to the public in an easy-to-read, simplified pamphlet or booklet form. Subjects range widely over the fields of archaeology, current international relations, social customs, trade, arts and politics, biographies of Shanghailanders (with a photograph of Chang Ping-lin, pub. July 1930, no no.), ancient music, and the World Red Swastika Society (no. 73, 1937). While catholic in scope, the scholarly level of the author-popularizers seems with some exceptions to have been less than first-rate.

9.8.10 KUWABARA Jitsuzō 桑原隲藏, **Tōyōshi setsuen** 東洋史說苑 (Collected writings on Oriental history), Kōbundō, 1927, 543 pp.

These articles by one of the leading sinologues of the day (professor at Kyōto University) begin with an historian's comment on the Republican Revolution (pp. 1–17, first printed in the **Ōsaka asahi shimbun** in 1912) and include other topical essays on the "Yellow Peril", modern Confucianism ("Shina no kokkyō mondai" 支那の國教問題, 1926, pp. 38–69), histories of the queue in China (pp. 315–334, originally printed in **Geimon**) and of the eunuchs (pp. 344–359), and more general discussions of Chinese temperament, in addition to ancient and medieval studies which were the author's specialty.

9.8.11 Shigakukai 史學會, ed., **Tōzai kōshōshi ron** 東西交涉史論 (Treatises on the history of East-West relations), Fuzambō, 1939, 2 vols., 1410 pp. (Fiftieth anniversary memorial volume).

The scholarly articles in these two handsome volumes deal mainly with pre-modern Chinese or with Japanese relations with the West. Relevant to modern China are the following: 1) Wada Sei, "Iwayuru 'Kōto rokujūshi ton' no mondai ni tsuite", pp. 1107–1146, reprinted in his volume **Tōashi ronsō** (1.1.16); 2) Shimmi Yoshiji, "Beikoku no Tōyō seisaku", pp. 1243–1292, (see 4.10.3); 3) Tabohashi Kiyoshi, "Giwakempiran to Nichi-Ro", pp. 1051–1106, (see 2.7.8); 4) Saitō Seitarō, "Rokoku no Tōa seisaku to rekkyō", pp. 1003–1050, (see 4.7.12). See also 1.1.13.

9.8.12 **Wada hakushi kanreki kinen Tōyōshi ronsō** 和田博士還曆記念 東洋史論叢 ("Oriental Studies presented to Sei [Kiyoshi] Wada...in celebration of His Sixtieth Birthday, 15 November 1950, by his pupils"), Dai Nippon Yūbenkai Kōdansha 大日本雄辯會講談社, 1951, 806+English summary 71 pp.

This impressive tribute to a leader among Japanese historians of China contains fifty articles on a great variety of subjects. Those by Messrs. Itano, Ichiko, Kanda, Konuma, Suzuki, Fujii, Masui, and Yazawa are noted by us elsewhere under the authors' names.

9.9 LEARNED JOURNALS

Note: Arranged alphabetically by title, these journals form an introductory core to the vast mass of Japanese periodical matter on China and Chinese studies. The Kyōto Institute's bibliography for 1946–50 (item 9.2.1), for example, lists some 220 Japanese periodicals of current academic interest in connection with Oriental history, from which we selected the following for examination. In general we list below only the periodicals which we have actually seen, together with an indication of the latest issues seen by us. A number of other periodicals from which we have cited articles are noted in earlier sections as sources of such articles, but are not listed here.

9.9.1 **Chūgoku bungaku geppō** (later **Chūgoku bungaku**) 中國文學月報 (Chinese literature monthly), ed. by Chūgoku Bungaku Kenkyūkai, pub. by Seikatsusha, monthly, no. 1 (March 1935)—no. 92 (March 1943); no. 93 (March 1946)—no. 105 (April-May 1948).

Edited consistently by Takeuchi Yoshimi 竹内好 until its wartime suspension and by Senda Kyūichi 千田九一 in the post-war period, and devoted mainly to short critical articles on modern Chinese literature of the late 19th and 20th centuries, this lively and interesting little journal also published translations, book reviews and studies of problems of translation. The final number of March 1943 終刊號 printed an index of contents arranged serially by issues. Of particular historical interest are a series of special numbers which include no. 26 (May 1937) on Wang Kuo-wei 王國維, no. 61 (May 1940) on Ts'ai Yuan-p'ei 蔡元培, no. 77 (Oct. 1941) to commemorate 30 years of the Chinese Republic, no. 83 (May 1942) on the Japanese study of spoken Chinese, including an article on the Tōa Dōbun Shoin at Shanghai, and no. 95 (May 1946) on the May Fourth Movement. The post-war issues are of course much concerned with the modern Chinese literature of social protest, especially with Lu Hsün, on whom there are several articles.

9.9.2 **Chūgoku kenkyū** 中國研究 ("The Chinese Research"), pub. at first by the Chūgoku Kenkyūjo ("Institute for China Research"), now by Gendai Chūgoku Gakkai 現代中國學會, at intervals of 1 to 6 months, ca. 100 pp. ea. issue, no. 1 (June 1947)—seen through no. 16 (Sept. 1952).

This institute organ is devoted mainly to articles which describe current developments in mainland China, mainly from official communist sources. While it touches many topics of current interest and presents a considerable amount of bibliography and interpretation, it contains few articles of academic historical research.

9.9.3 **Geimon** 藝文, pub. by the Kyōto Bungakukai 京都文學會 with offices in the Faculty of Letters, Kyōto Imperial University, monthly, vol. 1 (1910)—vol. 22, no. 3 (May 1931).

A learned journal covering the whole range of philosophy, literature, fine arts and archaeology, East and West. Its few articles on modern China relate mainly to the classical tradition in the late Ch'ing period. The last number (22.3) is a general index arranged (1) chronologically, (2) by authors' names, and (3) by topics.

9.9.4 **Jimbun kagaku** 人文科學 ("Quarterly Journal of Humanistic Science"), pub. by the Kyōto [Teikoku] Daigaku Jimbunkagaku Kenkyūjo ("Research Institute of Humanistic Science, Kyōto Imperial University"), ca. twice a year, 1. 1–2 (July 1946)—concluded with 2. 4 (July 1948).

This small post-war organ of the Kyōto Institute, though transitory, published useful research on the modern period in China. In general its materials seem briefer, more popular, less formal and definitive than those of the contemporary Kyōto journal, **Tōhō gakuhō, Kyōto**. This does not make them less interesting. Compare the somewhat similar relationship between

Tōyō bunka kenkyūjo kiyō and **Tōyō bunka,** both now published by the Institute for Oriental Culture, Tōkyō University.

9.9.5 Keizai ronsō 經濟論叢, pub. at first by the Kyōto Hōgakukai 京都法學會 (Kyōto society for legal science) and thereafter by Kyōto [Teikoku] Daigaku Keizai Gakkai 京都 (帝國) 大學經濟學會 (Society for economics, Kyōto [Imperial] University), monthly, 1.1 (July 1915)—seen through vol. 56 (June 1943), ... the most recent number confirmed, 71.6 (1953).

This organ of the Economics Department of the Faculty of Law (later, of the Faculty of Economics) of the Kyōto Imperial University published rather few articles on modern China, but did include writings by scholars such as Ojima Sukema.

9.9.6 Kokusaihō gaikō zasshi 國際法外交雜誌 ("Revue de droit international et diplomatique", later "The Journal of International Law and Diplomacy"), pub. by the Kokusaihō Gakkai 國際法學會 ("L'Association de droit international", later "The Association of International Law"), with offices in the Faculty of Law, the University of Tōkyō, monthly, 1.1 (Feb. 1902, entitled **Kokusaihō zasshi** 國際法雜誌 ("Revue de droit international")—renamed **Kokusaihō gaikō zasshi** from 11.1 (Oct. 1912, the first year of Taishō)—last issue seen 52.3 (June 1953).

This is a representative organ for Japanese scholars in international law and diplomacy (contemporary and historical), published by an influential scholarly association, closely connected with the Foreign Office and aided by grants from The Carnegie Endowment (from 1912 down to before the Pacific War). It has reflected Japan's political vicissitudes and also published many legal and diplomatic analyses or comments on contemporary China problems. Its recent volumes are noteworthy as publishing substantial historical research on modern China.

9.9.7 Mantetsu chōsa geppō 滿鐵調査月報 (South Manchurian Railway Company Research Department monthly), Dairen: pub. by the SMR, monthly, Sept. 1931—vol. 24, no. 2 (Feb. 1944).

This justly famous and important journal was preceded by a similar monthly under two different names: **Chōsa jihō** 調査時報 (Research reports), first published in April 1922, which ran to 100 numbers 號 and was renamed **Man-Mō jijō** 滿蒙事情 (Conditions in Manchuria and Mongolia) beginning with no. 101 of March 1930. Publication continued under this name through no. 119, the title being changed to **Mantetsu chōsa geppō** from Sept. 1931 with vol. 11, no. 9 (whole number 120). In fact, the journal thus dates from 1922. An index for 1931-1941 (whole nos. 120-243) was published in 22.1 (Jan. 1942), 200-244, arranged by subjects but without indices of authors or titles.

For our purposes in the present bibliographical survey we have selected relatively few items from this leading repository of Japanese research on China, for several reasons. (1) A full listing of pertinent items in **Mantetsu chōsa geppō** would greatly enlarge this present volume, and yet any serious Western student of the modern Chinese economy will have to examine its files in any case, without relying on bibliographic aids such as the present one. (2) A good many of its articles, such as those by Shimizu Morimitsu 清水盛光 or Amano Motonosuke 天野元之助, were later combined in books we have noted elsewhere. (3) The great bulk of the research reports deal with current conditions as studied in the field or from contemporary sources, rather than through library research.

We have therefore tried to confine our selections, which are nevertheless numerous, to articles of *historical research* or of direct interest to historians, including a number of items which open with brief historical introductions. A description of a sample issue may serve to indicate the value of this journal: vol. 14, no. 12 for December 1934 has three "research" articles, on the theoretical and historical status of silver in the finances of leading countries, on the expenditures of the Ch'ing government (see our entry, 7. 13. 2), and on wheat in Heilungkiang, respectively. These are followed by six items of factual "source material", on Manchurian hemp production, drought in China, usury in Hopei, the significance of the Chinese Communists' attack on rich peasants in Kiangsi (8 pp.), 22 fascist-type publications in Shanghai (listed and described), and a translation from Magyar's **Chinese peasant economy** (continued). Next are "miscelleneous notes on current events", political, social, economic, etc. (including the Chinese Soviet constitution), pp. 203-242; Manchoukuo customs figures in tables; a chronology of recent events, 7 pp.; recent laws and regulations, 9 pp.; concluded by 14 pp. of statistics and 36 pp. of international bibliography. Major articles are carefully organized with tables of contents. In general they deal with all aspects of the Chinese and especially Manchurian economy, and many aspects of social organization and political developments. Contributions are mainly from the research staff members in Dairen, Peiping and Shanghai.

9. 9. 8 Nippon oyobi Nipponjin 日本及日本人 (Japan and the Japanese), pub. by Seikyōsha 政教社, fortnighty, beginning under the title of **Nipponjin** 日本人 in April 1888, renamed **Nippon oyobi nipponjin** from January 1907,—concluded with No. 1309 (Feb. 1945)—post-war republication by Nippon Shimbunsha 日本新聞社, monthly, 1. 1 (Sept. 1950)....

Although devoted to Japanese and worldwide politics, art and letters generally, this popular journal had a good many able comments on current developments in revolutionary China by specialists such as Inaba Kunzan (Iwakichi), and Ichimura Sanjirō. It should have useful resource value for historical research, but we have not gone through it systematically.

9. 9. 9 Orientalica オリエンタリカ, ed. by Tōyōshi gakkai ("La société d'histoire de l'extrême-orient") with offices in the Faculty of Letters of Tōkyō University, vol. 1 (1948), pub. by Daichi Shobō 大地書房—— suspended with vol. 2 (1949), pub. by Kanrin Shoin 翰林書院·

This competent scholarly periodical, contributed to mainly by graduates of the Oriental history section in the Faculty of Letters of Tōkyō University, unfortunately faded away after only two numbers, which include a few noteworthy articles.

9. 9. 10 Rekishigaku kenkyū 歷史學研究 ("The Journal of Historical

Studies," formerly "Zeitschrift der Gesellschaft für Geschichtsforschung"),
ed. by the society 會 for the same ("The Historical Science Society") and
pub. by Iwanami Shoten (formerly pub. by the society itself), monthly,
vol. 1, no. 1 (Nov. 1933)—14.5 (Dec. 1944); suspended 1945; resumed
with no. 號 122 (June 1946)—. (Seen to no. 162).

This journal presented a rather general academic type of materials in its
early years but since the war has become more outspokenly a channel for
explicitly Marxist interpretations of history. Consequently its pages have
been particularly full of theoretical discussions. Since its inauguration, this
periodical has been, so-to-speak, an organ of the opposition party and younger
generation among historians, as compared for example with the **Shigaku zasshi**
which has been closely connected with the academic authority of the Tōkyō
Imperial University. Note the special memorial issue, **Ajia no henkaku アジ
アの變革** ("Transform of Asia"), 150 (Mar. 1951), 94 pp., with articles on phases
of modern history which we have listed separately (2.4.13; 2.7.10; 5.2.4).
See also comment under our item 5.1.4.

9.9.11 Seikyū gakusō 青丘學叢, pub. for the Seikyū Gakukai in Keijō
by Ōsakayagō Shoten, Keijō (Seoul), Korea, quarterly, no. 1 (Aug. 1930)
—seen through no. 29 (Aug. 1937).

This learned journal printed the work of professors at Keijō Imperial
University and other Japanese scholars at the Korean capital, mostly relating
to Korean history but with occasional reference to China.

9.9.12 Shakai kagagu kenkyū 社會科學研究 ("Journal of Social Science"),
ed. by the "Institute for Social Science" 社會科學研究所, Tōkyō University,
pub. by Hakujitsu Shoin 白日書院 (later by Nippon Hyōronsha, and then
by Yūhikaku), quarterly with many exceptions, no. 1 (Feb. 1948)—4.1
(Nov. 1952).

A journal devoted mainly to articles in political science and economics
concerning the West or Japan, with rather little attention to China.

9.9.13 Shakai keizai shigaku 社會經濟史學 ("The Socio-Economic History," "The Quarterly Journal of the Socio-Economic History Society"),
pub. for the society by Nippon Hyōronsha, Iwanami Shoten, Shinkigensha
新紀元社, and others, monthly from 1.1 (May 1931), later quarterly, and
bimonthly from 17.1 (April 1951)—seen through 18.6 (May 1953).

A scholarly journal largely devoted to pre-modern Japanese history, with
articles also on pre-modern Western and Chinese history but very little on
modern China. The few articles on China, however, are of high quality, and
are more numerous in later numbers.

9.9.14 Shichō 史潮 ("The Journal of History"), ed. by Ōtsuka Shigaku-
kai 大塚史學會 ("The Ōtsuka Historical Science Society"), Tōkyō Kyōiku
Daigaku 東京敎育大學 ("The Tōkyō University of Education",) formerly
Tōkyō Bunrika Daigaku 東京文理科大學 ("The Tōkyō University of

Literature and Science"), three times a year, 1.1 (Feb. 1931)—seen through no. 47 (Dec. 1952). Publication was suspended from 1944 to 1950.

This well established journal has rather few articles on modern China.

9.9.15 Shien 史苑 ("A Quarterly Journal of History"), pub. by Rikkyō Daigaku Shigakukai 立敎大學史學會 ("The Historical Society, Rikkyo [St. Paul's] University"), Tōkyō, quarterly, 2.2 (May 1929)—seen through 15.2 (April 1943).

Covering all areas and times, this journal seldom touches modern China.

9.9.16 Shien 史淵 ("Journal of History"), pub. by Kyūshū Shigakukai 九州史學會 ("The Historical Society of Kyūshū University"), Fukuoka, bimonthly but irregular, 1930—seen to no. 56 (Mar. 1953).

This excellent journal has rather little on modern China.

9.9.17 Shigaku zasshi 史學雜誌 ("Zeitschrift für Geschichtswissenschaft", later "The Journal of Historical Science"), edited by Shigakukai 史學會 ("Institution of Historical Science"), with offices in the Tōkyō Daigaku Bungakubu 東京大學文學部 ("Faculty of Letters, University of Tōkyō"), monthly, 1892—seen through 62.9 (Sept. 1953).

This is one of the oldest and most important Japanese historical journals for all fields, and it naturally has not published a great deal on modern China. Articles are usually brief, solid and authoritative. The 60-year memorial index, **Shigaku zasshi sōmokuroku, ippen—rokujippen** 總目錄1編—60編 (Yamakawa 山川 pub. co., 1952, 96 pp.) lists the contents, including book reviews, by title and author (4197 items), and supersedes the 40-year index, **Shigaku zasshi sōsakuin** 總索引 (Fuzambō, 1932, 198 pp.).

9.9.18 Shikan 史觀, ed. and pub. by Waseda Daigaku Bungakubu 早稻田大學文學部 (Faculty of Letters, Waseda University) and later by the ibid. Bungakukai (Literary Society), two or three times a year, no. 1 (Nov. 1931)—seen through no. 38 (Oct. 1952).

Number 38 of this learned journal has a general index to all previous numbers. Rather little has been published on modern China.

9.9.19 Shin Chūgoku 新中國 (New China), pub. by Jitsugyō no Nippon-sha, monthly, no. 1 (March 1946)—no. 19, the last number (Jan. 1948).

An inexpensive, 30 to 50-page journal, edited by Sanetō Keishū 實藤惠秀, containing cultural-political articles on contemporary China of an often leftward political orientation together with articles of a scholarly character on the bibliography or personalities of modern history, especially Chinese liberal leaders. Annual indices are provided in nos. 9 and 19. Special numbers of interest include no. 11 (Mar. 1947) on "the third force" and Democratic League (Min-chu t'ung-meng 民主同盟) and its leaders, and no. 18 (Nov. 1947) on the Chinese language, with articles by Niida Noboru (pp. 3–10) on "the ethical character of the Chinese as seen in verbal expressions"—an interesting analysis of the Chinese technique and psychology of verbal abuse; and by Yoshikawa Kōjirō on "the Chinese language and dictionaries" (pp. 11–21).

9.9.20 Shina 支那 ("The Shina, The China Review"), pub. by the Tōa Dōbunkai, monthly, beginning from 1910—concluded probably with 36.1 (Jan. 1945).

Published by an association in Tōkyō connected, as sponsors, with the Tōa Dōbun Shoin at Shanghai and including officials, journalists, and business men as members, this journal provided a regular channel for information and informed comment on the current Chinese scene and international issues affecting it. While scholars contributed to its pages, it seldom carried research articles in the academic sense.

9.9.21 Shina kenkyū 支那研究, ed. and pub. by Tōa Dōbun Shoin Shina Kenkyūbu 東亞同文書院支那研究部 (China research section of the T'ung-wen College), monthly, no. 1 (Aug. 1920)—concluded with no. 62 (Mar. 1942) and a special number (May 1942).

This organ of the Tōa Dōbun Shoin in Shanghai was the regular outlet for its energetic scholars and compilers, although not its only publication. Most of its materials relate to the current scene and they are often based on field investigation of aspects of the Chinese economy. Much of this research seems to have been weak on the side of social science analysis but it usually displayed a strong historical sense. A number of the leading China specialists now active in Japan, like Otake Fumio, Miyashita Tadao, Uchida Naosaku or Ueda Toshio made their early contributions in this journal. Note that special numbers were issued (sometimes outside the regular sequence), e.g. no. 18 (reprinted Feb. 1930) is a special issue on all aspects of Shanghai to commemorate ten years of the research department's work (804 pp.).

9.9.22 Shinagaku 支那學 ("Sinology"), ed. by the Shinagakusha 支那學社 and pub. by Kōbundō Shobō in Kyōto, monthly, later quarterly, 1.1 (Sept. 1920)—concluded with 12.5 (Aug. 1947).

A principal organ for Chinese classical, philosophical and historical studies, with frequent contributions from Naitō Torajirō, Kuwabara Jitsuzō and other leaders in research at Kyōto. Aside from articles of Yano Jin'ichi, however, and a few by Ojima Sukema, the unclassical modern period is relatively neglected.

9.9.23 Shirin 史林 ("The Shirin or the Journal of History"), pub. by Shigaku Kenkyūkai 史學研究會 ("The Historical Society"), with offices in the Faculty of Letters, Kyōto (Imperial) University, 1916–April 1950, quarterly; from 33.3 (May 1950)—monthly or bi-monthly for a short period.

Corresponding to **Shigaku zasshi** in Tōkyō, this well known journal has represented the work of historians, archaeologists and geographers at Kyōto University in all fields, Japanese, Western and Asiatic. Owing to the strength of the Kyōto scholars in archaeology and ancient history, East and West, not much has appeared in **Shirin** on modern China (except articles of Dr. Yano Jin'ichi which are included in his later books) until recently. There is an index to the first 20 volumes, **Shirin sōmoku sakuin** 史林總目索引, Kyōto: Naigai 內外 pub. and printing co., 1935, 126 pp., with a rather extensive listing of contents by subjects.

9.9.24 Tenri daigaku gakuhō 天理大學學報 ("Bulletin of Tenri University"), pub. by Tenri University, Tambaichi, Nara Prefecture, occasional, vol. 1. no. 1 (May 1949)—seen to vol. 4, no. 2 (Nov. 1952).

A journal of high quality devoted to studies of literature and literary history, including problems of translation, literary criticism, education, drama, and philosophy. A leading journal in this field. The sponsoring institution, formerly called Tenri Gaikokugo Gakkō 天理外國語學校, has long had an established reputation as a language school, sponsored by the Tenrikyō 天理教, one of the most influential popular (non-Buddhist) religious sects in Japan.

9.9.25 Tōa 東亞, pub. by Tōa Keizai Chōsakyoku, monthly, 1.1 (May 1928)—last issue seen, 18.2 (Feb. 1945).

This well-informed journal of news and comment on the current Chinese scene, rather comparable to **Shina,** printed popular articles by scholars like Tachibana Shiraki, Hanabusa Nagamichi, and Momose Hiromu and is a useful repository of day-to-day information on problems, events, negotiations, tendencies and prospects in China and Manchuria and other parts of Asia. Articles on current Chinese politics and on economic developments during the 1930s seem particularly useful, particularly the accounts of KMT policy.

9.9.26 Tōagaku 東亞學 ("Orientalica"), pub. by Nikkō Shoin, semiannually, no. 1 (Sept. 1939)—no. 9 (June 1944).

This journal should not be confused with **Orientalica,** 9.9.9, which is best known by that name. **Tōagaku,** although a commercial venture, published a relatively high quality of scholarly articles during the war years.

9.9.27 Tōa jimbun gakuhō 東亞人文學報, pub. for the the Kyōto Teikoku Daigaku Jimbun Kagaku Kenkyūjo 京都帝國大學人文科學研究所 by Kōbundō, Kyōto, quarterly (at first), 1.1 (Mar. 1941)—concluded with 4.2 (Mar. 1945).

This organ of the abovementioned institute was chiefly devoted to studies of modern China, historical, sociological, economic and legal. After disappearing at the end of the war, it was virtually revived under the title of **Jimbun gakuhō** 人文學報 ("Journal of Humanistic Science", no. 1, Dec. 1950—), as a principal organ of the institute to which, however, the members of only its Japanese and Western sections seem to contribute articles.

9.9.28 Tōa keizai kenkyū 東亞經濟研究 ("Revue d'Economie Politique d'Extrême-Orient"), pub. by the society for the same, with offices "à l'Ecole Supérieure de Commerce de Yamaguchi" 山口經濟專門學校, quarterly or three times a year, no. 2 (Nov. 1917)—seen through 28.3 (Mar. 1945), the last number being 29.1–2.

This journal published both articles of journalism on current economic developments and policy problems concerning China, and also articles of research on early and recent China by scholars like Yano Jin'ichi, Inaba Iwakichi, and Amano Motonosuke. The latter, however, are usually to be found in subsequent books, while the former are rather insubstantial. A good deal was also published in this journal on Manchuria or Manchoukuo and a certain amount on Korea, Indochina and Asiatic Russia, constant attention being centered on the problems of Sino-Jananese relations and trade.

9.9.29 Tōa keizai ronsō 東亞經濟論叢 (Collected essays on East Asian economics), Kyōto, pub. by the Tōa Keizai Kenkyūjo 東亞經濟研究所 (East Asia Research Institute) with offices in the Faculty of Economics, Kyōto Imperial University, quarterly (or occasional), 1.1 (Feb. 1941)— 3.3–4 (Sept., Dec. 1943, pub. July 1944).

A wartime journal of social science research studies produced at Kyōto University. **Tōa keizai sōsho** 東亞經濟叢書 was a wartime pamphlet series consisting of articles reprinted from the above journal, mainly in 1941–43, arranged according to a general plan in 14 sections 輯 on various aspects of the East Asian economy—economic theory, thought, and history; natural resources, mining, industry, finance, etc., and dealing with Southeast Asian countries as well as China. Often these were brief second-hand essays rather than research reports; a number of the more substantial items have been noted under their authors, with citation of their original appearance in **Tōa keizai ronsō,** although without the original page numbers.

9.9.30 Tōa kenkyū shohō 東亞研究所報 (Reports of the East Asia Research Institute), pub. by the Institute, bimonthly, no. 1, May 1939—no. 29, Aug. 1944.

The Tōa Kenkyūjo was set up on the basis of government funds and some private grants and placed under civilian management in order to mobilize Japanese scholarship as an academic aid to wartime policy formation, and eventually came to employ several hundred research and clerical personnel. The results of some of its projects are noted elsewhere in this volume—e.g. for the study of foreign investments in China, 7.18.1–4; field research in North China villages in cooperation with the SMR research department, (see 8.1.1 among others); or the rule of alien peoples over China, 3.3.1. This journal also published research reports but was mainly devoted to making available the results of foreign scholarship through translations, book reports, and extensive listings of bibliography.

9.9.31 Tōa mondai 東亞問題, pub. by Seikatsusha, monthly, 1.1 (Mar. 1939)—last issue seen, 5.2 (June 1943); the last number was published in Sept. 1944.

Though limited to some four years of publication, this journal served during that time as an outlet for the less formal but serious writings of leading scholars like Katō Shigeshi, Makino Tatsumi, and Miyazaki Ichisada, to name only a few of the many scholarly contributors. Their articles, while seldom documented research, maintained a level somewhat higher and more individualistic in content than was the case with more popular journals on China like **Tōa** and **Shina,** and make this journal of importance for modern studies.

9.9.32 Tōa ronsō 東亞論叢, pub. by Bunkyūdō Shoten 文求堂書店, semi-annually (at first), no. 1 (July 1939)—no. 5 (Oct. 1941)—concluded with no. 6 (April 1948).

These volumes published by a leading dealer in Chinese books cover the historical and contemporary scene in East Asia, but are mainly on modern and present-day China (including Manchuria). Most of these articles are of a high level of scholarship.

9.9.33 Tōhōgaku 東方學 ("Eastern Studies"), pub. by the Tōhō Gakkai 東方學會 ("Institute of Eastern Culture"), semi-annually, no. 1 (Mar. 1951)—seen through no. 6 (June 1953).

This new journal presents the best results of academic sinology and accordingly ranges over many periods and subjects, archaeological, literary, and historical, ancient and modern. Not much on modern China; there are useful reviews of Far Eastern studies in Japan by Wada Hironori 和田博德 (second son of Prof. Wada Sei).

9.9.34 Tōhō gakuhō, Kyōto 東方學報 ("Journal of Oriental Studies"), pub. by Tōhō Bunka Gakuin Kyōto Kenkyūjo and three other publishers in succession, now by Kyōto Daigaku Jimbunkagaku Kenkyūjo, more or less semi-annually, no. 1 (Mar. 1931)—seen through no. 22 (Feb. 1953).

Unlike its twin publication in Tōkyō, which was begun at the same time but concluded in 1944, this journal has been maintained as a principal organ of the Jimbunkagaku Kenkyūjo ("Institute of Humanistic Science"), the Oriental section of which has now, in a general way and after various vicissitudes, succeeded to the position (and library) of the Tōhō Bunka Gakuin Kyōto Kenkyūjo ("Academy of Oriental Culture, Kyōto Institute").

9.9.35 Tōhō gakuhō, Tōkyō 東方學報 ("Journal of Oriental Studies"), pub. by the Tōhō Bunka Gakuin Tōkyō Kenkyūjo ("The Academy of Oriental Culture, Tōkyō Institute"), more or less annually, no. 1 (Mar. 1931)—concluded with no. 15, part 2 (Aug. 1944).

Like its sister publication in Kyōto, this journal was devoted to sinology in the proper sense, with only occasional reference to modern China.

9.9.36 Tōkō 東光, pub. by Kōbundō, no. 1 (Aug. 1947)—seen through no. 7 (Jan. 1949). The journal concluded with no. 9.

This rather thin journal carried on for a time, in a somewhat popularized fashion, as the successor of **Shinagaku.**

9.9.37 Tōyō bunka 東洋文化 ("Oriental Culture"), pub. by Tōyō Gakkai 東洋學會 ("Society of Oriental Culture") of the Tōkyō Daigaku Tōyō Bunka Kenkyūjo ("Institute for Oriental Culture, University of Tokyo"), no. 1, Jan. 1950—every three to six months (no. 11, Nov. 1952), ca. 100 pp. ea. issue.

This institute organ publishes usually three or four scholarly articles, one or more long book reviews and occasional reports of scholarly discussions. Contributors are not confined to the Institute members and articles usually concern topics of broad significance or methodological value connected with China, Japan, India, or other areas of East Asia. This journal has the special merit of being interdisciplinary, interested in both historical and social science studies of the modern period. **Tōyō bunka** is the direct successor to **Tōyō bunka kenkyū,** and unconnected with the pre-war journal **Tōyō bunka,** which was a popular rather than academic publication.

9. 9. 38 Tōyō bunka kenkyū 東洋文化研究 ("The Oriental Culture Review"), pub. by Tōyō Gakkai 東洋學會 ("Society of Oriental Culture"), with offices in the Institute for Oriental Culture, Tōkyō University, irregular, no. 1 (Oct. 1944)—concluded with no. 11 (May 1949).

This was the principal organ of the Institute, in a rather popularized form, and was devoted to modern and contemporary East Asia, mainly China. Some of its articles are noteworthy as representing aspects of recent scholarly trends in Japan. It was succeeded by **Tōyō bunka,** 9. 9. 37.

9. 9. 39 Tōyōbunka kenkyūjo kiyō 東洋文化研究所紀要 (Memoirs of the Institute for Oriental Culture), pub. at first by Nikkō Shoin for the Tōkyō Teikoku Daigaku Tōyō Bunka Kenkyūjo (Institute for Oriental Culture, Tōkyō Imperial University), now by the Institute, no. 1 (Dec. 1943)—seen through no. 4 (Mar. 1953).

This annual volume serves, in effect, as a successor to the **Tōhō gakuhō, Tōkyō.**

9. 9. 40 Tōyō gakuhō 東洋學報 ("Reports of the Investigation of Oriental Society", "Reports of the Oriental Society"), pub. for the Tōyōkyōkai Chōsabu 東洋協會調查部, later by the Tōyō Gakujutsu Kyōkai 學術, by Sanyō Shobō 三養書房, by Kunitachi Shoin 國立書院, and now again by the Tōyō Gakujutsu Kyōkai, three times a year (later at longer intervals), 1. 1 (Jan. 1911)—seen through 35. 1 (Sept. 1952).

A major repository of studies in archaeology and premodern history of the mainland but offering rather little on modern China.

9. 9. 41 Tōyōshi kenkyū 東洋史研究 ("Revue des études de l'histoire de l'Extrême-Orient", "The Journal of Oriental Researches"), pub. for the Tōyōshi Kenkyūkai at Kyōto University with offices in the museum of the Faculty of Letters, first by Mannensha 萬年社, now by the Tōyōshi Kenkyūkai ("The Society of Oriental Researches"), bimonthly, 1. 1 (Oct. 1935)—seen through 12. 3 (March 1953).

Representing a major center of Chinese historical research, this journal has dealt largely with pre-modern times, although its post-war content has dealt more extensively with the modern period.

INDEXES

GENERAL INDEX

This is a single alphabetic list of all authors, editors, and compilers, and titles of books and articles, in romaji according to the Hepburn system. We have added subject headings for special points of content. Since materials are arranged above according to content, however, we have tried to avoid a duplicate indication of content in this index, which is purely supplementary to the text above.

CHARACTER INDEX OF AUTHORS' NAMES

For Western scholars reared in alphabetic writing systems, the mystery of Japanese personal names is by turns incredible, fascinating and exasperating. Some at first refuse to believe it. The fact remains that the sounds of personal names of Japanese scholars are normally a matter of conjecture to other scholars, whose conjectures may be, and not infrequently are, incorrect. The bearer of any name may choose between the native Japanese *kun* and borrowed Chinese *on* readings of it, or between variants of either one. While most given names, like surnames, have a conventional pronunciation, exceptions are always possible and intellectuals seem to enjoy them. Momose Hiromu is not named Hiroshi; Katō Shigeshi is not named Shigeru; it is incorrect to call Professor Otake either Kotake or Odake, though some do. The late Dr. Kuwabara is named Jitsuzo in the Harvard-Yenching index of 1933 (item 9.2.5), Zitsuzō in the same for 1940 (9.2.6), Shitsuzō in the KBS bibliography for 1934 (9.2.4), but Kuwahara Shitsuzō in the same for 1935 and 1933. (We understand Shitsuzō is correct but Jitsuzō appears on his published work in English.)

Since the sound of a given name is seldom written down in Japan, except by faddists for *romaji*, this built-in ambiguity does not impede the onward rush of scholarship. For Western students, who must transliterate Japanese characters into *romaji*, it seems deliberately unkind. In the case of Chinese names, the pronuciation (if one uses an agreed upon dictionary) can be worked out. As long as the sinological fraternity stick together in using the Wade-Giles system, ambiguity of sound (though not of characters) can be avoided. But for Japanese names, dictionaries are little help, while Japanese directories and who's who volumes avoid the issue by remaining safely inside the Japanese language. This leaves the Western student to fight the battle of nomenclature helpless and alone. In our experience the best way to ascertain the pronunciation of a personal name is to ask its owner, his heirs or his classmates.

This index follows the order of characters in the index to R. H. Mathews, *A Chinese-English Dictionary* (Shanghai, 1931), but adds purely Japanese characters in the order of the Ueda *Daijiten*. Readings have been checked with numerous individuals and with those given in the KBS bibliographies (9.2.4), and it is our belief at the present time that the following names are properly pronounced as indicated.

一 1					
一	又	正	雄	Ichimata Masao	
三	上	正	利	Mikami Masatoshi	
三	上	義	夫	Mikami Yoshio	
三	上	諦	聽	Mikami Taichō	
三	國	谷	宏	Mikuniya Hiroshi	
三	嶋	一		Mishima Hajime	

三	村	啓	吉	Mimura Keikichi
三	枝	茂	智	Saegusa Shigetomo
三	浦	周	行	Miura Hiroyuki
三	田 村	泰	助	Mitamura Taisuke
上	原		蕃	Uehara Shigeru
上	原	重	美	Uehara Shigemi
上	坂	西	三	Kōsaka Torizō
上	村	鎮	威	Kamimura Shizui

321

CHARACTER INDEX OF AUTHORS' NAMES

叉 29		
及 川 恒 忠	Oikawa Tsunetada	

口 30	
古 賀 元 吉	Koga Motokichi
向 山 寛 夫	Mukōyama Hiroo
吉 岡 義 豐	Yoshioka Gihō
吉 川 勝 治	Yoshikawa Katsuji
吉 川 幸 次 郎	Yoshikawa Kōjirō
吉 川 重 藏	Yoshikawa Shigezō
吉 田 金 一	Yoshida Kin'ichi
吉 野 作 造	Yoshino Sakuzō
周 藤 吉 之	Sudō Yoshiyuki
和 田 博 德	Wada Hironori
和 田 清	Wada Sei (incorrectly Kiyoshi)

口 31	
園 田 一 龜	Sonoda Kazuki
園 田 庸 次 郎	Sonoda Yōjirō
團 藤 重 光	Dandō Shigemitsu

土 32	
寺 谷 修 三	Teratani Shūzō
坂 本 一 郎	Sakamoto Ichirō
坂 本 是 忠	Sakamoto Koretada
坂 野 正 高	Banno Masataka
堀 江 義 廣	Horie Yoshihiro
堀 江 英 一	Horie Eiichi
塚 本 善 隆	Tsukamoto Zenryū
塙 作 樂	Hanawa Sakura
增 井 經 夫	Masui Tsuneo
增 淵 龍 夫	Masubuchi Tatsuo

夕 36	
外 山 軍 治	Toyama Gunji

大 37	
大 上 末 廣	Ōkami Suehiro
大 久 保 莊 太 郎	Ōkubo Sōtarō
大 井 專 三	Ōi Senzō
大 塚 久 雄	Ōtsuka Hisao
大 塚 令 三	Ōtsuka Reizō
大 島 晉	Ōshima Susumu
大 平 善 梧	Ōhira Zengo
大 村 欣 一	Ōmura Kin'ichi
大 熊 眞	Ōkuma Tadashi
大 竹 秀 男	Ōtake Hideo
大 谷 健 夫	Ōtani Takeo
大 谷 孝 太 郎	Ōtani Kōtarō
大 隈 重 信	Ōkuma Shigenobu
天 海 謙 三 郎	Amagai Kenzaburō

天 野 元 之 助	Amano Motonosuke
太 田 英 一	Ōta Eiichi
奥 平 武 彦	Okudaira Takehiko
奥 田 乙 次 郎	Okuda Otojirō

女 38	
姜 在 彦	Kan, Z. E. (Kyō Zai-gen)

宀 40	
宇 佐 美 誠 次 郎	Usami Seijirō
宇 垣 一 成	Ugaki Issei
宇 野 精 一	Uno Seiichi
安 田 薰	Yasuda Kaoru
安 達 生 恒	Adachi Ikutsune
安 部 健 夫	Abe Takeo
安 藤 彦 太 郎	Andō Hikotarō
安 藤 鎭 正	Andō Shizumasa
安 齋 庫 治	Anzai Kuraji
守 屋 美 都 雄	Moriya Mitsuo
宗 像 金 吾	Munekata Kingo
宮 下 忠 雄	Miyashita Tadao
宮 內 季 子	Miyauchi Kishi
宮 崎 其 二	Miyazaki Sonoji
宮 崎 市 定	Miyazaki Ichisada
宮 崎 正 義	Miyazaki Masayoshi
宮 崎 滔 天 (寅藏)	Miyazaki Tōten (Torazō)
宮 崎 龍 介	Miyazaki Ryūsuke
宮 川 尙 志	Miyagawa Hisayuki
宮 本 通 治	Miyamoto Michiharu
宮 澤 俊 義	Miyazawa Toshiyoshi
宮 脇 賢 之 介	Miyawaki Kennosuke
實 藤 惠 秀	Sanetō Keishū

小 42	
小 倉 芳 彦	Ogura Yoshihiko
小 堀 巖	Kobori Iwao
小 宮 義 孝	Komiya Yoshitaka
小 島 昌 太 郎	Kojima Shōtarō
小 島 祐 馬	Ojima Sukema
小 川 平 二	Ogawa Heiji
小 川 環 樹	Ogawa Tamaki
小 早 川 欣 吾	Kobayakawa Kingo
小 林 又 三	Kobayashi Matazō
小 林 澄 兄	Kobayashi Sumie
小 林 胖 生	Kobayashi Bansei
小 林 英 三 郎	Kobayashi Eizaburō
小 椋 廣 勝	Ogura Hirokatsu
小 沼 正	Konuma Tadashi
小 澤 三 郎	Ozawa Saburō
小 畑 龍 雄	Obata Tatsuo
小 竹 文 夫	Otake Fumio
小 野 則 秋	Ono Noriaki

CHARACTER INDEX OF AUTHORS' NAMES

CHARACTER INDEX OF AUTHORS' NAMES

CHARACTER INDEX OF AUTHORS' NAMES

CHARACTER INDEX OF AUTHORS' NAMES

CHARACTER INDEX OF AUTHORS' NAMES

329

INDEX OF NAMES AND ABBREVIATIONS
FREQUENTLY CITED

This is a single alphabetic list of abbreviations and names of publishers, associations, and institutions which appear more or less frequently and for which characters are therefore given here rather than being repeated throughout the text.